T0305798

Methods of Geometry in the Theory of Partial Differential Equations

Principle of the Cancellation of Singularities

Other World Scientific Titles by the Author

Applied Analysis: Mathematics for Science, Technology, Engineering
Third Edition
ISBN: 978-981-12-5735-3

Chemotaxis, Reaction, Network: Mathematics for Self-Organization
ISBN: 978-981-323-773-5

Mean Field Theories And Dual Variation:
A Mathematical Profile Emerged in the Nonlinear Hierarchy
ISBN: 978-90-78677-14-7

Methods of Geometry in the Theory of Partial Differential Equations

Principle of the Cancellation of Singularities

Takashi Suzuki
Osaka University, Japan

World Scientific

NEW JERSEY · LONDON · SINGAPORE · BEIJING · SHANGHAI · HONG KONG · TAIPEI · CHENNAI · TOKYO

Published by

World Scientific Publishing Co. Pte. Ltd.

5 Toh Tuck Link, Singapore 596224

USA office: 27 Warren Street, Suite 401-402, Hackensack, NJ 07601

UK office: 57 Shelton Street, Covent Garden, London WC2H 9HE

Library of Congress Control Number: 2024000476

British Library Cataloguing-in-Publication Data
A catalogue record for this book is available from the British Library.

METHODS OF GEOMETRY IN THE THEORY OF PARTIAL
DIFFERENTIAL EQUATIONS
Principle of the Cancellation of Singularities

Copyright © 2024 by World Scientific Publishing Co. Pte. Ltd.

All rights reserved. This book, or parts thereof, may not be reproduced in any form or by any means, electronic or mechanical, including photocopying, recording or any information storage and retrieval system now known or to be invented, without written permission from the publisher.

For photocopying of material in this volume, please pay a copying fee through the Copyright Clearance Center, Inc., 222 Rosewood Drive, Danvers, MA 01923, USA. In this case permission to photocopy is not required from the publisher.

ISBN 978-981-12-8789-3 (hardcover)
ISBN 978-981-12-8790-9 (ebook for institutions)
ISBN 978-981-12-8791-6 (ebook for individuals)

For any available supplementary material, please visit
https://www.worldscientific.com/worldscibooks/10.1142/13721#t=suppl

Printed in Singapore

Preface

To the memory of my parents

Mathematical models are used to describe the real world, and their analysis induces new predictions filled with unexpected phenomena.
A huge number of insights have been derived in a variety of scientific fields, while some principles that ensure these achievements have been noticed very recently.

This monograph focuses on one of them arising in the theory of partial differential equations (PDE); cancellation of singularities caused by symmetric interaction of many particles, skew-symmetric interaction of multiple species, and variations of physical quantities under the flow. These profiles are described by the language of geometry, particularly, that of global analysis (GA), dealing with the evolution of geometric objects and relaxation of internal forces described by differential forms and frames.

Five objects are selected in this monograph. They are widely spreaded, but are connected strongly because of common geometric backgrounds; evolution of geometric quantities subject to thermo-dynamical laws, models of multi-species in biology, interface vanishing in $d - \delta$ system formulated by differential forms, fundamental equations of mathematical physics derived from the transformation theory, and domain deformation associated with the problems in engineering and physics.

This monograph describes them in three parts. Part I deals with the evolution of geometric objects. In Chapter 1, we confirm the classical theory on curves and surfaces in $\mathbf{R}^3$. Chapter 2 and Chapter 3, are then devoted to the models provided with quantized blowup mechanis observed in physics because of their common geometric background; that is, Boltzmann Poisson (BP) equation in statistical mechanics and Smoluchowski

Poisson (SP) equation in non-equilibrium thermo-dynamics. Then we turn to the 2D normalized Ricci flow (2D-NRF) in Chapter 4 arising in geometry in comparison with BP and SP equations in mathematical physics.

Part II is concerned on the cancellation of singularities described by differential forms. First, Chapter 5 deals with the system of multiple species in biology provided with the Poisson structure. A class of skew-symmetric Lotka Volterra (LV) system is detected, realizing foliations made by periodic orbits in phase space, and then integrable systems are formulated usinig differential forms. Second, Chapter 6 is devoted to the interface vanishing of $d - \delta$ system including non-stationary Maxwell equations.

Part III is the theory of transformations, significantly related to the theories of function spaces, elliptic regularity, tensor analysis, and differential geometry. Chapter 7 is devoted to the elliptic regularity on Lipschitz domains, where frame on the boundary is used as a fundamental tool of analysis. Chapter 8 is devoted to Louville's theorems in transformation theory concerning the first and the second volume and area derivatives. These formulae induce numerical schemes to the free boundary problem in engineering. They also provide with mathematical justifications of several physical models. Chapter 9 is on the Hadamard variational formula of the Green function. The first and the second formulae are given rigorously under general perturbation of domains. Chapter 10, finally, deals with the Hadamard variation of eigenvalues, particularly, characterization of the first and the second unilateral derivatives, their unilateral continuity, and their smooth rearrangements.

In these three parts flexible use of the coordinates is made totally; those of the Euler and the Lagrange, both in individual and dependent variables. This approach reveals hidden principles in the real world through the study of PDE, by means of the languages of geometry, particularly GA. The author noticed the role of GA very recently in the theory of PDE although he has been involved by both of them for a long term. He thanks Dr. Nikos Kavallaris, Dr. Ken'ichi Nagasaki, Dr. Hiroshi Ohtsuka, Dr. Takasi Senba, Dr. Takuya Tsuchiya, Dr. Kazuo Watanabe, and Dr. Yoshio Yamada, for useful discussions and collaborations.

Osaka, Japan

November, 2023
Takashi Suzuki

Contents

Differential Forms and Singularities 187

PART 1

Evolution of Geometric Objects

Chapter 1

Curves and Surfaces in $\mathbf{R}^3$

Several phenomena of cancellation of singularities arising in the theory of partial differential equations are described in the languages of geometry. Typical examples are observed in the evolution of geometric objects; associated with symmetric interactions of physical particles. This chapter is concerned with the classical theory of curves and surfaces in $\mathbf{R}^3$. Its target is two-fold; to present the notions of curvature (§1.1–§1.4) and differential forms (§1.5). The former induces geometric problems studied in Part I, while the latter provides several tools to clarify the mechanism of singularity cancellation through the interaction of multiple species in Part II. Then we describe fundamental concept used in Part III, that is, the tangent and cotangent spaces, frames, covariant derivatives, and connections (§1.6–§1.8). Finally, we formulate an elliptic equation studied in Chapter 2 (§1.9–§1.10).

1.1 Surfaces

A surface is a set of points in $\mathbf{R}^3$ represented by two parameters,

$$x(u, v) = \begin{pmatrix} x_1(u, v) \\ x_2(u, v) \\ x_3(u, v) \end{pmatrix}, \ (u, v) \in \Omega.$$

This $x = x(u, v)$ is a smooth mapping from Ω to $\mathbf{R}^3$, where $\Omega \subset \mathbf{R}^2$ is the parameter region and $x(\Omega) = \mathcal{M}$.

Let the distance between two points on $\mathcal{M}$,

$$x(u, v), \ x(u + \Delta u, v + \Delta v)$$

be Δs, and the area of the parallelogram made by

$$a = x(u + \Delta u, v) - x(u, v), \quad b = x(u, v + \Delta v) - x(u, v)$$

3

be ΔS. Using

$$x(u + \Delta u, v + \Delta v) - x(u, v)$$
$$= x_u(u, v)\Delta u + x_v(u, v)\Delta v + o\left(\sqrt{\Delta u^2 + \Delta v^2}\right),$$

we have

$$(\Delta s)^2 = |x(u + \Delta u, v + \Delta v) - x(u, v)|^2$$
$$= |x_u|^2 \Delta u^2 + 2x_u \cdot x_v \Delta u \Delta v + |x_v|^2 \Delta v^2$$
$$+ o\left(\Delta u^2 + \Delta v^2\right), \qquad (1.1)$$

where $| \cdot |$ and $\cdot$ denote the length and inner product of vectors in $\mathbf{R}^3$, that is,

$$|a| = \sqrt{a_1^2 + a_2^2 + a_3^2}, \quad a \cdot b = a_1 b_1 + a_2 b_2 + a_3 b_3$$

defined for

$$a = \begin{pmatrix} a_1 \\ a_2 \\ a_3 \end{pmatrix}, \quad b = \begin{pmatrix} b_1 \\ b_2 \\ b_3 \end{pmatrix},$$

and

$$x_u = x_u(u, v) = \frac{\partial x}{\partial u}(u, v), \quad x_v = x_v(u, v) = \frac{\partial x}{\partial v}(u, v).$$

Equality (1.1) is represented by the *first fundamental form,*

$$ds^2 = E du^2 + 2F du dv + G dv^2, \qquad (1.2)$$

with

$$E = |x_u|^2, \quad F = x_u \cdot x_v, \quad G = |x_v|^2$$

standing for the *first fundamental quantities.*

Similarly, we obtain

$$\Delta S = |a \times b|$$
$$= |x_u(u, v) \times x_v(u, v)| \Delta u \Delta v + o\left(\Delta u^2 + \Delta v^2\right) \qquad (1.3)$$

by

$$a = x(u + \Delta u, v) - x(u, v) = x_u(u, v)\Delta u + o(\Delta u)$$
$$b = x(u, v + \Delta v) - x(u, v) = x_v(u, v)\Delta v + o(\Delta v),$$

where $\times$ denotes the outer product in $\mathbf{R}^3$:

$$\begin{pmatrix} a_1 \\ a_2 \\ a_3 \end{pmatrix} \times \begin{pmatrix} b_1 \\ b_2 \\ b_3 \end{pmatrix} = \begin{pmatrix} a_2 b_3 - a_3 b_2 \\ a_3 b_1 - a_1 b_3 \\ a_1 b_2 - a_2 b_1 \end{pmatrix}.$$

In fact, $a \times b$ dnotes the vector perpendicular to a, b in the direction of right-hand with the length equal to ΔS. This operation satisfies the axiom

$$a \times b = -(b \times a)$$
$$(\alpha a) \times b = \alpha(a \times b)$$
$$a \times (b + c) = (a \times b) + (a \times c)$$

for $a, b, c \in \mathbf{R}^3$ and $\alpha \in \mathbf{R}$. Then, equality (1.3) induces the *area element*

$$dS = |x_u \times x_v| \, dudv.$$

Since the vectors

$$x_u(u, v), \ x_v(u, v)$$

are tangent to the curves on $\mathcal{M}$ where v and u are constants, the vector

$$n = \frac{x_u \times x_v}{|x_u \times x_v|}$$

is the unit vector perpendicular to $\mathcal{M}$. Henceforth, we use

$$dx = x_u du + x_v dv$$

to write the first fundamental form as

$$ds^2 = dx \cdot dx = |x_u|^2 \, du^2 + 2x_u \cdot x_v dudv + |x_v|^2 \, dv^2. \qquad (1.4)$$

We have, on the other hand,

$$\begin{aligned}
|x_u \times x_v|^2 &= (x_u \times x_v) \cdot (x_u \times x_v) \\
&= (x_u \cdot x_u)(x_v \cdot x_v) - (x_u \cdot x_v)^2 \\
&= EG - F^2
\end{aligned}$$

by

$$(a \times b)(c \times d) = (a \cdot c)(b \cdot d) - (a \cdot d)(b \cdot c)$$

and hence

$$dS = \sqrt{EG - F^2} dudv, \ n = \frac{x_u \times x_v}{\sqrt{EG - F^2}}.$$

Then the *vector area element* is defined by

$$ndS.$$

Integration by parts of the function with multiple variables is described by the formulae of Gauss and Stokes (§1.5, §6.3). Vector area element, on the other hand, is used in Liouville's transformation theory in accordance with the Hodge operator to differential forms (§1.5, §8.7).

1.2   Curves

A curve in three-dimensional space is parametrized by the *length parameter*, denoted by s:

$$\mathcal{C} = \{x(s)\} \subset \mathbf{R}^3, \quad \left|\frac{dx}{ds}\right| = 1.$$

Then

$$t = dx/ds$$

is the unit *tangent vector*, and the plane containing $P = x(s)$ and is perpendicular to t is called the *normal plane*.

We take P_1, P_2 on $\mathcal{C}$ near P and let π be the plane made by these three points, P_1, P_2, and P. As

$$P_1, P_2 \to P,$$

this π converges to a plane orthogonal to the normal plane, called the *osculating plane*. The intersection of normal and osculating planes forms a line on which we take a unit vector n, called the *principal normal vector*.

Near P the curve $\mathcal{C}$ is approximated by a circle on the osculating plane. If its radius is denoted by ρ and the direction of n is taken counter-clockwise to t, it holds that

$$\frac{dt}{ds} = \frac{1}{\rho}n$$

with a scalar $1/\rho$, called the *curvature* of $\mathcal{C}$ at P, by

$$\frac{dt}{ds} \cdot t = 0$$

derived from

$$t \cdot t = |t|^2 = 1.$$

This ρ stands for the radius of the approximate circle mentioned above. The tangent and the principal normal vectors are perpendicular to each other. Then the *bi-principal normal vector* is defined by

$$b = t \times n.$$

The direction of b changes as $\mathcal{C}$ twists, and this rate is detected by the *torsion*, denoted by τ. To clarify this role, note that $b = t \times n$ implies

$$\frac{db}{ds} = \frac{dt}{ds} \times n + t \times \frac{dn}{ds}$$

$$= \frac{1}{\rho}n \times n + t \times \frac{dn}{ds} = t \times \frac{dn}{ds}$$

by

$$n \times n = 0.$$

We have also

$$b \cdot \frac{db}{ds} = 0$$

by

$$b \cdot b = 1,$$

and hence

$$\frac{db}{ds}$$

is perpendicular to both t and b.

Regarding these facts, we define the *torsion* τ by

$$\frac{db}{ds} = -\tau n$$

including its sign. It holds also that

$$n = b \times t,$$

which implies

$$\frac{dn}{ds} = \frac{db}{ds} \times t + b \times \frac{dt}{ds}$$
$$= -\tau n \times t + b \times \frac{1}{\rho} n = \tau b - \frac{1}{\rho} t$$

by

$$n \times t = -b, \quad b \times n = -t.$$

We thus end up with the *Frenet Serret formula,*

$$\frac{dt}{ds} = \frac{1}{\rho} n$$
$$\frac{dn}{ds} = -\frac{1}{\rho} t + \tau b$$
$$\frac{db}{ds} = -\tau n,$$

where ρ, τ, and s are *curvature radius, torsion,* and *length parameter,* respectively. These quantities are used to describe the status of higher-dimensional manifolds as sectional curvatures (§7.3).

1.3   Curves on Surfaces

Recall $x = x(u, v)$, a parameter representation of the surface $\mathcal{M}$, and let $\mathcal{C}$ be a curve on $\mathcal{M}$. This curve $\mathcal{C}$ is represented by

$$(u, v) = (u(s), v(s)),$$

or

$$x = x(u(s), v(s)),$$

using the length parameter s of $\mathcal{C}$.

The unit tangential vector of $\mathcal{C}$ is given by

$$t = \frac{dx}{ds} = x_u \frac{du}{ds} + x_v \frac{dv}{ds} \tag{1.5}$$

and hence

$$\left| \frac{dt}{ds} \right| = 1.$$

The line element $d\sigma^2$ on $\mathcal{C}$ is given by

$$d\sigma^2 = Edu^2 + 2Fdudv + Gdv^2 = (Eu'(s)^2 + 2Fu'(s)v'(s) + Gv'(s)^2)ds^2$$

$$= \left| \frac{dx}{ds}(u(s), v(s)) \right|^2 ds^2 = |t|^2 ds^2 = ds^2$$

by (1.2) in §1.1.

It holds also that

$$\frac{dt}{ds} = \frac{1}{\rho_{\mathcal{C}}} n_{\mathcal{C}},$$

where $\rho_{\mathcal{C}}$ and $n_{\mathcal{C}}$ denote the curvature radius and the principal normal unit vector of $\mathcal{C}$, respectively. If $\psi_{\mathcal{C}}$ denotes the angle between n and $n_{\mathcal{C}}$, then it follows that

$$\cos \psi_{\mathcal{C}} = n \cdot n_{\mathcal{C}},$$

or

$$\frac{\cos \psi_{\mathcal{C}}}{\rho_{\mathcal{C}}} = n \cdot \frac{dt}{ds},$$

where

$$n = \frac{x_u \times x_v}{|x_u \times x_v|}$$

stands for the unit normal vector of $\mathcal{M}$.

Since (1.5) implies

$$\frac{dt}{ds} = x_{uu} \left(\frac{du}{ds} \right)^2 + 2x_{uv} \left(\frac{du}{ds} \right) \left(\frac{dv}{ds} \right)$$
$$+ x_{vv} \left(\frac{dv}{ds} \right)^2 + x_u \frac{d^2u}{ds^2} + x_v \frac{d^2v}{ds^2}$$

it holds that

$$n \cdot \frac{dt}{ds} = n \cdot \left\{ x_{uu} \left(\frac{du}{ds} \right)^2 + 2x_{uv} \left(\frac{du}{ds} \right) \left(\frac{dv}{ds} \right) + x_{vv} \left(\frac{dv}{ds} \right)^2 \right\}$$

by

$$n \cdot x_u = n \cdot x_v = 0. \tag{1.6}$$

Then the *second fundamental quantities* are defined by

$$L = x_{uu} \cdot n = -x_u \cdot n_u$$
$$M = x_{uv} \cdot n = -x_u \cdot n_v = -x_v \cdot n_u$$
$$N = x_{vv} \cdot n = -x_v \cdot n_v, \tag{1.7}$$

because of (1.6).

Writing

$$\frac{\cos \psi_C}{\rho_C} = L \left(\frac{du}{ds} \right)^2 + 2M \frac{du}{ds}\frac{dv}{ds} + N \left(\frac{dv}{ds} \right)^2$$
$$= \frac{L du^2 + 2M\,dudv + N dv^2}{E du^2 + 2F\,dudv + G dv^2},$$

furthermore, we call

$$L du^2 + 2M\,dudv + N dv^2$$

the *second fundamental form*. It holds that

$$L du^2 + 2M\,dudv + N dv^2 = -dn \cdot dx \tag{1.8}$$

by (1.7), where

$$dn = n_u du + n_v dv, \quad dx = x_u du + x_v dv.$$

Equality (1.8) corresponds to (1.4),

$$ds^2 = dx \cdot dx,$$

for the first fundamental form. Here,

$$\eta = dx$$

stands for the arbitrary vector in $\mathcal{M}$, while dn is regarded as a tangential derivative of n on $\mathcal{M}$ as in

$$\frac{\partial n}{\partial \xi}, \quad \xi \in \mathcal{M}.$$

We thus regard the second fundamental form as a bi-linear form on the tangential space of $\mathcal{M}$ in §1.6, that is,

$$\mathcal{B}(\xi, \eta) = -\frac{\partial n}{\partial \xi} \cdot \eta, \quad \xi, \eta \in \mathcal{M}. \tag{1.9}$$

The second fundamental form on hyper-surfaces arises in the second variational formula of the Green function concerning domain perturbations (§9.11).

1.4 Curvatures

Using

$$k = dv/du,$$

the direction of the infinitesimal vector dx, we have

$$\frac{\cos \psi_C}{\rho_C} = \frac{L + 2Mk + Nk^2}{E + 2Fk + Gk^2}. \tag{1.10}$$

Here,

$$L, \ M, \ N, \ E, \ F, \ G$$

are determined by $P \in \mathcal{M}$, while

$$k = \frac{dv}{du} = \frac{dv/ds}{du/ds}$$

is determined by the tangential vector t of $\mathcal{C}$, as well as P, because of

$$t = \frac{dx}{ds} = \frac{du}{ds}x_u + \frac{dv}{ds}x_v.$$

Let $\prod$ be the plane containing P made by the tangential vector t of $\mathcal{C}$ and the normal vector n of $\mathcal{M}$, and let $\mathcal{C}'$ be the curve on $\mathcal{M}$ cut by $\prod$. Since the center of curvature $\mathcal{C}'$ at P is on $\prod$ it holds that either

$$(\psi_{C'}, \rho_{C'}) = (0, R)$$

or

$$(\psi_{C'}, \rho_{C'}) = (\pi, -R),$$

where R denotes the curvature of $\mathcal{C}'$ at $P \in \mathcal{M}$.

The right-hand side of (1.10), on the other hand, is determined by P and t so that we have

$$\frac{\cos \psi_C}{\rho_C} = \frac{\cos \psi_{C'}}{\rho_{C'}} = \frac{1}{R} \tag{1.11}$$

at P. This $1/R$ is called the *normal curvature* of $\mathcal{M}$ at P with the direction t. Fixing P, we seek t such that the normal curvature attains the minimum or the maximum. In these extremal cases, t and $1/R$ are called the *principal direction* and the *principal curvature*, respectively.

They are actually obtained by

$$\frac{d}{dk} \left(\frac{1}{R} \right) = \frac{d}{dk} \cdot \frac{L + 2Mk + Nk^2}{E + 2Fk + Gk^2} = 0,$$

or, equivalently,

$$\frac{1}{R}(F + Gk) = M + Nk. \tag{1.12}$$

Then, since

$$\frac{1}{R}(E + 2Fk + Gk^2) = L + 2Mk + Nk^2$$

it holds that

$$\frac{1}{R}(E + Fk) = L + Mk. \tag{1.13}$$

We have, on the other hand,

$$k = -\frac{F/R - M}{G/R - N}$$

again by (1.12) and then

$$\left(\frac{F}{R} - M \right)\left(\frac{F}{R} - M \right) - \left(\frac{E}{R} - L \right)\left(\frac{G}{R} - N \right) = 0 \tag{1.14}$$

follows from (1.13). Thus, the principal curvatures $1/R_1$ and $1/R_2$ are the solutions to (1.14).

Writing (1.14) as

$$(EG - F^2)\frac{1}{R^2} - (GL + EN - 2FM)\frac{1}{R} + LN - M^2 = 0, \tag{1.15}$$

we obtain

$$2H \equiv \frac{1}{R_1} + \frac{1}{R_2} = \frac{GL + EN - 2FM}{EG - F^2}$$

$$K \equiv \frac{1}{R_1} \cdot \frac{1}{R_2} = \frac{LN - M^2}{EG - F^2}. \tag{1.16}$$

These quantities

$$2H = \frac{1}{R_1} + \frac{1}{R_2}, \quad K = \frac{1}{R_1} \cdot \frac{1}{R_2} \tag{1.17}$$

are called the *mean curvature* and the *Gaussian (or scalar) curvature* of $\mathcal{M}$ at P, respectively. Then it follows that

$$2H = \text{tr } A, \quad K = \det A, \tag{1.18}$$

where

$$A = \begin{pmatrix} E & F \\ F & G \end{pmatrix}^{-1} \begin{pmatrix} L & M \\ M & N \end{pmatrix} \tag{1.19}$$

is called the *Weingarten matrix*.

If the Gaussian curvature is positive, the surface is convex at this place. Similarly, if it is negative it constitutes a saddle. The Gaussian curvature is determined by the first fundamental quantities as is shown in §1.8.

Equation (1.14) on $1/R$ takes an equivalent form on k by (1.12) and (1.13), that is,

$$FL - EM + (GL - EN)k + (GM - FN)k^2 = 0, \tag{1.20}$$

which provides the principal directions k_1 and k_2 as the solution. These directions are associated with the infinitesimal vectors

$$dx_1 = x_u du_1 + x_v dv_1$$
$$dx_2 = x_u du_2 + x_v dv_2$$

through $k_1 = dv_1/du_1$ and $k_2 = dv_2/du_2$, which implies

$$dx_1 \cdot dx_2 = E du_1 du_2 + F(du_2 dv_1 + du_1 dv_2) + G dv_1 dv_2$$
$$= (E + F(k_1 + k_2) + G k_1 k_2)\, du_1 du_2. \tag{1.21}$$

We obtain, on the other hand,

$$k_1 + k_2 = -\frac{GL - EN}{GM - FN}, \quad k_1 k_2 = \frac{FL - EM}{GM - FN}$$

by (1.20) and hence

$$E + F(k_1 + k_2) + G k_1 k_2 = \frac{1}{GM - FN}$$
$$\cdot \{(GM - FN)E - F(GL - EN) + G(FL - EM)\} = 0.$$

It thus holds that $dx_1 \cdot dx_2 = 0$ by (1.21), and, therefore, the principal directions are perpendicular to each other.

Curvature is a fundamental quantities of geometric objects. It appears in the study of point vortices (§2.5) and domain perturbations (§9.11).

1.5   Differential Forms

Each scalar function

$$f = f(x_1, x_2, x_3) \tag{1.22}$$

in $x = (x_1, x_2, x_3) \in \mathbf{R}^3$ is called a 0-*form*. Then 1-, 2-, 3- *forms* are given by

$$f dx_1 + g dx_2 + h dx_3,$$
$$f dx_2 \wedge dx_3 + g dx_3 \wedge dx_1 + h dx_1 \wedge dx_2,$$
$$f dx_1 \wedge dx_2 \wedge dx_3,$$

respectively. Here, the *wedge product* $\wedge$ is an operation satisfying

$$dx_i \wedge dx_j = -dx_j \wedge dx_i, \tag{1.23}$$

which results in

$$dx_i \wedge dx_i = 0.$$

Given 1-forms

$$\omega_1 = \sum_{i=1}^{3} P_i dx_i, \ \omega_2 = \sum_{i=1}^{3} Q_i dx_i,$$

we define

$$\omega_1 \wedge \omega_2 = \sum_{i=1}^{3} P_i Q_i \ dx_i \wedge dx_j = (P_2 Q_3 - P_3 Q_2) \ dx_2 \wedge dx_3$$
$$+ (P_3 Q_1 - P_1 Q_3) \ dx_3 \wedge dx_1 + (P_1 Q_2 - P_2 Q_1) \ dx_1 \wedge dx_2,$$

regarding (1.23). For 1- and 2-forms denoted by

$$\omega = \sum_{i=1}^{3} P_i dx_i \tag{1.24}$$

and

$$\theta = Q_1 dx_2 \wedge dx_3 + Q_2 dx_3 \wedge dx_1 + Q_3 dx_1 \wedge dx_2, \tag{1.25}$$

respectively, we have

$$\omega \wedge \theta = (P_1 dx_1 + P_2 dx_2 + P_3 dx_3)$$
$$\wedge (Q_1 dx_2 \wedge dx_3 + Q_2 dx_3 \wedge dx_1 + Q_3 dx_1 \wedge dx_2)$$
$$= (P_1 Q_1 + P_2 Q_2 + P_3 Q_3) \ dx_1 \wedge dx_2 \wedge dx_3$$

similarly.

The *outer derivative* is denoted by d. It acts on 0-, 1-, and 2-forms denoted by f, ω, and θ in (1.22), (1.24), and (1.25) as

$$df = \frac{\partial f}{\partial x_1} dx_1 + \frac{\partial f}{\partial x_2} dx_2 + \frac{\partial f}{\partial x_3} dx_3,$$

$$d\omega = dP \wedge dx_1 + dQ \wedge dx_2 + dR \wedge dx_3$$
$$= \left(\frac{\partial R}{\partial x_2} - \frac{\partial Q}{\partial x_3} \right) dx_2 \wedge dx_3 + \left(\frac{\partial P}{\partial x_3} - \frac{\partial R}{\partial x_1} \right) dx_3 \wedge dx_1$$
$$+ \left(\frac{\partial Q}{\partial x_1} - \frac{\partial P}{\partial x_2} \right) dx_1 \wedge dx_2,$$

and

$$d\theta = dP \wedge dx_2 \wedge dx_3 + dQ \wedge dx_3 \wedge dx_1 + dR \wedge dx_1 \wedge dx_2$$
$$= \left(\frac{\partial P}{\partial x_1} + \frac{\partial Q}{\partial x_2} + \frac{\partial R}{\partial x_3} \right) dx_1 \wedge dx_2 \wedge dx_3,$$

respectively. Then it follows that

$$d^2 = 0.$$

Given 1-forms ω_1 and ω_2, furthermore, we obtain

$$d(\omega_1 \wedge \omega_2) = d\omega_1 \wedge \omega_2 - \omega_1 \wedge d\omega_2. \tag{1.26}$$

The *divergence formula* of Gauss is represented by

$$\iiint_V d\theta = \iint_{\partial V} \theta,$$

where $V \subset \mathbf{R}^3$ is a bounded domain with smooth boundary ∂V, θ is a 2-form, and

$$dx_i \wedge dx_j, \quad i \neq j,$$

is a projection of dS to the plane $x = x_k$, $k \neq i, j$, with (i, j, k) oriented in a right-handed system.

The *Hodge operator* $*$ is defined for (1.22), (1.24), and (1.25) by

$$*f = f \, dx_1 \wedge dx_2 \wedge dx_3,$$
$$*\omega = P \, dx_2 \wedge dx_3 + Q \, dx_3 \wedge dx_1 + R \, dx_1 \wedge dx_2,$$

and

$$*\theta = P \, dx_1 + Q \, dx_2 + R \, dx_3,$$

respectively. We put also

$$*(f \, dx_1 \wedge dx_2 \wedge dx_3) = f,$$

which ensures $** = Id$. If dS and $n = (n_1, n_2, n_3)^T$ denote the area element and the outer normal unit vector on $\mathcal{M}$, respectively, it holds that

$$dx_2 \wedge dx_3 = n_1 dS, \; dx_3 \wedge dx_1 = n_2 dS, \; dx_1 \wedge dx_2 = n_3 dS$$

and hence

$$ndS = (*dx_1, *dx_2, *dx_3)^T. \tag{1.27}$$

Let

$$\varphi : D \subset \mathbf{R}^2 \;\rightarrow\; \mathcal{M} \subset \mathbf{R}^3$$

be a parametric representation of the surface $\mathcal{M}$, denoted by

$$\varphi(u, v) = x \equiv \begin{pmatrix} x_1(u, v) \\ x_2(u, v) \\ x_3(u, v) \end{pmatrix}.$$

Given a smooth function $f : \mathcal{M} \rightarrow \mathbf{R}$, its *pull-back* is defined by

$$\varphi^* f = f \circ \varphi : D \rightarrow \mathbf{R}.$$

The pull-backs of 1- and 2-forms on $\mathcal{M}$ in the forms of (1.24) and (1.25), respectively, are given by

$$\varphi^* \omega = \varphi^* P \cdot d(\varphi^* x_1) + \varphi^* Q \cdot d(\varphi^* x_2) + \varphi^* R \cdot d(\varphi^* x_3)$$

and

$$\varphi^* \theta = \varphi^* P \cdot d(\varphi^* x_2) \wedge d(\varphi^* x_3) + \varphi^* Q \cdot d(\varphi^* x_3) \wedge d(\varphi^* x_1) \\ + \varphi^* R \cdot d(\varphi^* x_1) \wedge d(\varphi^* x_2),$$

respectively.

Then we obtain the following theorems.

Theorem 1.1. *If f, ω, and θ are 0-, 1-, and 2-forms, respectively, it holds that*

$$\varphi^*(df) = d(\varphi^* f), \quad \varphi^*(d\omega) = d(\varphi^* \omega), \quad \varphi^*(d\theta) = d(\varphi^* \theta).$$

Theorem 1.2. *If ω and θ are 1 and 2-forms, respectively, it holds that*

$$\iint_{\mathcal{M}} \theta = \iint_D \varphi^* \theta, \quad \int_{\partial \mathcal{M}} \omega = \int_{\partial D} \varphi^* \omega.$$

Green's formula on $D \subset \mathbf{R}^2$ indicates

$$\iint_D d(\varphi^* \omega) = \int_{\partial D} \varphi^* \omega,$$

where ∂D is oriented counter-clockwise. Then the *Stokes formula* on $\mathcal{M}$

$$\iint_{\mathcal{M}} d\omega = \int_{\partial \mathcal{M}} \omega$$

arises by Theorems 1.1 and 1.2.

Theory of differential forms has the origin in differentiation of functions of multiple variables. Later we use it to describe cancellation of singularities by skew-symmetric interaction of multiple species (§5.10) and interface vanishing of some components of multi-valued solution to fundamental equations of physics (§6.7).

1.6   Tangent and Cotangent Spaces

Let

$$\mathcal{M} = \bigcup_{d=1}^{s} U_\alpha$$

be a global covering of a surface, and

$$\varphi_\alpha : D_\alpha \subset \mathbf{R}^2 \to U_\alpha \subset \mathcal{M}$$

be a *local chart*. Hence

$$\varphi_\beta^{-1} \circ \varphi_\alpha : \tilde{D}_\alpha = \varphi_\alpha^{-1}(U_\beta \cap U_\alpha) \subset D_\alpha \to \varphi_\beta^{-1} \circ \varphi_\alpha(\tilde{D}_\alpha) \subset D_\beta$$

is a *diffeomorphism*, which means that a continuously differentiable bijection with its inverse mapping, provided that

$$U_\beta \cap U_\alpha \neq \emptyset. \tag{1.28}$$

Then, each $(u_\alpha, v_\alpha) \in U_\alpha$ is called the local coordinate.

If (1.28) arises and

$$(u_\alpha, u_\alpha) \in D_\alpha, \ (u_\beta, u_\beta) \in D_\beta$$

are *local coordinates*, we have

$$u_\beta = u_\beta(u_\alpha, v_\alpha), \quad v_\beta = v_\beta(u_\alpha, v_\alpha),$$

which induces

$$du_\beta = \frac{\partial u_\beta}{\partial u_\alpha} du_\alpha + \frac{\partial u_\beta}{\partial v_\alpha} dv_\alpha, \quad dv_\beta = \frac{\partial v_\beta}{\partial u_\alpha} du_\alpha + \frac{\partial v_\beta}{\partial v_\beta} dv_\beta.$$

The vector spaces of 1-forms generated by $\{du_\alpha, dv_\alpha\}$ and $\{du_\beta, dv_\beta\}$ are identified with this relation as in

$$\{du_\alpha, dv_\alpha\} \sim \{du_\alpha, dv_\alpha\}.$$

Then the *cotangent space* of $\mathcal{M}$ at $x \in U_\alpha$ is defined as a *quotient vector space* by this equivalence:

$$T_x^*\mathcal{M} = \{du_\alpha, du_\beta\}/\sim .$$

Then

$$T^*\mathcal{M} = \bigcup_{x \in \mathcal{M}} T_x^*\mathcal{M}$$

is called the *cotangent bundle.*

The *dual basis* of $\{du_\alpha, dv_\alpha\}$, denoted by

$$\left\{ \frac{\partial}{\partial u_\alpha}, \frac{\partial}{\partial v_\alpha} \right\},$$

is defined by

$$\left\langle du_a, \frac{\partial}{\partial u_b} \right\rangle = \delta_{ab}, \quad a, b = \alpha, \beta.$$

Using the transformation of variables,

$$\frac{\partial}{\partial u_\alpha} = \frac{\partial u_\beta}{\partial v_\alpha} \frac{\partial}{\partial u_\beta} + \frac{\partial v_\beta}{\partial v_\alpha} \frac{\partial}{\partial v_\beta}, \quad \frac{\partial}{\partial v_\alpha} = \frac{\partial u_\beta}{\partial v_\alpha} \frac{\partial}{\partial u_\beta} + \frac{\partial v_\beta}{\partial v_\alpha} \frac{\partial}{\partial v_\beta},$$

we introduce the *tangent space* of $\mathcal{M}$ at $x \in U_\alpha$ by

$$T_x\mathcal{M} = \left\{ \frac{\partial}{\partial u_\alpha}, \frac{\partial}{\partial v_\alpha} \right\}/\sim,$$

and then

$$T(M) = \bigcup_{x \in \mathcal{M}} T_x\mathcal{M}$$

is called the *tangential bundle.*

Through the local chart

$$\varphi : (u, v) \in D \subset \mathbf{R}^2 \mapsto \vec{x} = \begin{pmatrix} x_1(u, v) \\ x_2(u, v) \\ x_3(u, v) \end{pmatrix} \in \mathcal{M} \subset \mathbf{R}^3,$$

the tangential vector field $X = \xi x_u + \eta x_v$ on $\mathcal{M}$ is dfined. It is indicated as

$$X \in \mathcal{X}(\mathcal{M}),$$

and the operation

$$Xf = \xi \frac{\partial}{\partial u}(\varphi^* f) + \eta \frac{\partial}{\partial v}(\varphi^* f)$$

is induced for $\varphi^* f = f \circ \varphi$, where $f, \xi, \eta : \mathcal{M} \to \mathbf{R}$ are 0-forms. Hence

$$\frac{\partial}{\partial u}, \ \frac{\partial}{\partial v} \in T\mathcal{M}$$

are identified with $x_u, x_v \in \mathcal{X}(\mathcal{M})$ by

$$x_u = \frac{\partial}{\partial u}(\varphi^* \cdot), \ x_v = \frac{\partial}{\partial v}(\varphi^* \cdot) \in \mathbf{R}^3$$

through the local chart.

General theory of manifolds is described in §5.1. Geometric quantities of objects are formulated there through general coordinates.

1.7   Frames and Covariant Derivatives

Given a surface $\mathcal{M} \subset \mathbf{R}^3$ and a point $x \in \mathcal{M}$, let

$$e_1, e_2 \in T_x\mathcal{M}, \ e_3 \in \mathbf{R}^3$$

be a right-handed ortho-normal system of unit vectors in $\mathbf{R}^3$. For the moment we regard each e_j as a vector valued 0-form. We call

$$\{e_1, e_2, e_3\} = \{e_1(x), e_2(x), e_3(x)\}$$

an ortho-normal *frame* if it is smooth in $x \in \mathcal{M}$. Then it holds that

$$dx = \omega^1 e_1 + \omega^2 e_2$$

as a vector-valued 1-form, and this $\{\omega^1, \omega^2\}$ is called the *dual frame*. Consequently, these ω^1 and ω^2 are scalar-valued 1-forms.

The vector valued 1-form de_j on $\mathcal{M}$ takes the form

$$de_j = \sum_{i=1}^{3} \omega_j^i e_i$$

for $j = 1, 2, 3$, and each ω_j^i is called a *connection form*. Connection forms are scalar-valued 1-forms on $\mathcal{M}$.

By $e_i \cdot e_j = \delta_{ij}$ we obtain

$$de_i \cdot e_j + e_i \cdot de_j = 0,$$

which implies

$$\omega_j^i + \omega_i^j = 0 \tag{1.29}$$

and hence

$$\omega_i^i = 0. \tag{1.30}$$

We thus end up with

$$d\begin{pmatrix} e_1 \\ e_2 \\ e_3 \end{pmatrix} = \begin{pmatrix} 0 & \omega_1^2 & \omega_1^3 \\ -\omega_1^2 & 0 & \omega_2^3 \\ -\omega_1^3 & -\omega_2^3 & 0 \end{pmatrix} \begin{pmatrix} e_1 \\ e_2 \\ e_3 \end{pmatrix}. \tag{1.31}$$

The first two and the last equations of (1.31) are called the *Gauss equation* and the *Weingarten equation*, respectively.

Since

$$d(\omega^i e_j) = (d\omega^i)e_j - \omega^i \wedge (de_j)$$

holds as in (1.26) in §1.5, there arises that

$$0 = d^2 x = d(\omega^1 e_1 + \omega^2 e_2)$$
$$= (d\omega^1)e_1 + (d\omega^2)e_2 - \omega^1 \wedge (de_1) - \omega^2 \wedge (de_2),$$

where

$$\omega^1 \wedge de_1 = \sum_{i=1}^{3}(\omega^1 \wedge \omega_1^i)e_i, \quad \omega^2 \wedge de_2 = \sum_{i=1}^{3}(\omega^2 \wedge \omega_2^i)e_i.$$

Then we obtain

$$d\omega^1 + \omega_2^1 \wedge \omega^2 = 0$$
$$d\omega^2 + \omega_1^2 \wedge \omega^1 = 0$$
$$\omega_1^3 \wedge \omega^1 + \omega_2^3 \wedge \omega^2 = 0$$

by (1.30). It follows also that

$$d\omega_j^i - \sum_{k=1}^{3}\omega_j^k \wedge \omega_k^i = 0$$

from

$$0 = d^2 e_j = d\left(\sum_{i=1}^{3}\omega_j^i e_i\right) = \sum_{i=1}^{3}(d\omega_j^i \wedge e_i - \omega_j^i \wedge de_i)$$

and

$$\sum_{i=1}^{3}\omega_j^i \wedge de_i = \sum_{i,k=1}^{3}\omega_j^i \wedge \omega_i^k e_k = \sum_{i,k=1}^{3}\omega_j^k \wedge \omega_k^i e_i.$$

Let

$$\mathcal{M} : \vec{x} = \vec{x}(u,v) \in \mathbf{R}^3, \quad (u,v) \in D \subset \mathbf{R}^2 \tag{1.32}$$

be a parametric representation of $\mathcal{M}$, and $e_3 = n$ be the unit normal vector on $\mathcal{M}$. Then we take a frame on $\mathcal{M}$ denoted by $\{e_1, e_2, e_3\}$, and regard e_1, e_2 as tangential vector fields on $\mathcal{M}$: $e_1, e_2 \in \mathcal{X}(\mathcal{M})$.

Given 0-forms $\xi^i : \mathcal{M} \to \mathbf{R}$, $i = 1, 2$, let

$$X = \sum_{i=1}^{2} \xi^i e_i \in \mathcal{X}(\mathcal{M}).$$

We obtain

$$dX = \sum_{i=1}^{2}((d\xi^i)e_i + \xi^i de_i) = \sum_{i=1}^{2}\{(d\xi^i)e_i + \xi^i \sum_{k=1}^{3} \omega_i^k e_k\},$$

using the connection forms ω_j^i in (1.29), and hence

$$dX = \sum_{i=1}^{2}\{d\xi^i + \sum_{k=1}^{2} \xi^k \omega_k^i\}e_i + \sum_{i=1}^{2} \xi^i \omega_i^3 e_3.$$

We call the first term on the right-hand side, denoted by ∇X, the *covariant derivative* of X:

$$\nabla X = \sum_{i=1}^{2}\{d\xi^i + \sum_{k=1}^{2} \xi^k \omega_k^i\}e_i. \qquad (1.33)$$

Frame on higher dimensional manifold is also formulated (§7.3), to induce Liouville's formulae on the deformation of domains (§8.3, §8.7).

1.8   Connections

Regarding (1.33), we put, for $Y \in \mathcal{X}(\mathcal{M})$ that

$$\nabla_X Y = \sum_{i=1}^{2}\{\langle Y, d\xi^i \rangle + \sum_{k=1}^{2} \xi^k \langle Y, \omega_k^i \rangle\}e_i \in \mathcal{X}(\mathcal{M}),$$

where $\langle\ ,\ \rangle$ denotes the pairing between vector field and 1-form. We thus reach the bilinear form

$$\nabla : (X, Y) \in \mathcal{X}(\mathcal{M}) \times \mathcal{X}(\mathcal{M}) \mapsto \nabla_X Y \in \mathcal{X}(\mathcal{M}), \qquad (1.34)$$

satisfying the axiom of *connection*,

$$\nabla_{fX} Y = f \nabla_X Y$$
$$\nabla_X (fY) = (Xf)Y + f \nabla_X Y$$
$$\nabla_X Y - \nabla_Y X = [X, Y] \equiv XY - YX,$$

where $X, Y \in \mathcal{X}(\mathcal{M})$ and f is 0-form on $\mathcal{M}$.

Using the embedding

$$\mathcal{M} \hookrightarrow \mathbf{R}^3,$$

we obtain

$$X(Y \cdot Z) = \nabla_X Y \cdot Z + Y \cdot \nabla_X Z, \quad X, Y, Z \in \mathcal{X}(\mathcal{M}) \qquad (1.35)$$

where $\cdot$ stands for the inner product in $\mathbf{R}^3$. Define the 0-form

$$g(X, Y) = X \cdot Y$$

on $\mathcal{M}$ for each $X, Y \in \mathcal{X}(\mathcal{M})$. Then we get

$$X(g(Y, Z)) = g(\nabla_X Y, Z) + g(Y, \nabla_X Z)$$

by (1.35). Such connection ∇ is called the *Levi Chivita connection* with respect to g.

Under the parametrization (1.32) in §1.7, we put

$$g_{ij} = g\left(\frac{\partial}{\partial x_i}, \frac{\partial}{\partial x_j}\right), \quad i, j = 1, 2,$$

identifying $\mathcal{X}(\mathcal{M})$ with $T\mathcal{M}$ for

$$x_1 = u, \ x_2 = v.$$

Then we call

$$\sum_{ij} g_{ij} dx_i dx_j$$

a *Riemann metric* on $\mathcal{M}$, and $(\mathcal{M}, g)$ a *Riemann surface*. These g_{ij} are actually the first fundamental quantities,

$$g_{11} = E, \quad g_{12} = g_{21} = F, \quad g_{22} = G, \qquad (1.36)$$

and it holds that

$$ds^2 = \sum_{ij} g_{ij} dx_i dx_j.$$

We define the *Cristoffel symbol* Γ_{ij}^k by

$$\nabla_{\frac{\partial}{\partial x_j}} \frac{\partial}{\partial x_i} = \sum_k \Gamma_{ij}^k \frac{\partial}{\partial x_k}.$$

Then it holds that

$$\Gamma_{ij}^k = \frac{1}{2} \sum_{\ell} g^{k\ell} \left(\frac{\partial g_{i\ell}}{\partial x_j} + \frac{\partial g_{j\ell}}{\partial x_i} - \frac{\partial g_{ij}}{\partial x_\ell}\right) \qquad (1.37)$$

for

$$(g^{ij}) = (g_{ij})^{-1}.$$

Connection on higher dimensional manifolds is used in accordance with the Poisson structure of biological models (§5.1, §5.5).

1.9   Fundamental Equation of Surfaces

In §1.9–§1.10 we describe geometric background of the Boltzmann Poisson equation studied in Chapter 2.

From (1.36), we obtain

$$\Gamma^1_{11} = \frac{GE_u - 2FF_u + FE_v}{2(EG - F^2)}, \quad \Gamma^1_{12} = \Gamma^1_{21} = \frac{GE_v - 2FG_u}{2(EG - F^2)}$$

$$\Gamma^1_{22} = \frac{2GF_v - GG_u - FG_v}{2(EG - F^2)}, \quad \Gamma^2_{11} = \frac{2EF_u - FF_v - FE_u}{2(EG - F^2)}$$

$$\Gamma^2_{12} = \Gamma^2_{21} = \frac{2EG_u - FF_v}{2(EG - F^2)}, \quad \Gamma^2_{22} = \frac{EG_v - 2FF_u + FG_v}{2(EG - F^2)}, \quad (1.38)$$

recalling $x_1 = u$, $x_2 = v$. Then it follows that

$$\Gamma^1_{11}E + \Gamma^2_{11}F = x_{uu} \cdot x_u, \quad \Gamma^1_{11}F + \Gamma^2_{11}G = x_{uu} \cdot x_v,$$

$$\Gamma^1_{12}E + \Gamma^2_{12}F = x_{uv} \cdot x_u, \quad \Gamma^1_{12}F + \Gamma^2_{12}G = x_{uv} \cdot x_v$$

$$\Gamma^1_{11}E + \Gamma^2_{11}F = x_{vv} \cdot x_u, \quad \Gamma^1_{22}F + \Gamma^2_{22}G = x_{uv} \cdot x_u,$$

which ensures the *Gauss formula*

$$x_{uu} = \Gamma^1_{11}x_u + \Gamma^2_{11}x_v + Ln$$

$$x_{uv} = \Gamma^1_{12}x_u + \Gamma^2_{12}x_v + Mn$$

$$x_{vv} = \Gamma^1_{22}x_u + \Gamma^2_{22}x_v + Nn. \quad (1.39)$$

Using the *Weingarten matrix* in (1.19),

$$A = \begin{pmatrix} E & F \\ F & G \end{pmatrix}^{-1} \begin{pmatrix} L & M \\ M & N \end{pmatrix} \equiv \begin{pmatrix} A^1_1 & A^1_2 \\ A^2_1 & A^2_2 \end{pmatrix},$$

on the other hand, we obtain the *Weingarten formula*

$$n_u = -A^1_1 x_u - A^2_1 x_v, \quad n_v = -A^1_2 x_u - A^2_2 x_v. \quad (1.40)$$

by (1.7) in §1.3.

Equalities (1.40) and (1.39) are summarized as the *fundamental equation of surfaces*,

$$X_u = X\Omega, \quad X_v = X\Lambda, \quad (1.41)$$

where

$$X = [x_u, x_v, n]$$

and

$$\Omega = \begin{pmatrix} \Gamma^1_{11} & \Gamma^2_{11} & -A^1_1 \\ \Gamma^2_{11} & \Gamma^2_{12} & -A^2_1 \\ L & M & 0 \end{pmatrix}, \quad \Lambda = \begin{pmatrix} \Gamma^1_{21} & \Gamma^1_{22} & -A^1_2 \\ \Gamma^2_{21} & \Gamma^2_{22} & -A^2_1 \\ M & N & 0 \end{pmatrix}. \quad (1.42)$$

By (1.42) we obtain

$$X_{uv} = (X\Omega)_v = X_v\Omega + X\Omega_v = X(\Lambda\Omega + \Omega_v)$$

and similarly,

$$X_{vu} = X(\Omega\Lambda + \Lambda_u),$$

which results in

$$\Omega_v - \Lambda_u = \Omega\Lambda - \Lambda\Omega \qquad (1.43)$$

because X is non-singular.

Equality (1.43) is composed of the *Theorem Egregium*,

$$K = \frac{E(E_vG_v - 2F_uG_v + G_u)^2}{4(EG - F^2)^2}$$
$$+ \frac{F(E_uG_v - E_vG_u - 2E_vF_v - 2F_uG_u + 4F_uF_v)}{4(EG - F^2)}$$
$$+ \frac{G(E_uG_u - 2E_uF_v + E_v^2)^2}{4(EG - F^2)} - \frac{E_{vv} - 2F_{uv} + G_{uu}}{2(EG - F^2)}$$

by (1.16), and the *Codazzi Mainardi equation*

$$L_v - M_u = L\Gamma_{12}^1 + M(\Gamma_{12}^2 - \Gamma_{11}^1) - N\Gamma_{11}^2$$
$$M_v - N_u = L\Gamma_{22}^1 + M(\Gamma_{22}^2 - \Gamma_{12}^1) - N\Gamma_{12}^2. \qquad (1.44)$$

1.10 Conformal Geometry

Parametrization (1.32) is called *conformal* if

$$|x_u|^2 = |x_v|^2 \equiv E = E(u, v), \quad x_u \cdot x_v = 0, \qquad (1.45)$$

everywhere in D. In this case it holds that

$$A = \frac{1}{E}\begin{pmatrix} L & M \\ M & N \end{pmatrix}$$

and hence

$$2H = \text{tr } A = \frac{L + N}{E}, \quad K = \det A = \frac{LN - M^2}{E^2}$$

by (1.18)–(1.19). There arises also that

$$E = G = e^{2\sigma}, \ F = 0$$

with $\sigma = \sigma(u, v)$. Then equalities (1.38) imply

$$\Gamma_{11}^1 = \frac{2\sigma_u e^{4\sigma}}{2e^{4\sigma}} = \sigma_u, \qquad \Gamma_{12}^1 = \Gamma_{21}^1 = \frac{2\sigma_v e^{4\sigma}}{2e^{4\sigma}} = \sigma_v,$$
$$\Gamma_{22}^1 = \frac{-2\sigma_u e^{4\sigma}}{2e^{4\sigma}} = -\sigma_u, \qquad \Gamma_{11}^2 = \frac{-2\sigma_v e^{4\sigma}}{2e^{4\sigma}} = -\sigma_v$$
$$\Gamma_{12}^2 = \Gamma_{21}^2 = \frac{2\sigma_u e^{4\sigma}}{2e^{4\sigma}} = \sigma_u, \ \Gamma_{22}^2 = \frac{2\sigma_v e^{4\sigma}}{2e^{4\sigma}} = \sigma_v.$$

It holds also that

$$A = e^{-2\sigma} \begin{pmatrix} L & M \\ M & N \end{pmatrix} = \begin{pmatrix} A_1^1 & A_2^1 \\ A_1^2 & A_2^2 \end{pmatrix}$$

and hence

$$\Omega = \begin{pmatrix} \sigma_u & \sigma_v & -e^{-2\sigma}L \\ -\sigma_v & \sigma_u & -e^{-2\sigma}L \\ L & M & 0 \end{pmatrix}, \quad \Lambda = \begin{pmatrix} \sigma_v & -\sigma_u & -e^{-2\sigma}M \\ \sigma_u & \sigma_v & -e^{-2\sigma}N \\ M & N & 0 \end{pmatrix}$$

by (1.42). Equality (1.41) now implies

$$\sigma_{uu} + \sigma_{vv} + e^{-2\sigma}(LN - M^2) = 0$$
$$L_v - M_u = \sigma_v(L + N)$$
$$N_u - M_v = \sigma_u(L + N),$$

and in particular,

$$K = \frac{LN - M^2}{E^2} = -e^{-2\sigma}(\sigma_{uu} + \sigma_{vv}) = -\frac{\Delta \log E}{2E} \tag{1.46}$$

by $E = e^{2\sigma}$.

From the Codazzi Mainardi equation (1.44), on the other hand, there arises that

$$E_v H = 2e^{2\sigma}\sigma_v H = 2E\sigma_v H = (L + N)\sigma_v = L_v - M_u$$
$$E_u H = 2e^{2\sigma}\sigma_u H = 2E\sigma_u H = (L + N)\sigma_u = N_u - M_v.$$

Then we obtain

$$\begin{aligned} M_v + \frac{1}{2}(L - N)_u &= N_u - E_u H + \frac{1}{2}(L - N)_u \\ &= -E_u H + \frac{1}{2}(L + N)_u \\ &= -E_u H + (EH)_u = EH_u \end{aligned} \tag{1.47}$$

and similarly,

$$\begin{aligned} M_u - \frac{1}{2}(L - N)_v &= L_v - E_v H - \frac{1}{2}(L - N)_v \\ &= -E_v H + \frac{1}{2}(L + N)_v \\ &= -E_v H + (EH)_v = EH_v. \end{aligned} \tag{1.48}$$

Now we use the complex variable

$$z = u + \sqrt{-1}v.$$

Since

$$\frac{\partial}{\partial z} = \frac{1}{2}\left(\frac{\partial}{\partial u} - \sqrt{-1}\frac{\partial}{\partial v}\right), \quad \frac{\partial}{\partial \bar{z}} = \frac{1}{2}\left(\frac{\partial}{\partial u} + \sqrt{-1}\frac{\partial}{\partial v}\right)$$

it holds that

$$
\begin{aligned}
EH_z &= \frac{1}{2}E(H_u - \sqrt{-1}H_v)\\
&= \frac{1}{2}\{M_v + \frac{1}{2}(L - N)_u - \sqrt{-1}(M_u - \frac{1}{2}(L - N)_v)\}\\
&= \frac{1}{4}\left\{(L - N - 2\sqrt{-1}M)_u + \sqrt{-1}(L - N - 2\sqrt{-1}M)_v\right\} = \frac{1}{2}\phi_{\bar{z}},
\end{aligned}
$$

where

$$\phi = L - N - 2\sqrt{-1}M.$$

A closed surface $\mathcal{M}$ of genous 1 with constant mean curvature H is realized by a doubly periodic conformal mapping

$$x = x(u, v) : \mathbf{R}^2 \to \mathbf{R}^3.$$

Then the above

$$\phi = L - N - 2\sqrt{-1}M$$

is holomorphic by

$$\phi_{\bar{z}} = 2EH_z = 0,$$

and therefore, is a constant by Liouville's theorem on bounded entire holomorphic functions.

Since

$$
\begin{aligned}
|\phi|^2 &= (L - N)^2 + 4M^2 = (L + N)^2 + 4(M^2 - LN)\\
&= 4E^2(H^2 - K)
\end{aligned}
$$

there arises that

$$\frac{\lambda^2}{4E^2} = H^2 - K = H^2 + \frac{\Delta \log E}{2E}, \quad \lambda = |\phi|,$$

and hence

$$\Delta \log E + 2EH^2 - \frac{\lambda^2}{2E} = 0. \tag{1.49}$$

Equations (1.46) and (1.49), concerned on the Gauss and mean curvatures, take the form of elliptic PDE. These equations are used to classify surfaces with prescribed curvatures in accordance with the complex function theory in §2.3.

Chapter 2

Static Recursive Hierarchy

This chapter is devoted to the study of equilibrium statistical mechanics of point vortices (§2.1–§2.2). It is associated with conformal geometry in Chapter 1, particularly, the case that K is constant in (1.46) in §1.10. We use, to this end, the Liouville integral (§2.3), which is a parametric representation of such surface by complex variables. This formula induces quantized blowup mechanism of the Boltzmann Poisson (BP) equation (§2.4–§2.5) because of its geometric feature of the nonlinearity associated with flat plane and round sphere. There arises also a physical insight of *recursive hierarchy* in the context of Onsager's theory on many point vortices. Thus, the location of blowup points of the family of solutions to the BP equation is formulated by the Hamiltonian of the original system of point vortices. This control of the Hamiltonian on the critical spots of the solution to nonlinear partial differential equations spreads widely, involving analytic, geometric, and physical phenomena, where the method of scaling is systematically used for their analysis (§2.6).

2.1 Point Vortices

Incompressible, non-viscous fluid is called the *ideal fluid*. Velocity

$$v = \begin{pmatrix} v_1(x,t) \\ v_2(x,t) \\ v_3(x,t) \end{pmatrix} \in \mathbf{R}^3, \ x = (x_1, x_2, x_3) \in \mathbf{R}^3,$$

and *pressure*

$$p = p(x,t) \in \mathbf{R}$$

in three space dimension, are subject to the *Euler equation of motion*

$$v_t + (v \cdot \nabla)v = -\nabla p, \ \nabla \cdot v = 0 \quad \text{in } \mathbf{R}^3 \times (0, T), \tag{2.1}$$

where physical parameters are put to be one. Under this flow, the particle $x_0 \in \mathbf{R}^3$ at $t = 0$ moves to $x = x(t) \in \mathbf{R}^3$ at $t = t$, satisfying

$$\frac{dx}{dt} = v(x, t), \quad x(0) = x_0. \tag{2.2}$$

Let $f = f(x, t) \in \mathbf{R}$ be a state quantity distributing in the space-time variables (x, t). Then the trajectory of the particle detected by this $f = f(x, t) \in \mathbf{R}$,

$$h(t) = f(x(t), t),$$

satisfies

$$\frac{dh}{dt} = f_t + v \cdot \nabla f \tag{2.3}$$

by (2.2). The right-hand side of (2.3), denoted by

$$\frac{Df}{Dt} = f_t + v \cdot \nabla f,$$

is called the *material derivative*, and the *acceleration* of this particle is defined by

$$\frac{dv}{dt}(x(t), t) = v_t + (v \cdot \nabla)v = \frac{Dv}{dt}. \tag{2.4}$$

Since the kinetic force acting on this particle is equal to the minus of the pressure gradient, the first equation of (2.1) is due to *Newton's equation of motion*,

$$m\ddot{x} = f,$$

where m, x, and f stand for the mass, position, and outer force, respectively.

Let the solution to (2.2) and the volume of the domain $\omega \subset \mathbf{R}^3$ be $x(t) = T_t x_0$ and $|\omega|$, respectively. Then, *Liouville's first volume derivative*, Theorem 8.2 in §8.3, guarantees

$$\frac{d}{dt}|T_t(\omega)|_{t=0} = \int_\omega (\nabla \cdot v)(x, 0) \, dx. \tag{2.5}$$

Since the left-side of (2.5) stands for the rate of volume dilation of the fluid in ω at $t = 0$, the second equation of (2.1) describes that the fluid is *incompressible*.

In the case of *rigid body*, rotation of the velocity v,

$$\nabla \times v = \begin{pmatrix} \frac{\partial v_3}{\partial x_2} - \frac{\partial v_2}{\partial x_3} \\ \frac{\partial v_1}{\partial x_3} - \frac{\partial v_3}{\partial x_1} \\ \frac{\partial v_2}{\partial x_1} - \frac{\partial v_1}{\partial x_2} \end{pmatrix}$$

is equal to the twice of its *angular velocity* (Chapter 2 of [Suzuki (2022a)]). We call this

$$\omega = \nabla \times v$$

the *vorticity* of the fluid. Equation (2.1) implies

$$\omega_t + (v \cdot \nabla)\omega = (\omega \cdot \nabla)v, \ \nabla \cdot v = 0 \quad \text{in } \mathbf{R}^3 \times (0, T). \tag{2.6}$$

For the *two-dimensional flow* described by

$$v = \begin{pmatrix} v_1(x_1, x_2, t) \\ v_2(x_1, x_2, t) \\ 0 \end{pmatrix} \tag{2.7}$$

it holds that

$$\omega = \nabla \times v = \begin{pmatrix} 0 \\ 0 \\ \frac{\partial v_2}{\partial x_1} - \frac{\partial v_1}{\partial x_2} \end{pmatrix},$$

which results in

$$(\omega \cdot \nabla)v = 0.$$

Then we use, without confusing, the symbol ω to indicate the two-dimensional scalar field

$$\omega = \frac{\partial v_2}{\partial x_1} - \frac{\partial v_1}{\partial x_2},$$

which satisfies

$$\omega_t + (v \cdot \nabla)\omega = 0, \ \nabla \cdot v = 0,$$

and therefore,

$$\omega_t + \nabla \cdot (v\omega) = 0, \ \nabla \cdot v = 0 \quad \text{in } \mathbf{R}^2 \times (0, T). \tag{2.8}$$

The second equation of (2.8), on the other hand, implies

$$\nabla \cdot v = \frac{\partial v_1}{\partial x_1} + \frac{\partial v_2}{\partial x_2} = 0. \tag{2.9}$$

In the whole space $\mathbf{R}^2$, this (2.9) guarantees the existence of the *stream function*

$$\psi = \psi(x_1, x_2, t) \in \mathbf{R}$$

such that

$$v_1 = \frac{\partial \psi}{\partial x_2}, \ v_2 = -\frac{\partial \psi}{\partial x_1},$$

and hence

$$v = \nabla^{\perp}\psi, \quad \nabla^{\perp} = \begin{pmatrix} \frac{\partial}{\partial x_2} \\ -\frac{\partial}{\partial x_1} \end{pmatrix}. \tag{2.10}$$

Then we obtain

$$\omega = \frac{\partial v_2}{\partial x_1} - \frac{\partial v_1}{\partial x_2} = -\Delta\psi$$

by (2.10).

System (2.8) is now reduced to the hyperbolic elliptic system of (ω, ψ),

$$\omega_t + \nabla \cdot (\omega\nabla^{\perp}\psi) = 0, \quad -\Delta\psi = \omega \quad \text{in } \mathbf{R}^2 \times (0, T), \tag{2.11}$$

called the *vorticity equation*. This (2.11) is provided with a similarity between the parabolic elliptic system called the *Smoluchowski Poisson equation* studied in Chapter 3.

If $\Omega \subset \mathbf{R}^2$ is a bounded domain with smooth boundary $\partial\Omega$, we impose the null normal velocity on the boundary:

$$v_t + (v \cdot \nabla)v = -\nabla p, \ \nabla \cdot v = 0 \text{ in } \Omega \times (0, T), \quad \nu \cdot v|_{\partial\Omega} = 0.$$

If Ω is simply-connected, there is a stream function $\psi = \psi(x, t)$ such that

$$v = \nabla^{\perp}\psi$$

by $\nabla \cdot v = 0$ in Ω, and then the boundary condition $\nu \cdot v = 0$ is reduced to

$$\psi = \text{constant} \quad \text{on } \partial\Omega. \tag{2.12}$$

We can assume that this constant is zero without loss of generality, and then there arises the vorticity equation

$$\omega_t + \nabla \cdot (\omega\nabla^{\perp}\psi) = 0, \quad -\Delta\psi = \omega \quad \text{in } \Omega \times (0, T), \quad \psi|_{\partial\Omega} = 0. \tag{2.13}$$

Let $G = G(x, x')$ be the *Green function* to the *Poisson equation*

$$-\Delta\psi = \omega \text{ in } \Omega, \quad \psi|_{\partial\Omega} = 0, \tag{2.14}$$

which satisfies

$$-\Delta_x G(x, x') = \delta_{x'}(dx), \ G(x, x')|_{x \in \partial\Omega} = 0, \quad x' \in \Omega,$$

where $\delta_{x'}(dx)$ is the delta function supported at $x = x'$. We then obtain

$$\omega_t + \nabla \cdot (\omega\nabla^{\perp}\psi) = 0,$$

$$\psi(\cdot, t) = \int_{\Omega} G(\cdot, x')\omega(x', t)dx' \text{ in } \Omega \times (0, T), \tag{2.15}$$

and

$$\nu \cdot \omega \nabla^{\perp} \psi \big|_{\partial \Omega} = 0.$$

System (2.15) takes a weak form based on the symmetry of the Green function,

$$G(x, x') = G(x', x). \tag{2.16}$$

Thus, each $\varphi \in C^1(\overline{\Omega})$ admits

$$\begin{aligned} \frac{d}{dt} \int_{\Omega} \varphi \omega \, dx &= \int_{\Omega} \omega \nabla^{\perp} \psi \cdot \nabla \varphi \, dx \\ &= \iint_{\Omega \times \Omega} \omega(x, t) \nabla_x^{\perp} G(x, x') \omega(x', t) \cdot \nabla \varphi(x) \, dx' dx \\ &= \frac{1}{2} \iint_{\Omega \times \Omega} \rho_{\varphi}^{\perp} \omega \otimes \omega \, dx dx', \end{aligned} \tag{2.17}$$

provided that $\omega(\cdot, t) \in L^{\infty}(\Omega)$, where

$$\omega \otimes \omega = (\omega \otimes \omega)(x, x', t) = \omega(x, t)\omega(x', t)$$

and

$$\rho_{\varphi}^{\perp} = \rho_{\varphi}^{\perp}(x, x') = \nabla_x^{\perp} G(x, x') \cdot \nabla \varphi(x) + \nabla_{x'}^{\perp} G(x, x') \cdot \nabla \varphi(x').$$

Here we use the interior regularity of the Green function,

$$G(x, x') = \Gamma(x - x') + K(x, x'), \quad K \in C^2(\overline{\Omega} \times \Omega) \cap C^2(\Omega \times \overline{\Omega}) \tag{2.18}$$

for

$$\Gamma(x) = \frac{1}{2\pi} \log \frac{1}{|x|},$$

which results in

$$\begin{aligned} \rho_{\varphi}(x, x')^{\perp} &= \nabla^{\perp} \Gamma(x - x') \cdot (\nabla \varphi(x) - \nabla \varphi(x')) \\ &+ (\nabla_x^{\perp} K(x, x') \cdot \nabla \varphi(x) + \nabla_{x'}^{\perp} K(x, x') \cdot \nabla \varphi(x')) \end{aligned}$$

if supp $\varphi \subset \Omega$. In fact, this $\Gamma(x)$ is the fundamental solution to $-\Delta$ in two space dimensions, satisfying

$$-\Delta \Gamma = \delta_0(dx),$$

which implies

$$-\Delta w = 0 \text{ in } \Omega, \quad w = -\Gamma(\cdot - x') \text{ on } \partial \Omega$$

for $x' \in \Omega$ and

$$w(x) = G(x, x') - \Gamma(x - x').$$

Since this $w = w(x)$ is smooth on $\overline{\Omega}$, there arises that (2.18) (Chapter 5 of [Suzuki (2005)]).

We have

$$\nabla^{\perp}\Gamma(x) = -\frac{1}{2\pi}\frac{x^{\perp}}{|x|^2}, \qquad x^{\perp} = \begin{pmatrix} x_2 \\ -x_1 \end{pmatrix}$$

for

$$x = \begin{pmatrix} x_1 \\ x_2 \end{pmatrix},$$

and therefore,

$$\rho_{\varphi}^{\perp} \in L^{\infty}(\Omega \times \Omega).$$

This cancellation of the singularity of $\rho_{\varphi}^{\perp}$ is a result of (2.16), the symmetry of the Green function to the Poisson equation (2.14), which comes from the *action reaction law*. Deriving this cancellation by (2.14) is called the *method of symmetrization* (Chapter 5 of [Suzuki (2005)]).

The weak formulation (2.17) admits ω satisfying

$$\omega \in C^1([0,T], L^1(\Omega)). \tag{2.19}$$

Now we confirm that $\omega(\cdot, t) \in L^1(\Omega)$ implies

$$\lim_{\varepsilon \downarrow 0} \iint_{|x-x'|<\varepsilon} \nabla^{\perp}\Gamma(x - x') \cdot (\nabla\varphi(x) - \nabla\varphi(x'))\omega \otimes \omega \, dxdx' = 0 \tag{2.20}$$

for

$$\varphi \in C_0^2(\Omega), \quad \varphi(x) = A(r), \quad r = |x - x_0|, \quad x_0 \in \Omega. \tag{2.21}$$

In fact, for such $\varphi(x)$ we have

$$\nabla^{\perp}\Gamma(x - x') \cdot (\nabla\varphi(x) - \nabla\varphi(x')) = -\frac{1}{2\pi}\nabla^2\varphi(\theta x + (1-\theta)x')[e, e^{\perp}]$$

for

$$0 < \theta < 1, \quad e = \frac{x - x'}{|x - x|},$$

and

$$\nabla^2 A(0) = A''(0)E,$$

where E denotes the unit matrix, and then the result follows from

$$e_* \cdot e_*^{\perp} = 0, \quad \forall e_* \in \mathbf{R}^2, \ |e_*| = 1.$$

We thus end up with the inequality derived from (2.13),

$$\frac{d}{dt}\int_\Omega \omega\varphi \; dx = \lim_{\varepsilon\downarrow 0}\frac{1}{2}\iint_{|x-x'|>\varepsilon}\rho_\varphi^\perp\omega\otimes\omega \; dxdx'$$

$$+\lim_{\varepsilon\downarrow 0}\frac{1}{2}\iint_{|x-x'|<\varepsilon}(\nabla_x^\perp K(x,x')\cdot\nabla\varphi(x)+\nabla_{x'}^\perp K(x,x')\cdot\nabla\varphi(x'))$$

$$\cdot\omega\otimes\omega \; dxdx' \tag{2.22}$$

valid to ω and φ satisfying (2.19) and (2.21), respectively.

The right-hand side of (2.22) converges even to the sum of *point vortices*,

$$\omega(dx,t) = \sum_{i=1}^{\ell}\alpha_i\delta_{x_i(t)}(dx), \quad \alpha_i \in \mathbf{R}, \; x_i(t)\in\Omega, \; 1\le i\le\ell. \tag{2.23}$$

In fact, fix $1\le i\le\ell$, and let $I=(a,b)$ be a time interval and taking $x_0\in\Omega$ and $0<R\ll 1$ such that

$$B_R(x_0)\cap\mathcal{S}_t = \{x_i(t)\}, \; t\in I, \tag{2.24}$$

for $\mathcal{S}_t = \{x_j(t)\mid 1\le j\le\ell\}$. Then, if φ in (2.21) satisfies

$$\operatorname{supp}\varphi\subset B_R(x_0),$$

we obtain

$$\alpha_i\frac{d\varphi}{dt}(x_i) = \alpha_i\sum_{j\neq i}\alpha_j\nabla_x^\perp G(x_i,x_j)\cdot\nabla\varphi(x_i)+\alpha_i^2\nabla_x^\perp K(x_i,x_i)$$

$$= \sum_{j\neq i}\alpha_i\alpha_j\nabla_x^\perp G(x_i,x_j)\cdot\nabla\varphi(x_i)+\frac{\alpha_i^2}{2}\nabla^\perp R(x_i)\cdot\nabla\varphi(x_i), \; t\in I$$

for $x_i = x_i(t)$ by (2.22), where

$$R(x) = \left[G(x,x')+\frac{1}{2\pi}\log|x-x'|\right]_{x'=x} = K(x,x)$$

stands for the *Robin function*. It then follows the *Kirchhoff equation*,

$$\alpha_i\frac{dx_i}{dt} = \frac{1}{2}\alpha_i^2\nabla^\perp R(x_i)+\sum_{j\neq i}\alpha_i\alpha_j\nabla_x^\perp G(x_i,x_j), \quad 1\le i\le\ell \tag{2.25}$$

by

$$\frac{d\varphi}{dt}(x_i) = \frac{dx_i}{dt}\cdot\nabla\varphi(x_i),$$

because x_0 and $I=(a,b)$ in (2.24) are arbitrary.

This (2.25) is the *Hamilton system*

$$\alpha_i\frac{dx_i}{dt} = \nabla^\perp H_\ell(x_1,\cdots,x_\ell), \quad 1\le i\le\ell,$$

or

$$\alpha_i \frac{dx_{i1}}{dt} = \frac{\partial H_\ell}{\partial x_{i2}}, \quad \alpha_i \frac{dx_{i2}}{dt} = -\frac{\partial H_\ell}{\partial x_{i1}}, \quad 1 \le i \le \ell, \tag{2.26}$$

where $x_i = (x_{i1}, x_{i2})$ and

$$H_\ell(x_1, \ldots, x_\ell) = \frac{1}{2} \sum_{i=1}^\ell \alpha_i^2 R(x_i) + \sum_{1 \le i < j \le \ell} \alpha_i \alpha_j G(x_i, x_j) \tag{2.27}$$

stands for the *Hamiltonian*.

2.2   Mean Field Limit

Micro-canonical statistical mechanics is concerned on the motion of mean field of classical particles subject to the Hamilton mechanics formulated in §5.2.

First, the Hamilton system

$$\frac{dq_i}{dt} = \frac{\partial H}{\partial p_i}, \quad \frac{dp_i}{dt} = -\frac{\partial H}{\partial q_i}, \quad 1 \le i \le N \tag{2.28}$$

is concerned on the *position* $(q_1, \ldots, q_N) \in \mathbf{R}^{3N}$ and the *momentum* $(p_1, \ldots, p_N) \in \mathbf{R}^{3N}$ in general coordinate, where

$$H = H(q_1, \ldots, q_N, p_1, \ldots, p_N)$$

is the associated Hamiltonian. This system (2.28) is provided with the conservation law

$$\frac{d}{dt} H(q(t), p(t)) = 0$$

and therefore, the set of orbits $\{\mathcal{O}\}$ formed by (2.28) is classified according to their energy $H = E$.

Each orbit $\mathcal{O} = \{(p(t), q(t))\}$ is a curve in the *phase space*

$$x = (p_1, \cdots, p_N, q_1, \cdots, q_N) \in \Gamma = \mathbf{R}^{6N},$$

while the hyper-surface

$$\Gamma_E = \{x \in \Gamma \mid H(x) = E\} \subset \Gamma$$

is provided with the *area element* $d\Sigma(E)$. Then the *micro-canonical measue* $\mu^{E,N}(dx)$ on Γ_E is defined by the *co-area formula*

$$dx = dE \cdot \frac{d\Sigma(E)}{|\nabla H|}$$

where

$$dx = dq_1 \cdots dq_N dp_1 \ldots dp_N$$

stands for the Lebesgue measure on $\mathbf{R}^{6N}$. We thus put

$$\mu^{E,N}(dx) = \frac{1}{W(E)} \cdot \frac{d\Sigma(E)}{|\nabla H|}, \quad W(E) = \int_{H=E} \frac{d\Sigma(E)}{|\nabla H|} \quad (2.29)$$

for each $E \in \mathbf{R}$. This $\mu^{E,N}$ detects the magnitude of the set of orbits with energy E, and $W(E)$ is called the *weight factor*.

If (2.28) describes the motion of gas molecules, *statistical ensembles* are associated with their thermo-dynamical features. In this case of the Hamilton system, the particles are put on the space without any interactions outside, which forms an *isolated system*.

The system of materially closed and thermo-dynamically open, on the other hand, is called a *closed system*. There, the second law of thermo-dynamics takes a different form from the entropy increasing. If temperature T is constant in such closed system, thus, the *free energy of Helmholtz* decreases. The *canonical ensemble* is defined as an equivalent class of the same temperature, and the *canonical measue* is defined for each T by the principle of a priori equal probabilities.

Heat bath is a standard method to induce the *canonical measure*, where Boltzmann's and thermo-dynamical relations

$$S = k \log W, \quad \frac{\partial S}{\partial E} = \frac{1}{T} \quad (2.30)$$

are applied with the *Boltzmann constant k*. It then holds that

$$\mu^{\beta,N}(dx) = \frac{e^{-\beta H} dx}{Z(\beta, N)}, \quad Z(\beta, N) = \int_\Gamma e^{-\beta H} dx, \quad (2.31)$$

where $\Gamma = \mathbf{R}^{6N}$ stands for the phase space, $E = H$ is the energy, and $\beta = (kT)^{-1}$ is called the *inverse temperature* (Chapter 9 of [Suzuki (2015b)]).

Order structures are observed during a long term in two-dimensional turbulence emerged from many point vortices. L. Onsager proposed a statistical mechanics based on the Hamilton system (2.26) ([Onsager (1949)]). Here, it is assumed that the vorticities in (2.23) are stable and their intensities α_i, $1 \le i \le \ell$, take the same value denoted by $\alpha > 0$:

$$\omega_\ell(dx) = \sum_{i=1}^{\ell} \alpha \delta_{x_i}(dx),$$

which results in

$$H(x_1, \cdots, x_\ell) = \frac{\alpha^2}{2} \sum_{i=1}^{\ell} R(x_i) + \alpha^2 \sum_{1 \le i < j \le \ell} G(x_i, x_j).$$

Hence we have $\Gamma = \Omega^\ell$ and the position of point vortices is indicated as $X_\ell = (x_1, \cdots, x_\ell) \in \Omega^\ell$.

Let

$$\mu_\ell = \mu_\ell(dx_1 \cdots dx_\ell)$$

be the distribution function of these point vortices. The principle of *a priori equal probabilities* ensures that the measure

$$\rho_{1,i}^\ell(dx_i) = \int_{\Omega^{\ell-1}} \mu_\ell(dx_1 \ldots dx_{i-1} dx_{i+1} \cdots dx_\ell)$$

is independent of $1 \le i \le \ell$. Assuming

$$\rho_{1,i}^\ell(dx_i) = \rho_{1,i}^\ell(x_i) dx_i,$$

we call

$$\rho_1^\ell(x) = \rho_{1,i}^\ell(x_i), \ x = x_i, \quad 1 \le i \le \ell$$

one-point reduced particle density function. Similarly, we call

$$\rho_k^\ell(x_1, \ldots, x_k) dx_1 \cdots dx_k = \int_{\Omega^{\ell-k}} \mu_\ell(dx_{k+1} \cdots dx_n)$$

the k-point reduced particle density function.

Phase space mean of $\omega_\ell(dx)$ is defined by

$$\langle \omega_\ell(dx) \rangle = \sum_{i=1}^{\ell} \int_{\Omega^\ell} \alpha \delta(x_i - x) \mu_\ell(dx_1 \ldots dx_\ell) = \ell \alpha \rho_1^\ell(x),$$

Letting $\tilde{E}$ and $\tilde{\beta}$ be the energy and temperature of this system of ℓ-point vortices, respectively, we take the limit $\ell \to \infty$ of this phase space mean under the scaling of

$$\alpha \ell = 1, \ \alpha^2 \ell^2 \tilde{E} = E, \ \alpha^2 \ell \tilde{\beta} = \beta.$$

Then it follows formally that the limit

$$\lim_{\ell \to \infty} \langle \omega_\ell(dx) \rangle = \rho(x) = \lim_{\ell \to \infty} \rho_1^\ell(x), \tag{2.32}$$

satisfies

$$\rho = \frac{e^{-\beta\psi}}{\int_\Omega e^{-\beta\psi} dx}, \ \psi = \int_\Omega G(\cdot, x') \rho(x') dx' \quad \text{in } \Omega \tag{2.33}$$

([Montgomery–Joyce (1974)]). These ψ and ρ are due to the stream function and the point vortex density, respectively, constituting a duality.

Using $v = \psi$ and $\lambda = -\beta$, actually, we can write (2.33) as

$$-\Delta v = \frac{\lambda e^v}{\int_\Omega e^v dx} \text{ in } \Omega, \quad v = 0 \text{ on } \partial\Omega. \tag{2.34}$$

This (2.34) is called the *Boltzmann Poisson equation*, because it splits in the Poisson part

$$-\Delta v = u \text{ in } \Omega, \quad v = 0 \text{ on } \partial\Omega$$

and the Boltzmann part due to (2.31),

$$u = \frac{\lambda e^v}{\int_\Omega e^v \, dx}.$$

The convergence (2.32) is justified if $\{\rho_k^\ell\}$ is uniformly bounded in k, and also if the solution to (2.34) is unique. These two requirements are valid, provided that $\beta = -\lambda > -8\pi$, which implies, simultaneously, *propagation of chaos* formulated by

$$\rho_k^\ell \rightharpoonup \rho^{\otimes k}, \quad \rho^{\otimes k}(x_1, \ldots, x_k) = \prod_{i=1}^{k} \rho(x_i)$$

in the sense of measures ([Caglioti–Lions–Marchioro–Pulvirenti (1992); Suzuki (1992)]).

From (2.30) it follows that

$$\beta = \frac{\partial}{\partial E} \log W(E). \tag{2.35}$$

The co-area formula implies

$$\Theta(E) \equiv \int_{H<E} dx_1 \ldots dx_n = \int_{-\infty}^{E} dE' \int_{H=E'} \frac{d\Sigma(E')}{|\nabla E'|}$$
$$= \int_{-\infty}^{E} W(E') \, dE',$$

and therefore, equality (2.35) means

$$\beta = \frac{\Theta''(E)}{\Theta'(E)}. \tag{2.36}$$

Since the mapping

$$E \mapsto \Theta(E)$$

is bounded and non-decreasing, it has a point of inflection. In particular, negative inverse temperature $\beta < 0$ can happen in the range of $E \gg 1$.

In this way, L. Onsager reached the conclusion that the ordered structure arises in the range of negative inverse temprature ([Onsager (1949)]). The *quantized blowup mechanism* of the Boltzmann Poisson equation, on the contrary, shows the formation of singular limits of (2.58) at

$$\lambda = 8\pi\ell, \quad \ell \in \mathbf{N},$$

of which singular points coincide with those of the Hamiltonian in (2.27) ([Nagasaki–Suzuki (1990a)]). Thus there emerges a *recursive hierarchy* in the sense that the point vortex Hamiltonian controls the mean field again with its quantized concentration (Theorem 2.1 in §2.4).

2.3   Liouville Integral

Insight towards the quantized blowup mechanism of the Boltzmann Poisson equation emerges from the complex geometry in §1.10 (Chapter 3 of [Suzuki (2020)]). In fact, equation (2.34) implies

$$-\Delta u = e^u \quad \text{in } \Omega \qquad (2.37)$$

for

$$u = v - \log \int_\Omega e^v dx,$$

where $\Omega \subset \mathbf{R}^2$ is a bounded domain. This (2.37), on the other hand, is also derived from (1.46) if K is a positive constant:

$$u = \log E + \log(2K).$$

There is an integral of equation (2.37) found by J. Liouville. Here we take the complex variable

$$z = x_1 + \sqrt{-1}x_2, \quad \bar{z} = x_1 - \sqrt{-1}x_2$$

for $x = (x_1, x_2) \in \Omega$, to introduce the function

$$s = u_{zz} - \frac{1}{2}u_z^2. \qquad (2.38)$$

From (2.37) it follows that

$$u_{z\bar{z}} = -\frac{1}{4}e^u,$$

and hence

$$s_{\bar{z}} = u_{zz\bar{z}} - u_z u_{z\bar{z}} = -\frac{1}{4}e^u u_z + \frac{1}{4}e^u u_z = 0,$$

which means that $s = s(z)$ in (2.38) is a *holomorphic function* of $z \in \Omega \subset \mathbf{C}$.

Regarding (2.38) as a *Riccati equation*, we see that

$$\phi = e^{-u/2}$$

satisfies

$$\phi_{zz} + \frac{1}{2}s\phi = 0. \tag{2.39}$$

Taking a point $x^* = (x_1^*, x_2^*) \in \Omega$, we define the fundamental system of solutions, denoted by $\{\phi_1, \phi_2\}$, to the linear ordinary equation (2.39) in z by

$$\left(\phi_1, \frac{\partial\phi_1}{\partial z}\right)\Big|_{z=z^*} = (1, 0), \quad \left(\phi_2, \frac{\partial\phi_2}{\partial z}\right)\Big|_{z=z^*} = (0, 1) \tag{2.40}$$

for $z^* = x_1^* + \imath x_2^*$.

These $\phi_1 = \phi_1(z)$ and $\phi_2 = \phi_2(z)$ are *analytic functions* of $z \in \Omega$, possibly multi-valued when Ω has a positive genus, while the relation

$$\phi \equiv e^{-u/2} = \overline{f}_1(\overline{z})\phi_1(z) + \overline{f}_2(\overline{z})\phi_2(z)$$

holds for some functions $\overline{f}_1$, $\overline{f}_2$ of $\overline{z}$. Then there arises that

$$f_1(z) = C_1\phi_1(z), \quad f_2(z) = C_2\phi_2(z) \tag{2.41}$$

for

$$C_1 = e^{-u/2}\Big|_{x=x^*}, \quad C_2 = \frac{\lambda}{8}e^{u/2}\Big|_{x=x^*} \tag{2.42}$$

if $x^* = (x_1^*, x_2^*) \in \Omega$ is a critical point of $u = u(x)$,

$$\nabla u(x^*) = 0, \tag{2.43}$$

where

$$f_1(z) = \overline{\overline{f}_1(\overline{z})}, \quad f_2(z) = \overline{\overline{f}_2(\overline{z})}.$$

By

$$W(\phi_1, \phi_2) \equiv \phi_1\phi_{2z} - \phi_{1z}\phi_2 = 1$$

it holds that

$$\overline{f}_1(\overline{z}) = W(\phi, \phi_2) = \phi\phi_{2z} - \phi_z\phi_2$$
$$\overline{f}_2(\overline{z}) = W(\phi_1, \phi) = \phi_1\phi_z - \phi_{1z}\phi.$$

These $\overline{f}_1(\overline{z})$, $\overline{f}_2(\overline{z})$ are independent of z, and therefore, we obtain

$$\overline{f}_1(\overline{z}) = \phi(z^*, \overline{z}), \quad \overline{f}_2(\overline{z}) = \phi_z(z^*, \overline{z}), \tag{2.44}$$

putting $z = z^*$.

Since $\phi = e^{-u/2}$ is real-valued it solves also

$$\phi_{\bar{z}\bar{z}} + \frac{1}{2}\bar{s}\phi = 0 \tag{2.45}$$

for $\bar{s}(\bar{z}) = \overline{s(z)}$, and so do $\overline{f}_1(\bar{z})$, $\overline{f}_2(\bar{z})$ in (2.44). A fundamental system of solutions to (2.45), furthermore, is given by $\{\overline{\phi}_1, \overline{\phi}_2\}$ for

$$\overline{\phi}_1(\bar{z}) = \overline{\phi_1(z)}, \quad \overline{\phi}_2(\bar{z}) = \overline{\phi_1(z)},$$

which satisfies

$$\left(\overline{\phi}_1, \frac{\partial\overline{\phi}_1}{\partial\bar{z}}\right)\bigg|_{\bar{z}=\bar{z}^*} = (1,0), \quad \left(\overline{\phi}_2, \frac{\partial\overline{\phi}_2}{\partial\bar{z}}\right)\bigg|_{\bar{z}=\bar{z}^*} = (0,1).$$

We thus end up with

$$\overline{f}_1(\bar{z}^*) = \phi(z^*, \bar{z}^*) = e^{-u/2}\bigg|_{x=x^*} = C_1,$$

$$\frac{\partial}{\partial\bar{z}}\overline{f}_1(\bar{z}^*) = \phi_{\bar{z}}(z^*, \bar{z}^*) = \frac{\partial}{\partial\bar{z}}e^{-u/2}\bigg|_{x=x^*} = 0$$

$$\overline{f}_2(\bar{z}^*) = \phi_z(z^*, \bar{z}^*) = \frac{\partial}{\partial\bar{z}}e^{-u/2}\bigg|_{x=x^*} = 0$$

and

$$\frac{\partial}{\partial\bar{z}}\overline{f}_2(\bar{z}^*) = \phi_{z\bar{z}}(z^*, \bar{z}^*) = \frac{1}{4}\Delta e^{-u/2}\bigg|_{x=x^*}$$

$$= -\frac{1}{8}e^{-u/2}\Delta u\bigg|_{x=x^*} = \frac{\lambda}{8}e^{u/2}\bigg|_{x=x^*} = C_2,$$

using (2.43). These relations imply (2.41), which results in the *Liouville integral*

$$e^{-u/2} = C_1|\phi_1|^2 + C_2|\phi_2|^2 \tag{2.46}$$

for C_1 and C_2 defined by (2.42).

Put

$$\psi_1 = C_1^{1/2}8^{-1/4}\phi_1, \quad \psi_2 = C_1^{-1/2}8^{1/4}\phi_2, \quad F = \psi_2/\psi_1.$$

Since

$$W(\psi_1, \psi_2) = W(\phi_1, \phi_2) = 1$$

it holds that

$$\left(\frac{1}{8}\right)^{1/2}e^{u/2} = \frac{1}{|\psi_1|^2 + |\psi_2|^2} = \frac{|F'|}{1 + |F|^2} \equiv \rho(F). \tag{2.47}$$

Here, F is a *meromorphic function* of $z \in \Omega \subset \mathbf{C}$ made by a quotient of two linearly independent solutions to (2.39). Hence it satisfies

$$\{F; z\} = -\frac{1}{2}s,$$

where

$$\{F; z\} = \frac{3}{4}\left(\frac{F''}{F'}\right)^2 - \frac{1}{2}\frac{F'''}{F'}$$

is the *Schwarzian derivative*.

The quantity

$$\rho(F) = \frac{|F'|}{1 + |F|^2} \tag{2.48}$$

stands for the *spherical derivative* of the conformal mapping $F : \Omega \to S^2$, where

$$S^2 \subset \mathbf{R}^3$$

denotes the Gauss sphere, that is, a round sphere with diameter 1. This property means

$$\rho(F) = \frac{d\sigma}{ds},$$

where the right-hand side stands for the ratio of standard line element $d\sigma$ in S^2 and that ds in Ω mapped by F.

More precisely, if $S^2 \subset \mathbf{R}^3$ is the round sphere with the south and the north poles at $(0, 0, 0)$ and $(0, 0, 1)$, respectively, and if

$$\tau : \mathbf{C} \cup \{\infty\} \to S^2$$

denotes the inverse *stereographic projection*, then under the conformal mapping

$$\overline{F} = \tau \circ F : \Omega \to S^2, \tag{2.49}$$

the standard metrics on $\Omega \subset \mathbf{C} \cong \mathbf{R}^2$ and S^2 denoted by $ds^2 = dx_1^2 + dx_2^2$ and $d\sigma^2$, respectively, are so related as

$$\frac{d\sigma}{ds} = \rho(F). \tag{2.50}$$

The immersed length of the image of $\partial\Omega$ and the immersed area of the image of Ω under $\overline{F}$, are, therefore, given by

$$\ell(\partial\Omega) = \int_{\partial\Omega} \rho(F)\, d\sigma, \quad m(\Omega) = \int_\Omega \rho(F)^2 dx, \tag{2.51}$$

respectively.

2.4   Coverings of the Sphere

The *Gel'fand equation* in two space dimension takes the form

$$-\Delta u = \lambda e^u \text{ in } \Omega, \quad u|_{\partial\Omega} = 0, \tag{2.52}$$

where $\Omega \subset \mathbf{R}^2$ is a bounded domain with smooth boundary $\partial\Omega$ and $\lambda > 0$ is a constant. It is a *nonlinear eigenvalue problem* to find u and λ, simultaneously.

The equation

$$-\Delta u = \lambda e^u \quad \text{in } \Omega$$

has the integral

$$(\frac{\lambda}{8})^{1/2} e^{u/2} = \rho(F)$$

similarly, where $\rho(F)$ is the spherical derivative (2.48) of a meromorphic function $F(z)$ of $z = x_1 + \sqrt{-1}x_2$ for $x = (x_1, x_2) \in \Omega$. Then, the boundary value problem (2.52) is reduced to finding a conformal mapping (2.49) such that

$$\frac{d\sigma}{ds}\Big|_{\partial\Omega} = (\frac{\lambda}{8})^{1/2},$$

where ds and $d\sigma$ denote the line elements on Ω and S^2, respectively.

Since $|S^2| = \pi$, this expression induces a *quantized blowup mechanism*, which exhibits

$$\int_\Omega \rho(F)^2 \, dx = \frac{1}{8} \int_\Omega \lambda e^v \, dx \equiv \frac{\Sigma}{8} \; \to \ell\pi, \; \ell \in \mathbf{N}$$

as the singularity arises to F. More precisely, ℓ-covering of S^2 is realized as $L \downarrow 0$ in $\overline{F}(\Omega)$, where L stands for the immersed length of $\partial\Omega$:

$$L = (\frac{\lambda}{8})^{1/2} |\partial\Omega|.$$

This property of covering is the first reason that 8π appears in the singular state of the Boltzmann Poisson equation (Theorem 2.1 in §2.5). Thus we can classify the behavior of the sequence $\{(\lambda_k, u_k)\}$ of the solution to (2.52) for $\lambda = \lambda_k$ and $u = u_k$, satisfying

$$\lim_{k\to\infty} \lambda_k = 0$$

as follows [Nagasaki–Suzuki (1990a)].

Remark 2.1. The second reason comes from the classification of the entire solution, (2.52) for $\Omega = \mathbf{R}^2$ (Theorem 2.6 in §2.6). The third reason is the scaling invariance of the dual Trudinger Moser inequality ((3.12) in §3.2).

First, passing to a subsequence, we obtain

$$\Sigma_k = \int_\Omega \lambda_k e^{u_k}\, dx \to 8\pi\ell \qquad (2.53)$$

for some $\ell = 0, 1, \cdots, +\infty$. If $\ell = 0$ and $\ell = +\infty$, second, it holds that

$$\lim_{k\to\infty} \|u_k\|_\infty = 0$$

and

$$\lim_{k\to\infty} u_k = +\infty \quad \text{locally uniformly in } \Omega,$$

respectively. If $0 < \ell < +\infty$, third, there arises that

$$\lim_{k\to\infty} \|u_k\|_\infty = +\infty.$$

In this case, it holds, furthermore, that

$$\mathcal{S} \subset \Omega, \quad \sharp \mathcal{S} = \ell,$$

where

$$\mathcal{S} = \{x_0 \in \overline{\Omega} \mid \exists x_k \to x_0,\ u_k(x_k) \to +\infty\} \qquad (2.54)$$

denotes the *blowup set* of $\{u_k\}$. If

$$\mathcal{S} = \{x_1^*, \cdots, x_\ell^*\},$$

then $x_* = (x_1^*, \cdots, x_\ell^*)$ is a critical point of the point vortex Hamiltonian,

$$H_\ell(x_1, \cdots, x_\ell) = \frac{1}{2}\sum_{j=1}^\ell R(x_j) + \sum_{1\le i<j\le\ell} G(x_i, x_j), \qquad (2.55)$$

and there arises that

$$\lim_{k\to\infty} u_k = u_0 \quad \text{locally uniformly in } \overline{\Omega}\setminus\mathcal{S}$$

for

$$u_0(x) = 8\pi\sum_{j=1}^\ell G(x, x_j^*). \qquad (2.56)$$

The above feature of the quantization of Σ_k as in (2.53) and the control of the location of the blowup points $\{x_1^*, \cdots, x_\ell^*\}$ by the point vortex Hamiltonian as in

$$\nabla_{x_i} H(x_1^*, \cdots, x_\ell^*) = 0, \ 1 \le i \le \ell \qquad (2.57)$$

is the property that we call *recursive hierarchy* in connection with Onsager's theory on point vortices. Since extremal states of the solution to several elliptic problems other than the Gel'fand equation are controlled by this Hamiltonian, its study reveals a unified structure of the solution to a class of elliptic problems, where $u_0(x)$ in (2.56) is viewed as the *singular limit* of the solution (§3.3 of [Suzuki (2020)]). In fact, the recursive hierarchy in $2D$ elliptic equation is robust under the perturbation of the nonlinearity $f(u) = e^u$, including the control of the Hamiltonian of the linearized stability and instability of u_k ([Suzuki (2022b); Sato–Suzuki (2023)] and the references therein).

2.5   Boltzmann Poisson Equation

The Boltzmann Poisson equation in (2.34),

$$-\Delta v = \frac{\lambda e^v}{\int_\Omega e^v dx} \text{ in } \Omega \subset \mathbf{R}^2, \quad v|_{\partial\Omega} = 0. \tag{2.58}$$

is an equivalent form of the Gel'fand equation (2.52) under the correspondence of

$$\lambda \leftrightarrow \frac{\lambda}{\int_\Omega e^v dx}.$$

Then the above result on the equation implies the following theorem.

Theorem 2.1 ([Nagasaki–Suzuki (1990a)]). *Let* $\Omega \subset \mathbf{R}^2$ *be a bounded domain with smooth boundary* $\partial\Omega$*, and* (λ_k, v_k)*,* $k = 1, 2, \cdots$*, be a sequence of the classical solutions to (2.58):*

$$-\Delta v_k = \frac{\lambda_k e^{v_k}}{\int_\Omega e^{v_k} dx} \text{ in } \Omega, \quad v_k|_{\partial\Omega} = 0, \tag{2.59}$$

satisfying

$$\lim_{k\to\infty} \lambda_k = \lambda_0 \in (0, +\infty), \quad \lim_{k\to\infty} \|v_k\|_\infty = +\infty. \tag{2.60}$$

Then, it holds that

$$\lambda_0 = 8\pi\ell, \quad \ell \in \mathbf{N} \equiv \{1, 2, \cdots\}.$$

Passing to a subsequence, there is a set of ℓ*-interior points* $x_1^*, \cdots, x_\ell^*$ *such that*

$$\nabla_{x_j} H_\ell(x_1^*, \cdots, x_\ell^*) = 0, \quad 1 \le j \le \ell. \tag{2.61}$$

Here, H_ℓ *is the* ℓ*-th point vortex Hamiltonian defined by (2.55), and*

$$\mathcal{S} = \{x_1^*, \cdots, x_\ell^*\}$$

coicides with the blowup set of $\{v_k\}$ *defined by*

$$\mathcal{S} = \{x_0 \in \overline{\Omega} \mid \exists x_k \to x_0, \ \lim_{k\to\infty} v_k(x_k) = +\infty\}.$$

It holds, furthermore, that

$$v_k \to v_0 \text{ locally uniformly in } \overline{\Omega} \setminus \mathcal{S} \tag{2.62}$$

for

$$v_0 = v_0(x) = 8\pi \sum_{j=1}^{\ell} G(x, x_j^*).$$

Here we provide a proof based on the complex function theory, using several technical tools such as the L^1 *elliptic estimate* [Brezis–Strauss (1973)] and the *method of moving planes* [Gidas–Ni–Nirenberg (1979)].

Lemma 2.1. *Let* $B \equiv \{|x| < 1\} \subset \mathbf{R}^n$, $n \geq 2$, *and* $v \in W^{1,p}(B)$ *for some* $1 < p < \infty$. *Assume, furthermore,*

$$\Delta v = 0 \quad in \ B \setminus \{0\}.$$

Then there is $a \in \mathbf{R}$ *such that* $v - aE$ *is harmonic in* B, *where*

$$E(x) = \begin{cases} |x|^{2-n}, & n > 2 \\ -\log|x|, & n = 2 \end{cases}$$

denotes the fundamental solution: $-\Delta E = \delta$.

Proof. Since the support of the distribution $\Delta v \in \mathcal{D}'(B)$ is contained in $\{0\}$, it holds that

$$\Delta v = \sum_{|\alpha| \leq k} C_\alpha D^\alpha \delta, \quad \exists k, \ \exists C_\alpha \in \mathbf{R}.$$

We take $\zeta \in C_0^\infty(B)$ satisfying

$$(-1)^{|\alpha|} D^\alpha \zeta(0) = C_\alpha, \quad |\alpha| \leq k,$$

to put $\zeta_\epsilon(x) = \zeta(x/\epsilon)$ for $0 < \epsilon \ll 1$. Then it holds that

$$-\int_B \nabla v \cdot \nabla \zeta_\varepsilon \, dx = \langle \zeta_\varepsilon, \Delta v \rangle_{\mathcal{D}, \mathcal{D}'} = \sum_{|\alpha| \leq k} C_\alpha^2 / \varepsilon^{|\alpha|}.$$

Since

$$\|\nabla \zeta_\varepsilon\|_\infty = O(\varepsilon^{-1})$$

implies

$$\|\nabla \zeta_\varepsilon\|_{p'} = o(\varepsilon^{-1}), \quad \frac{1}{p'} + \frac{1}{p} = 1,$$

we have

$$\int_B \nabla v \cdot \nabla \zeta_\varepsilon \, dx = o(\varepsilon^{-1}), \quad \varepsilon \downarrow 0$$

by $\nabla v \in L^p(B)$, which implies $C_\alpha = 0$ for $|\alpha| \geq 1$. Hence the result follows. $\qquad \square$

Proof of Theorem 2.1. Since L^1 norm of the right-hand side of (2.59) is bounded, the elliptic estimate guarantees

$$\|v_k\|_{W^{1,q}} = O(1), \quad 1 \le q < 2 = \frac{n}{n-1} \tag{2.63}$$

by $n = 2$ ([Brezis–Strauss (1973)]). Now we apply the method of moving planes to (2.59).

First, we take the case that Ω is convex ([De Figueiredo–Lions–Nussbaum (1982)]). Since $v_k > 0$ in Ω, there is a family of simplicities with uniform shape, denoted by

$$A = \{T\},$$

such that $v_k(x)$ takes the maximum in $T \in A$ at its vertex which is closest to $\partial\Omega$, and furthermore,

$$\omega \equiv \Omega \cap \hat{\omega} \subset \bigcup_{T \in A} T,$$

where $\hat{\omega}$ is an open set containing $\partial\Omega$ ([Gidas–Ni–Nirenberg (1979)]).

This family $A = \{T\}$, furthermore, is determined by Ω, independent of the nonlinearity. We call this property the *monotone decreasing* of v_k near $\partial\Omega$. Since

$$\|v_k\|_1 = O(1)$$

follows from (2.63), this monotone decreasing property implies

$$\|v_k\|_{L^\infty(\omega)} = O(1). \tag{2.64}$$

For general Ω, we take a disc outscribing $\partial\Omega$. Then, regarding its center the origin, we apply the *Kelvin transformation*

$$T : x \mapsto y = x/|x|^2, \quad v(y) = |x|^{n-2}u(x),$$

which results in

$$\Delta v_y = |x|^{n+2} \Delta_x u.$$

Due to $n = 2$, we have the monotonicity of v_k near $\partial\Omega$, and obtain (2.64) again by

$$f(v) = \frac{\lambda e^v}{\int_\Omega e^v dx} \ge 0$$

([Gidas–Ni–Nirenberg (1979)]).

By (2.64) and local elliptic regularity near the boundary to (2.59), we obtain the uniform boundedness of any partial derivatives of $\{v_k\}$ near $\partial\Omega$. In particular, the holomorphic functions

$$s_k = v_{kzz} - \frac{1}{2}v_{kz}^2, \ k = 1, 2, \cdots,$$

of $z = x_1 + \sqrt{-1}x_2$ are uniformly bounded near $\partial\Omega$, and then the classical *maximum principle* and *Montel's theorem* on holomorphic functions guarantee that the family

$$\{s_k(z)\}$$

takes a subsequence, denoted by the same symbol, such that

$$\lim_{k\to\infty} s_k = s_0 \quad \text{locally uniformly in } \Omega.$$

Using

$$u_k = v_k + \log \lambda_k - \log \int_\Omega e^{v_k} dx, \tag{2.65}$$

we have

$$-\Delta u_k = e^{u_k} \quad \text{in } \Omega \tag{2.66}$$

and

$$s_k = u_{kzz} - \frac{1}{2}u_{kz}^2, \tag{2.67}$$

where the argument in the previous section is applicable.

Let $x_k = (x_{1k}, x_{2k}) \in \Omega$ be the maximum point of $v_k(x)$, and let

$$\{\varphi_{1k}(z), \varphi_{2k}(z)\}$$

be the fundamental system of solutions to

$$\varphi_{zz} + \frac{1}{2}s_k(z)\varphi = 0$$

satisfying

$$\left(\varphi_{1k}, \frac{\partial\varphi_{1k}}{\partial z}\right)\bigg|_{z=z_k^*} = (1, 0), \quad \left(\varphi_{2k}, \frac{\partial\varphi_{2k}}{\partial z}\right)\bigg|_{z=z_k^*} = (0, 1)$$

for

$$z_k^* = x_{1k} + \sqrt{-1}x_{2k}.$$

Then, we obtain

$$e^{-u_k/2} = \tilde{c}_k|\varphi_{1k}|^2 + \frac{\tilde{c}_k^{-1}}{8}|\varphi_{2k}|^2 \tag{2.68}$$

for

$$\tilde{c}_k = e^{-u_k(x_k)/2}.$$

From (2.65), equality (2.68) means

$$e^{-v_k/2} = c_k|\varphi_{1k}|^2 + \frac{\sigma_k c_k^{-1}}{8}|\varphi_{2k}|^2 \qquad (2.69)$$

for

$$c_k = e^{-v_k(x_k)/2}, \quad \sigma_k = \frac{\lambda_k}{\int_\Omega e^{v_k}\,dx}.$$

Passing to a subsequence, we have

$$\lim_{k\to\infty} x_k = x_0^* \equiv (x_{10}^*, x_{20}^*) \in \Omega$$

by (2.64). Then we define the fundamental system of solutions

$$\{\varphi_{10}(z), \varphi_{20}(z)\}$$

to

$$\varphi_{zz} + \frac{1}{2}s_0(z)\varphi = 0$$

by

$$\left(\varphi_{10}, \frac{\partial\varphi_{10}}{\partial z}\right)\Big|_{z=z_0^*} = (1,0), \quad \left(\varphi_{20}, \frac{\partial\varphi_{20}}{\partial z}\right)\Big|_{z=z_0^*} = (0,1),$$

where

$$z_0^* = x_{10}^* + \sqrt{-1}x_{20}^*.$$

It thus holds that

$$\lim_{k\to\infty} \varphi_{1k} = \varphi_{10}, \ \lim_{k\to\infty} \varphi_{2k} = \varphi_{20} \quad \text{locally uniformly in } \Omega.$$

In (2.69) we have

$$v_k(x_k) = \|v_k\|_\infty \to +\infty$$

by (2.60), and therefore,

$$\lim_{k\to\infty} c_k = 0. \qquad (2.70)$$

Passing to a subsequence, we have

$$\lim_{k\to\infty} \sigma_k c_k^{-1} = \gamma > 0 \qquad (2.71)$$

by (2.64), (2.69), and (2.70). The blowup set $\mathcal{S}$ of $\{v_k\}$ thus coincides with the zero set of φ_{20} in Ω.

Since zeros of the analytic function $\varphi_{20}(z)$ do not take an accumulating point in Ω, the set $\mathcal{S}$ is finite. There arises also (2.62) for $v_0 = v_0(x)$ defined by

$$e^{-v_0/2} = \gamma |\varphi_{20}|^2.$$

By the elliptic regularity, this convergence is valid up to their derivatives of any order. We have also

$$\sigma_k \to 0$$

by (2.70) and (2.71), which implies

$$\lim_{k \to \infty} \int_\Omega e^{v_k} dx = +\infty. \tag{2.72}$$

Taking the limit in (2.59), we obtain

$$-\Delta v_0 = 0 \text{ in } \Omega \setminus \mathcal{S}, \quad v_0 = 0 \text{ on } \partial\Omega.$$

This $v_0(x)$ takes each element in

$$\mathcal{S} = \{x_1^*, \cdots, x_\ell^*\}$$

as an isolated singular point. By Lemma 2.1 we obtain

$$a_j \in \mathbf{R}, \quad 1 \le j \le \ell,$$

for which

$$u_0(x) = v_0(x) + \sum_{j=1}^\ell a_j \log |x - x_j^*| \tag{2.73}$$

is harmonic in Ω.

Now we make $k \to \infty$ in (2.67), to obtain

$$s_0 = v_{0zz} - \frac{1}{2} v_{0z}^2.$$

This $s_0(z)$ is thus a holomorphic function in z. The singularities on the right-hand side at

$$z = x_{1j}^* + \sqrt{-1} x_{2j}^*, \quad x_j^* = (x_{1j}^*, x_{2j}^*)$$

are, therefore, removable. By (2.73), we take an expansion of the right-hand side.

Vanishing of the pole $z = z_j^*$ of the second and the first orders implies $a_j = 4$, which is equivalent to $m_j = 8\pi$, and (2.61), respectively. In fact, it holds that

$$s_0 \equiv u_{0zz} - u_{0z} = (w_{jzz} - \frac{1}{2} w_{jz}^2) + (H_{jzz} - \frac{1}{2} H_{jz}^2) - w_{jz} H_{jz}$$

$$= \frac{a_j/2 - a_j^2/8}{(z - \kappa_j)^2} + (H_{jzz} - \frac{1}{2} H_{jz}^2) + \frac{a_j}{2(z - \kappa_j)} H_{jz}$$

by

$$\frac{\partial}{\partial z} \log |z| = \frac{\partial}{\partial z} \frac{1}{2} \log(z\bar{z}) = \frac{1}{2} \frac{\bar{z}}{z\bar{z}} = \frac{z}{2}.$$

□

2.6   Method of Scaling

Scaling invariance is observed commonly in fundamental equations of physics. In the case of (2.37), or,

$$-\Delta v = e^v,$$

if $v(x)$ is a solution and $\mu > 0$ is a constant, then

$$v^\mu(x) = v(\mu x) + 2\log \mu \tag{2.74}$$

satisfies the same equation, although the domains where they are defined are different. This scaling invariance causes a lack of compactness of the family of (approximate) solutions, and this mechanism is clarified by *blowup analysis*, the ingredients of which are summarized as follows (§5.3 of [Suzuki (2020)]):

(1) scaling invariance of the equation
(2) classification of the entire solution
(3) control at infinity of the rescaled solution
(4) hierarchical argument

Here we describe the outcomes of this method applied to the Boltzmann Poisson equation (Chapter 12 of [Suzuki (2015b)]). The first theorem is derived from the pre-scaled analysis, but is free from the boundary condition and deals with the nonhomogeneous coefficient case. In the setting of Theorem 2.1, it is applicable to

$$u_k = v_k - \log \sigma_k, \quad \sigma_k = \frac{\lambda_k}{\int_\Omega e^{v_k}\, dx}.$$

The singular limit $v_0(x)$ in (2.62) induces

$$-\Delta v_0 = 8\pi \sum_{j=1}^{\ell} \delta_{x_j^*}(dx), \tag{2.75}$$

which stands for the limit of

$$V_k(x)e^{u_k}\, dx, \quad V_k(x) = 1$$

in the sensese of measures.

Theorem 2.2 ([Brezis–Merle (1991)]). *Let $\Omega \subset \mathbf{R}^2$ be a bounded domain and $v_k = v_k(x)$, $k = 1, 2, \ldots$, be a sequence of solutions to*

$$-\Delta v_k = V_k(x)e^{v_k}, \ 0 \le V_k(x) \le C \quad in \ \Omega$$

$$\int_\Omega e^{v_k}\, dx \le C. \tag{2.76}$$

Then, passing to a subsequence, there arises the following alternatives:

(1) $\{v_k\}$ *is locally uniformly bounded in* Ω.

(2) $v_k \to -\infty$ *locally uniformly in* Ω.

(3) *There is a finite set* $\mathcal{S} = \{x_j^*\} \subset \Omega$ *and* $m_j \geq 4\pi$ *such that*

$$v_k \to -\infty \quad locally\ uniformly\ in\ \Omega \setminus \mathcal{S}$$

$$V_k(x)e^{v_k}\,dx \rightharpoonup \sum_j m_j \delta_{x_j^*}(dx) \quad in\ \mathcal{M}(\Omega). \tag{2.77}$$

Furthermore, this $\mathcal{S}$ *is the blowup set of* $\{v_k\}$ *in* Ω:

$$\mathcal{S} = \{x_0 \in \Omega \mid \exists x_k \to x_0,\ \lim_{k\to\infty} v_k(x_k) = +\infty\}.$$

If $\{V_k(x)\}$ in (2.76) is compact in the set of continuous functions in Ω, the value m_j is so quantized as 8π times integer.

Theorem 2.3 ([Li–Shafrir (1994)]). *In the third case of the above theorem there arises that*

$$m_j = 8\pi n_j, \quad n_j \in \mathbf{N}, \tag{2.78}$$

provided that

$$\lim_{k\to\infty} V_k = V \quad locally\ uniformly\ in\ \Omega$$

for some $V = V(x) \in C(\Omega)$.

The integer n_j in (2.78) indicates the number of sub-collapses which make up a collision at x_j^*. Thus, the boundary condition in Theorem 2.1 excludes the *collision of sub-collapses* at each blowup point x_j^* in (2.75).

Theorem 2.3 is localized as follows, where $B = B_R(0)$ and $B_r = B_r(0)$ for $0 < r < R$.

Theorem 2.4 ([Li–Shafrir (1994)]). *If*

$$-\Delta v_k = V_k(x)e^{v_k} \ in\ B$$

$$0 \leq V_k = V_k(x) \in C(\overline{B}), \quad \lim_{k\to\infty} V_k = V \ uniformly\ on\ \overline{B}$$

$$\lim_{k\to\infty} \max_{\overline{B}} v_k = +\infty, \quad \lim_{k\to\infty} \max_{\overline{B}\setminus B_r} v_k = -\infty, \quad 0 < \forall r < R$$

$$\lim_{k\to\infty} \int_B V_k(x)e^{v_k}\,dx = \alpha, \quad \int_B e^{v_k}\,dx \leq C,$$

it holds that $\alpha \in 8\pi\mathcal{N}$.

Collision of sub-collapses, $\alpha = 8\pi\ell$, $\ell \geq 2$, as in (2.75), is actually excluded by a local boundary condition. There arises, furthermore, a local uniform behavior of $\{v_k(x)\}$ as follows.

Theorem 2.5 ([Li (1999)]). *It holds that*

$$\alpha = 8\pi \tag{2.79}$$

in the previous theorem, provided that

$$\max_{\partial B} v_k - \min_{\partial B} v_k \le C, \quad \|\nabla V_k\|_\infty \le C, \tag{2.80}$$

and furthermore,

$$\left| v_k(x) - \log \frac{e^{v_k(0)}}{\left(1 + \frac{V_k(0)}{8} e^{v_k(0)} |x|^2\right)^2} \right| \le C, \quad \forall k, \ x \in B. \tag{2.81}$$

Theorem 2.4 is proven by blowup anlaysis. We take $x_k \in B$ satisfying

$$v_k(x_k) = \|v_k\|_\infty, \ x_k \to 0,$$

to put

$$\tilde{v}_k(x) = v_k(\delta_k x + x_k) + 2\log \delta_k, \ \delta_k = e^{-v_k(x_k)/2} \to 0.$$

It holds that

$$-\Delta \tilde{v}_k = V_k(\delta_k x + x_k)e^{\tilde{v}_k}, \ \tilde{v}_k \le \tilde{v}_k(0) = 0 \text{ in } B(0, \tfrac{R}{2\delta_k})$$

$$\int_{B(0,\frac{R}{2\delta_k})} e^{\tilde{v}_k} dx \le C_0,$$

and Theorem 2.2 is applicable to this $\{\tilde{v}_k\}$.

Thus, $\{\tilde{v}_k\}$ is locally uniformly bounded in $\mathbf{R}^2$, and passing to a subsequence, we have

$$\lim_{k\to\infty} \tilde{v}_k = \tilde{v} \quad \text{locally uniformly in } \mathbf{R}^2 \tag{2.82}$$

for $\tilde{v} = \tilde{v}(x)$, satisfying

$$-\Delta \tilde{v} = V(0)e^{\tilde{v}}, \ \tilde{v} \le \tilde{v}(0) = 0 \text{ in } \mathbf{R}^2, \quad \int_{\mathbf{R}^2} e^{\tilde{v}} dx \le C_0$$

by the elliptic regularity. Then we obtain $V(0) > 0$, and assume

$$0 < a \le V_k(x) \le b < +\infty, \quad x \in B$$

without loss of generality, with a, b independent of k. It holds also that

$$\tilde{v}(x) = \log \frac{1}{(1 + \frac{V(0)}{8}|x|^2)^2}, \quad \int_{\mathbf{R}^2} V(0)e^{\tilde{v}} dx = 8\pi$$

by the following theorem of the classification of the *scaling limit*.

Theorem 2.6 ([Chen–Li (1991)]). *If*

$$-\Delta v = e^v \text{ in } \mathbf{R}^2, \quad \int_{\mathbf{R}^2} e^v \, dx < +\infty, \tag{2.83}$$

it holds that

$$v(x) = \log\left\{\frac{8\mu^2}{(1 + \mu^2 |x - x_0|^2)^2}\right\}, \quad \exists x_0 \in \mathbf{R}^2, \ \exists \mu > 0, \tag{2.84}$$

and hence

$$\int_{\mathbf{R}^2} e^v \, dx = 8\pi. \tag{2.85}$$

Equality (2.85) is the second reason why 8π is the unit of quantization in the singular state of the Boltzmann Poisson equation. Because of the convergence (2.82), collision of sub-collapses arises at $x = 0$, which makes $\alpha = 8\pi\ell$ with $\ell \geq 2$. Control at infinity of $\{\tilde{v}_k\}$, therefore, is essential in the proof of Theorem 2.5. Originally, method of moving planes is used for this purpose [Li (1999)], while the other argument detects their behavior in the intermediate region [Lin (2007)]. There is, however, a simple argument for exclusion of the collision of sub-collapses, (2.79), using the method of symmetrization developed in Chaptere 3 [Naito–Suzuki (2008)].

Equation (2.83) is called the *Liouville equation*. Combination of the methods of scaling and symmetrization are systematically used in the study of a dynamical model of the Boltzmann Poisson equation. That is the Smoluchowski Poisson equation in Chapter 3, where dynamics of sub-collapses of the blowup solution in finite time, and that of collapses of the blowup solution in infinite time, are clarified.

Chapter 3

Kinetic Recursive Hierarchy

This chapter is devoted to the quantized blowup mechanism realized in the Smoluchowski Poisson equation in two space dimension. This mechanism is observed in three levels of the temporal state; stationary, in finite time, and in infinite time. The stationary state is realized as the Boltzmann Poisson equation in the field variable, where the recursive hierachy is observed in Chapter 2 (Theorem 2.1 in §2.5). The Smoluchowski Poisson equation, sometimes mentioned in accordance with the simplified system of chemotaxis, is concerned on the motion of the mean field of many point vorticities in relaxation time, that is, from quasi-equilibrium to equilibrium. This physical background is the origin of recursive hierarchy observed in the kinetic level, realized as the collapse and sub-collapse dynamics in blowup in infinite time and in finite time, respectively. These dynamics are actually controlled by the point vortex Hamiltonian in Chapter 2 in the prescaled and the resclaed forms (Theorem 3.4 in §3.10 and Theorem 3.8 in §3.12).

In this chapter, first, we formulate the Smoluchowski Poisson equation in accordance with the transport theory and thermo-dynamical laws. Noting its scaling invariance, the critical dimension $n = 2$ and the critical mass 8π are detected (§3.1). Its criticality is confirmed from the view point of real analysis, the Trudinger Moser inequality (§3.2), and then quantized blowup mechanism both in finite and infinite time is formulated (§3.3). Story of the proof of these properties are described (§3.4–§3.5), and then the detailed proof is given for the whole domain $\mathbf{R}^2$ (§3.6–§3.9) and the bounded domain (§3.10–§3.14). Finally, we show that the total set of stationary solutions induces this quantized blowup mechanism, summarized as the *potential of self-organization* in the context of a simplified system of chemotaxis (§3.15).

3.1   Smoluchowski Poisson Equation

The *Smoluchowski Poisson equation*

$$u_t = \nabla \cdot (\nabla u - u\nabla v) \text{ in } \Omega \times (0, T)$$

$$\left.\frac{\partial u}{\partial \nu} - u\frac{\partial v}{\partial \nu}\right|_{\partial\Omega} = 0, \quad u|_{t=0} = u_0(x) \geq 0$$

$$-\Delta v = u, \quad v|_{\partial\Omega} = 0 \tag{3.1}$$

is subject to the thermo-dynamical law of *closed systems* (Chapter 5 of [Suzuki (2015b)]), that is, conservation of total mass and decrease of Helmholtz's free energy under the constant temperature, where $\Omega \subset \mathbf{R}^n$ is a bounded domain with smooth boundary $\partial\Omega$. If $u_0 = u_0(x)$ is smooth, there is a unique classical solution $u = u(\cdot, t)$ local in time, and henceforh its maximal existence time is denoted by $T = T_{\max} > 0$.

Physically, $u = u(x, t)$ and $v = v(x, t)$ stand for the particle density and the field distribution, respectively. These particles are suject to diffusion and self-attractive forces, and therefore, the flux (§8.5) is given by

$$j = -\nabla u + u\nabla v.$$

Here, the first term $-\nabla u$ stands for the diffusion, while the second term is a multiplication of the particle mass and its velocity due to the above described self-attractive features, called the *chemotaxis* term in theoretical biology (Chapter 2 of [Suzuki (2005)]). Hence system (3.1) is composed of the equation of conservation law (§8.5),

$$u_t + \nabla \cdot j = 0, \quad \nu \cdot j|_{\partial\Omega} = 0$$

and the Poisson equation

$$-\Delta v = u, \quad v|_{\partial\Omega} = 0.$$

This system is invariant under the *self-similar transformation*,

$$u_\mu(x, t) = \mu^2 u(\mu x, \mu^2 t), \ v_\mu(x, t) = v(\mu x, \mu^2 t), \quad \mu > 0, \tag{3.2}$$

as a consequence of the quadratic nonlinearity $u\nabla v$. There arise also non-negativity preserving, $u = u(x, t) \geq 0$, more strongly, the positivity

$$u(x, t) > 0, \quad (x, t) \in \overline{\Omega} \times (0, T)$$

unless $u_0 \equiv 0$, and total mass conservation

$$\|u(\cdot, t)\|_1 = \|u_0\|_1 \equiv \lambda. \tag{3.3}$$

From this structure, $n = 2$ is regarded as a critical dimension of this system for the existence of the solution global in time, because there arises that

$$\|u_\mu(\cdot, t)\|_1 = \mu^{2-n}\|u(\cdot, t)\|_1$$

in (3.2).

System (3.1) is a fundamental equation in non-equilibrium statistical mechanics, proposed in the context of astrophysics [Sire–Chavanis (2002)]. Hence this model is provided with the decreasing of *Helmholtz's free energy*,

$$\mathcal{F}(u) = \int_\Omega u(\log u - 1) \; dx - \frac{1}{2}\langle(-\Delta)^{-1}u, u\rangle, \tag{3.4}$$

where $v = (-\Delta)^{-1}u$ indicates

$$-\Delta v = u \text{ in } \Omega, \quad v|_{\partial\Omega} = 0. \tag{3.5}$$

This decreasing property is described by

$$\frac{d}{dt}\mathcal{F}(u) = -\int_\Omega u|\nabla(\log u - v)|^2 dx \le 0, \tag{3.6}$$

which may be derived by writing (3.1) as a *model (B) equation* associated with (3.4),

$$u_t = \nabla \cdot u\nabla \delta\mathcal{F}(u), \; u > 0 \text{ in } \Omega \times (0, T), \quad \frac{\partial}{\partial\nu}\delta\mathcal{F}(u)\bigg|_{\partial\Omega} = 0 \tag{3.7}$$

(Chapter 4 of [Suzuki (2015b)]). In fact we obtain

$$\frac{d}{ds}\mathcal{F}(u + sw)\bigg|_{s=0} = (\log u - (-\Delta)^{-1}u, w),$$

where $(\cdot, \cdot)$ denotes the L^2 inner product, and then there arises that (3.7) under the identification

$$\delta\mathcal{F}(u) = \log u - (-\Delta)^{-1}u \quad \text{in } L^2(\Omega).$$

Helmholtz's free energy H is the fundamental quantity which governs the kinetic evolution of thermo-dynamically colsed systems. It is given by

$$H = E - TS,$$

where E, T, S stand for the inner energy, temperature, and entropy, respectively (Chapter 5 of [Suzuki (2015b)]), and equation (3.4) is subject to this formula exactly, where

$$E = -\frac{1}{2}\langle(-\Delta)^{-1}u, u\rangle, \quad S = -\int_\Omega u(\log u - 1) \; dx,$$

and $T = 1$.

The model (B) equation (3.7) realizes, as a result, the total mass conservation (3.3) and the free energy decreasing (3.6). In fact, from (3.7), it follows that (3.3) and (3.6) as

$$\frac{d}{dt}\int_{\Omega} u\,dx = \int_{\partial\Omega} \nu \cdot (u\nabla\delta\mathcal{F}(u))\,dS = 0$$

and

$$\frac{d}{dt}\mathcal{F}(u) = \langle u_t, \delta\mathcal{F}(u)\rangle = -\int_{\Omega} u|\nabla\delta\mathcal{F}(u)|^2 dx \leq 0.$$

By (3.3), (3.5), and (3.6), on the other hand, the stationary state of (3.1) is formulated as

$$u > 0, \ \log u - v = \text{constant}, \ \|u\|_1 = \lambda \qquad (3.8)$$

and

$$-\Delta v = u \text{ in } \Omega, \quad v|_{\partial\Omega} = 0. \qquad (3.9)$$

It follows that

$$u = \frac{\lambda e^v}{\int_{\Omega} e^v dx} \qquad (3.10)$$

from (3.8), and then we obtain the Boltzmann Poisson equation

$$-\Delta v = \frac{\lambda e^v}{\int_{\Omega} e^v dx} \text{ in } \Omega, \quad v|_{\partial\Omega} = 0 \qquad (3.11)$$

by (3.9), of which quantized blowup mechanism and recursive hierarchy are observed in Chapter 2. Then the scaling (3.2) in the kinetic level induces that in the static level (2.74).

The correspondence between v and u is exactly the duality of field function and particle distribution, called the *Toland duality* (Chapter 3 of [Suzuki (2015b)]), and the scaling (3.2) is the origin of the blowup analysis developed in §2.6. In more details, first, equation (3.10) implies

$$u = e^{\tilde{v}}$$

for

$$\tilde{v} = v + \log\lambda - \int_{\Omega} e^v dx.$$

Then we define $\tilde{v}^{\mu}(x)$ by

$$u_{\mu} = e^{\tilde{v}^{\mu}}, \ u_{\mu}(x) = \mu^2 u(\mu x).$$

Then it follows that

$$\tilde{v}^{\mu}(x) = \log u_{\mu}(x) = \log u(\mu x) + 2\log\mu = \tilde{v}(\mu x) + 2\log\mu,$$

or (2.74) for $\tilde{v}(x)$.

3.2   Trudinger Moser Inequality

Dual Trudinger Moser inequality valid in two space dimension is a fundamental real-analytic feature of the above $\mathcal{F}(u)$, that is,

$$\inf\{\mathcal{F}(u) \mid u \geq 0, \ \|u\|_1 = 8\pi\} > -\infty. \tag{3.12}$$

As $n = 2$ is the critical dimension, thus $\lambda = 8\pi$ is the critical mass in (3.1).

This critical mass is again detected by the scaling (3.2). To confirm this property, we take the case of $\Omega = \mathbf{R}^2$, to put

$$\mathcal{F}_*(u) = \int_{\mathbf{R}^2} u(\log u - 1)dx - \frac{1}{2}\langle \Gamma * u, u \rangle, \quad u = u(x) \geq 0,$$

using the fundamental solution to $-\Delta$ in two space dimension, that is,

$$\Gamma(x) = \frac{1}{2\pi} \log \frac{1}{|x|}.$$

Then it follows that

$$\mathcal{F}_*(u_\mu) = (2\lambda - \frac{\lambda^2}{4\pi}) \log \mu + \mathcal{F}_*(u), \quad u_\mu(x) = \mu^2 u(\mu x),$$

where

$$\mu > 0, \ \lambda = \|u\|_1.$$

Hence

$$\inf\{\mathcal{F}_*(u) \mid u \geq 0, \ \|u\|_1 = \lambda\} > 0$$

is valid to $\lambda > 0$ only if $\lambda = 8\pi$ is the case. This property is the third reason that 8π is the critical mass for the blowup of the solution to the Boltzmann Poisson equation, the stationary state of the Smoluchowski Poisson equation. This criticality is kept to Smoluchowski Poisson equation as is clarified in this chapter (Theorem 3.3 in §3.6, Theorem 3.4 in §3.10, and Theorem 3.7 in §3.12).

Inequality (3.12) is the dual form of the *Trudinger Moser inequality*

$$\inf J_{8\pi}(v) > -\infty, \quad v \in H_0^1(\Omega), \tag{3.13}$$

where

$$J_\lambda(v) = \frac{1}{2}\|\nabla v\|_2^2 - \lambda \log \int_\Omega e^v \, dx.$$

In fact, there is a *Toland duality* indicated as

$$L|_{v=(-\Delta)^{-1}u} = \mathcal{F}(u), \quad L|_{u=\frac{\lambda e^v}{\int_\Omega e^v \, dx}} = J_\lambda(v)$$

for

$$\mathcal{F}(u) = \int_{\Omega} u(\log u - 1)dx - \frac{1}{2}\langle(-\Delta)^{-1}u, u\rangle.$$

Here,

$$L(u, v) = \int_{\Omega} u(\log u - 1) \, dx + \frac{1}{2}\|\nabla v\|_2^2 - \langle v, u\rangle \qquad (3.14)$$

is the *Lagrange function* defined for $u \in L\log L(\Omega)$ and $v \in H_0^1(\Omega)$. It constitutes of the paring between $L\log L$ and BMO,

$$\langle v, u\rangle = \int_{\Omega} uv \, dx, \qquad (3.15)$$

derived from $H_0^1(\Omega) \hookrightarrow L\log L(\Omega)$ and

$$(-\Delta)^{-1}u \in BMO(\Omega), \quad u \in L^1(\Omega)$$

valid to $n = 2$ (§3.4 of [Suzuki (2015b)]). It then holds that

$$\inf\{L(u, v) \mid u \geq 0, \|u\|_1 = \lambda, \, v \in H_0^1(\Omega)\}$$
$$= \inf\{\mathcal{F}(u) \mid u \geq 0, \|u\|_1 = \lambda\}$$
$$= \inf\{J_\lambda(v) \mid v \in H_0^1(\Omega)\}, \qquad (3.16)$$

which results in the equivalence of (3.12) and (3.13) (Chapter 3 of [Suzuki (2015b)]).

There are several versions of the Trudinger Moser inequality. First, *Chang Yang's inequality* is indicated as

$$\log\left(\frac{1}{|\Omega|}\int_{\Omega} e^v dx\right) \leq \frac{1}{8\pi}\|\nabla v\|_2^2 + \frac{1}{|\Omega|}\int_{\Omega} vdx + K, \quad v \in H^1(\Omega), \quad (3.17)$$

where K is a constant determined by $\Omega \subset \mathbf{R}^2$, a bounded domain with smooth boundary [Chang–Yang (1988)]. Inequality (3.13), on the other hand, is a consequence of the *Moser Onofri inequality*,

$$\log\left(\frac{1}{|\Omega|}\int_{\Omega} e^v dx\right) \leq \frac{1}{16\pi}\|\nabla v\|_2^2 + K, \quad v \in H_0^1(\Omega) \qquad (3.18)$$

([Moser (1971); Onofri (1982)]). This (3.18) holds for any bounded domain $\Omega \subset \mathbf{R}^2$, and the optimal constant $K = 1$ is confirmed [Nagai–Senba–Suzuki (2001)]. Then $K = 2$ is expected to be optimal in (3.17).

The difference of the constants 8π and 16π in (3.17) and (3.18) comes from the behavior of the minimizing sequences of these inequalities. In fact, this sequence forms a delta function (collapse) inside Ω in (3.18), while it concentrates on $\partial\Omega$ in (3.18). More precisely, the collapse mass made by these minimizing sequences is so quantized as 8π, while 4π is realized as a collapse mass if this collapse is formed on the boundary (§3.15). Hence the constant 8π in the right-hand side of (3.17) is reduced more if Ω has corners [Chang–Yang (1988)].

This Trudinger Moser inequality in two space dimension is associated with the Soblev and Morrey embeddings,

$$W_0^{1,p}(\Omega) \hookrightarrow \begin{cases} L^{\frac{np}{n-p}}(\Omega),\ 1 \le p < n \\ C^{1-\frac{n}{p}}(\Omega),\ p > n, \end{cases}$$

where $\Omega \subset \mathbf{R}^n$ is a bounded domain and $W_0^{1,p}(\Omega)$ denotes the closure of $C_0^\infty(\Omega)$ in $W^{1,p}(\Omega)$, the set of p-integrable functions including their distributional derivatives of the first order.

In the critical case $p = n$, therefore, $W_0^{1,p}(\Omega)$ is embedded into the *Orlicz space* ([Pohozaev (1965); Trudinger (1967)]). One of the sharp forms assures a constant $C > 0$ such that

$$\int_{S^2} w\, dS = 0,\ \|\nabla w\|_2 \le 1 \quad\Rightarrow\quad \int_{S^2} e^{4\pi w^2}\, dS \le C, \tag{3.19}$$

where S^2 is the two-dimensional unit sphere and dS is the surface element ([Moser (1971)]

If $v \not\equiv 0$ satisfies

$$\int_{S^2} v\, dS = 0, \tag{3.20}$$

then we obtain $v \notin \mathbf{R}$ and

$$\int_{S^2} e^{4\pi w^2}\, dS \le C, \quad w = \frac{v}{\|\nabla v\|_2}.$$

Since

$$v = w \|\nabla v\|_2 \le 4\pi w^2 + \frac{1}{16\pi} \|\nabla v\|_2^2$$

it follows that

$$\int_{S^2} e^v dS \le C \cdot \exp\left(\frac{1}{16\pi} \|\nabla v\|_2^2\right),$$

which implies

$$\log\left(\frac{1}{4\pi} \int_{S^2} e^v\, dS\right) \le \frac{1}{16\pi} \|\nabla v\|_2^2 + K \tag{3.21}$$

for $v \in H^1(S^2)$ satisfying (3.20), where

$$K = \log\left(C/(4\pi)\right).$$

Given $v \in H^1(S^2)$, we take

$$v - \frac{1}{4\pi} \int_{S^2} v \, dS$$

for v in (3.21), to obtain

$$\log \left(\frac{1}{4\pi} \int_{S^2} e^v dS \right) \leq \frac{1}{16\pi} \|\nabla v\|_2^2 + \frac{1}{4\pi} \int_{S^2} v \, dS + K.$$

The best constant K in this inequality is $K = 0$, which results in

$$\log \left(\frac{1}{4\pi} \int_{S^2} e^v dS \right) \leq \frac{1}{16\pi} \|\nabla v\|_2^2 + \frac{1}{4\pi} \int_{S^2} v \, dS$$

([Onofri (1982); Hong (1987)]).

Generally, if (Ω, dS) is a Riemann surface without boundary, it holds that

$$\inf\{J_{8\pi}(v) \mid v \in H^1(\Omega), \int_\Omega v \, dS = 0\} > -\infty \qquad (3.22)$$

for

$$J_\lambda(v) = \frac{1}{2}\|\nabla v\|_2^2 - \lambda \left(\int_\Omega e^v dS \right)$$

([Fontana (1993)]). Then, the Toland duality (3.16) ensures the dual Fontana inequality

$$\inf\{\mathcal{F}(u) \mid u \geq 0, \ \|u\|_1 = 8\pi\} > -\infty, \qquad (3.23)$$

where $L(u, v)$ the Lagrangian functional defined by (3.14)–(3.15) and $\mathcal{F}(u)$ is the associated free energy defined by (3.4):

$$L(u, v) = \int_\Omega u(\log u - 1) \, dS + \frac{1}{2}\|\nabla v\|_2^2 - \langle v, u \rangle$$

$$\mathcal{F}(u) = \int_\Omega u(\log u - 1) \, dS - \frac{1}{2}\langle (-\Delta)^{-1}u, u \rangle$$

for

$$v \in H^1(\Omega), \quad 0 \leq u \in L \log L(\Omega).$$

Differently from (3.5), here, equality

$$v = (-\Delta)^{-1}u$$

means

$$-\Delta v = u - \frac{1}{|\Omega|} \int_\Omega u \, dS, \quad \int_\Omega v \, dS = 0 \qquad (3.24)$$

(Chapter 3 of [Suzuki (2015b)]). Inequality (3.23) plays a fundamental role in the study of the $2D$ normalized Ricci flow in Chapter 4.

Inequality (3.22), similarly, follows from the other form of the Trudinger Moser inequality [Fontana (1993)],

$$v \in H^1(\Omega), \int_\Omega v \, dS = 0, \|\nabla v\|_2 \le 1 \quad \Rightarrow \quad \int_\Omega e^{4\pi v^2} \, dS \le C, \qquad (3.25)$$

which is a generalization of (3.19) for $\Omega = S^2$. An immediate consequence of (3.25) is that any $K > 0$ admits $C(K) > 0$ such that

$$v \in V = H^1(\Omega), \|v\|_V \le K \quad \Rightarrow \quad \|e^{|v|}\|_1 \le C(K), \qquad (3.26)$$

where

$$\|v\|_V = (\|v\|_2^2 + \|\nabla v\|_2^2)^{1/2}.$$

Property (3.26) is also used in the study of normalized Ricci flow in two space dimension (2D–NRF) in Chapter 4.

3.3 Quantized Blowup Mechanism

This section is devoted to the description of the quantized blowup mechanism to (3.1). This mechanism is noticed by the blowup threshold of the total mass $\lambda = \|u_0\|_1$. Based on the study of free energy and the second moment, this value $\lambda = 8\pi$ is realized as the threshold for the blowup in finite time of the solution [Biler–Hilhorst–Nadzieja (1994); Nagai (1995); Nagai–Senba–Yoshida (1997); Biler (1998); Gajewski–Zacharias (1999); Senba–Suzuki (2001)].

Thus, if $\lambda = \|u_0\|_1 > 8\pi$ holds with a concentration at some interior point of $u_0 = u_0(x)$, there arises that $T = T_{\max} < +\infty$, while $\lambda < 8\pi$ ensures

$$T = +\infty, \quad \|u(\cdot, t)\|_\infty \le C. \qquad (3.27)$$

Establishing the quantized blowup mechanism and Hamiltonian control of (3.1), first, we show that blowup in finite time exhibits *formation of collapses* with quantized mass. Hence $T < +\infty$ in (3.1) implies $\mathcal{S} \subset \Omega$, $\sharp \mathcal{S} < +\infty$, and

$$u(x,t)dx \rightharpoonup \sum_{x_0 \in \mathcal{S}} m(x_0)\delta_{x_0}(dx) + f(x)dx \quad \text{in } \mathcal{M}(\overline{\Omega}) = C(\overline{\Omega})' \qquad (3.28)$$

as $t \uparrow T$, where

$$m(x_0) \in 8\pi \mathbf{N}, \quad 0 < f = f(x) \in L^1(\Omega) \cap C(\overline{\Omega} \setminus \mathcal{S}),$$

and

$$\mathcal{S} = \{x_0 \in \overline{\Omega} \mid \exists x_k \to x_0, \ \exists t_k \uparrow T, \ u(x_k, t_k) \to +\infty\} \tag{3.29}$$

denotes the blowup set (Theorem 3.4 in §3.10).

So far, the method of asymptotic expansion has been applied to pick up several profiles of blowup solution in finite time; that is, a radially symmetric blowup pattern with bounded free energy (Remark 3.12 in §3.9, [Herrero–Velázquez (1996)]) and collision of sub-collapses (Remark 3.13 in §3.9, [Luckhaus–Sugiyama–Velázquez (2012)]). These profiles are actually consistent to the behavior of general solutions (Theorem 3.4 and Theorem 3.5 in §3.10). Significance of these theorems lies in *collapse mass quantization*,

$$m(x_0) \in 8\pi \mathbf{N} \tag{3.30}$$

in (3.28) (Theorem 3.4 in §3.10). This quantization heritages that of the family of solutions to the Boltzmann Poisson equation (3.11) in §3.1, where each collapse takes the normalized mass 8π and location of these collapses is controlled by the point vortex Hamiltonian defined by (2.55) in §2.4, that is the recursive hierarchy. This recursive hierarchy, however, is involved by the space-in-time scaling and an anti-gradient flow (§3.10–§3.11).

More precisely, each blowup point of (3.1) is composed of a finite sum of *sub-collapses*, the rescaled, translated solutions to the stationary problem on the whole space, that is the Liouville equation (2.83) in §2.6. A collapse composed of one sub-collapse is said to be simple. Otherwise, it is multiple, provided with the *collision* of sub-collapses. In this terminology, a family of blowing-up stationary solutions exibits always simple collapses; collision of sub-collapses indicated by $m(x_0) = 8\pi\ell$, $\ell \geq 2$, in (3.1), actually, was missed in the first approach of the author [Suzuki (2005)].

Above stated mechanism of formation of collapses in (3.1), however, is different from that in (2.76) in §2.6 without boundary condition. In fact, in contrast that Theorem 2.4 classifies the mechanism of formation of sub-collapses at the origin, with possibly different concentration rates, in (3.1), the movement of sub-collapses is subject to the Hamiltonian on the whole space, that is, (3.60)–(3.61) in §3.4. In particular, they are formed in a common rate.

The profile of the solution, blowup in infinite time in (3.1), on the other hand, is rather more similar to that of the family of stationary solutions formulated to (3.11) described in Theorem 2.1 in §2.5. This profile is actually observed in the dynamics of collapses formed in infinite time. Theorem 3.7 in §3.12 thus says that

$$T = +\infty, \quad \limsup_{t\uparrow+\infty} \|u(\cdot,t)\|_\infty = +\infty \tag{3.31}$$

occurs only if

$$\lambda = \|u_0\|_1 \in 8\pi\mathbf{N} \tag{3.32}$$

and there is a critical point

$$x_* = (x_1^*, \cdots, x_\ell^*)$$

of $H = H(x)$, $x = (x_1, \cdots, x_\ell)$, defined by

$$H(x_1, \cdots, x_\ell) = \frac{1}{2}\sum_{j=1}^{\ell} R(x_j) + \sum_{1\leq i<j\leq\ell} G(x_i, x_j) \tag{3.33}$$

for $\lambda = 8\pi\ell$, that is, (2.57) in §2.4. These profiles of blowup of the solution in infinite time govern the dynamics of (3.1) essentially, in accordance with the shape of domain and also the concentration of initial value (§7.5.4 of [Suzuki (2020)]).

Totally, the singular limit (2.61)–(2.62) in §2.5, emerged from the total set of stationary solutions, control the dynamics of the uniformly bounded classiacal solution global in time as well as the blowup mechanisms of the soluition both in infinite time and finite time. This property is called *nonlinear spectral mechanics*, because the stationary state takes the form of a nonlinear eigenvalue problem (3.11). It is also called the *potential of self-organization*, because blowup with a conservative quantity, realized in the closed system of thermo-dynamics, exihibits the profile of *bottom up self-organization* (Chapter 1 of [Suzuki (2005)])

3.4 Blowup in Finite Time

Here we describe the story of the proof of collapse mass quantization for solutions to (3.1) blowup in finite time. This proof is executed in §3.5–§3.9 and §3.10–§3.11 for the cases that $\Omega = \mathbf{R}^2$ and that $\Omega \subset \mathbf{R}^2$ is a bounded domain with smooth boundary $\partial\Omega$, respectively. Here we concentrate on the latter case.

First, formation of collapse, (3.28) in §5.13 as $t \uparrow T < +\infty$, is a consequence of ε-*regularity*, which means the existence of $\varepsilon_0 > 0$ such that

$$\lim_{R\downarrow 0}\limsup_{t\uparrow T}\|u(\cdot,t)\|_{L^1(\Omega\cap B(x_0,R))} < \varepsilon_0 \quad \Rightarrow \quad x_0 \notin \mathcal{S}, \tag{3.34}$$

and the *monotonicity formula*

$$\left|\frac{d}{dt}\int_\Omega u\varphi\,dx\right| \leq C(\lambda)\|\nabla\varphi\|_{C^1} \tag{3.35}$$

valid to

$$\varphi \in C^2(\overline{\Omega}), \ \left.\frac{\partial\varphi}{\partial\nu}\right|_{\partial\Omega} = 0$$

(§3.6). In fact, inequality (3.35) improves (3.34) to

$$x_0 \in \mathcal{S} \quad \Rightarrow \quad \lim_{R\downarrow 0}\liminf_{t\uparrow T}\|u(\cdot,t)\|_{L^1(\Omega\cap B(x_0,R)} \geq \varepsilon_0, \tag{3.36}$$

which implies $\sharp\mathcal{S} < +\infty$ and also (3.28),

$$u(x,t)dx \rightharpoonup \sum_{x_0\in\mathcal{S}} m(x_0)\delta_{x_0}(dx) + f(x)dx \quad \text{in } \mathcal{M}(\overline{\Omega}), \tag{3.37}$$

with

$$m(x_0) \geq \varepsilon_0, \quad 0 \leq f = f(x) \in L^1(\Omega) \cap C(\overline{\Omega}\setminus\mathcal{S}) \tag{3.38}$$

by (3.3) ([Senba–Suzuki (2001)]).

Property (3.34) is a standard consequence of the combination of Gagliardo Nirenberg inequality in two space dimensions, semigroup estimate, and total mass control (3.3), while the monotonicity formula (3.35) is a consequence of the *weak form* of (3.1),

$$\frac{d}{dt}\int_\Omega u\varphi\,dx = \int_\Omega u\Delta\varphi\,dx + \frac{1}{2}\iint_{\Omega\times\Omega}\rho_\varphi(x,x')u\otimes u\,dxdx', \tag{3.39}$$

where

$$u\otimes u = u(x,t)u(x',t)$$

and

$$\rho_\varphi(x,x') = \nabla\varphi(x)\cdot\nabla_x G(x,x') + \nabla\varphi(x')\cdot\nabla_{x'}G(x,x'). \tag{3.40}$$

Equality (3.39) with (3.40) is derived by the *method of symmetrization* in §2.1, using the symmetry of Green's function indicated by (2.16) in §2.1,

$$G(x',x) = G(x,x').$$

Here we mention that this profile of symmetry concerned on the Poisson part of (3.1) follows from the action reaction law in classical mechanics, and results in the self-adjointness of the associated elliptic operator (Chapter 2 of [Suzuki (2005)]). This symmetry induces a *cancellation of singularities* caused by the self-interaction of many particles and is used to derive the Kirchhoff equation (2.25) in §2.1, where u stands for their distribution.

More precisely, behavior of $G = G(x, x')$ on $(\Omega \cap B(x_0, R)) \times (\Omega \cap B(x_0, R))$ for $0 < R \ll 1$ and $x_0 \in \partial\Omega$, is controlled by the fundamental solution

$$\Gamma(x) = \frac{1}{2\pi} \log \frac{1}{|x|} \qquad (3.41)$$

and the conformal mapping $X : \Omega \cap B(x_0, R) \to \mathbf{R}_+^2$ as in §3.10. Then we obtain

$$\rho_\varphi \in L^\infty(\Omega \times \Omega) \qquad (3.42)$$

and hence (3.35). A careful analysis, however, is required to prove (3.42) because $\rho_\phi(x, x')$ is not continuous at $x = x'$ ([Suzuki (2013)]).

A *weak solution* is then defined, based on the weak form (3.39) ([Senba–Suzuki (2002a)]). Let

$$\mathcal{Y} = \{\varphi \in C^2(\overline{\Omega}) \mid \left.\frac{\partial\varphi}{\partial\nu}\right|_{\partial\Omega} = 0\}$$
$$\mathcal{X}_0 = \{\rho_\varphi + \psi \mid \varphi \in \mathcal{Y}, \ \psi \in C(\overline{\Omega} \times \overline{\Omega}),\}$$

and $\mathcal{X}$ be the closure in $L^\infty(\Omega \times \Omega)$ of $\mathcal{X}_0$. Then we say that

$$0 \leq \mu = \mu(dx, t) \in C_*([0, T], \mathcal{M}(\overline{\Omega}))$$

is a weak solution to (3.1) if there is $\mathcal{N} = \mathcal{N}(\cdot, t) \in L_*^\infty([0, T], \mathcal{X}')$, called the *multiplicated operator* satisfying the following properties:

(1) The mapping

$$t \in [0, T] \mapsto \langle \varphi, \mu(dx, t) \rangle$$

is absolutely continuous for any $\varphi \in \mathcal{Y}$.

(2) It holds that

$$\frac{d}{dt}\langle \varphi, \mu \rangle = \langle \Delta\varphi, \mu \rangle + \frac{1}{2}\langle \rho_\varphi, \mathcal{N}(\cdot, t) \rangle \quad \text{a.e. } t \in [0, T].$$

(3) There arises that

$$\mathcal{N} \geq 0, \quad \mathcal{N}|_{C(\overline{\Omega} \times \overline{\Omega})} = \mu \otimes \mu.$$

Here are several direct consequences. First, it holds that

$$\mu(\overline{\Omega}, t) = \mu(\overline{\Omega}, 0) \equiv \lambda, \quad 0 \le t \le T.$$

Second, we obtain

$$\left| \frac{d}{dt} \langle \varphi, \mu(dx, t) \rangle \right| \le C(\lambda) \|\nabla \varphi\|_{C^1} \tag{3.43}$$

for $\varphi \in \mathcal{Y}$ similarly to (3.35). Third, since the above $\mathcal{X}$ is separable, there is a *generation of the weak solution* with the aid of (3.43). Thus we obtain the following theorem.

Theorem 3.1 ([Senba–Suzuki (2002a)]). *If*

$$\mu_k(dx, t) \in C_*([0, T], \mathcal{M}(\overline{\Omega})), \quad k = 1, 2, \cdots,$$

is a sequence of weak solutions associated with the multiplicated operators

$$\mathcal{N}_k \in L_*^\infty([0, T], \mathcal{X}'), \quad k = 1, 2, \cdots,$$

satisfying

$$0 \le \mu_k(\overline{\Omega}, t) \le C, \quad \|\mathcal{N}_k(\cdot, t)\|_{\mathcal{X}'} \le C, \quad 0 \le t \le T, \ k = 1, 2, \cdots, \tag{3.44}$$

then there is a subsequence, denoted by the same symbol, such that

$$\mu_k(dx, t) \rightharpoonup \mu(dx, t) \qquad in \ C_*([0, T], \mathcal{M}(\overline{\Omega}))$$
$$\mathcal{N}_k(\cdot, t) \rightharpoonup \mathcal{N}(\cdot, t) \qquad in \ L_*^\infty([0, T], \mathcal{X}'),$$

where $\mu(dx, t)$ a weak solution to (3.1) with the multiplicated operator $\mathcal{N}(\cdot, t)$.

Any classical solution $u = u(x, t)$, next, is identified with the weak solution by

$$\mu(dx, t) = u(x, t)dx,$$

provided with the associated multiplicated operator

$$\mathcal{N}(\cdot, t) = u(x, t) \otimes u(x', t) \ dx dx'.$$

Given a family of classical solutions

$$u_k = u_k(x, t), \quad k = 1, 2, \cdots,$$

defined on $\overline{\Omega} \times [0, T]$, then, the condition (3.44) is satisfied if

$$\|u_{0k}\|_1 \le C, \quad k = 1, 2, \cdots,$$

where $u_{0k} = u_k|_{t=0} \ge 0$.

Finally, *blowup criterion* for classical solutions derived from the *second moment* [Biler–Hilhorst–Nadzieja (1994); Nagai (1995)] is also efficient by its proof. Hence (3.45) below is a criterior of the *instant blowup* of the weak solution, where $\varphi = \varphi_{x_0, R} \in \mathcal{Y}$ is a cut-off function:

$$0 \le \varphi \le 1, \quad \varphi = 0 \text{ on } \overline{\Omega} \cap B(x_0, R)^c, \quad \varphi = 1 \text{ on } \overline{\Omega} \cap B(x_0, R/2).$$

Theorem 3.2 ([Senba–Suzuki (2002a)]). *There is no weak solution*

$$\mu(dx, t) \in C_*([0, T], \mathcal{M}(\overline{\Omega}))$$

to (3.1) for any $T > 0$, *if*

$$\mu(\{x_0\}, 0) > 8\pi, \quad \lim_{R \downarrow 0} \frac{1}{R^2} \left\langle |x - x_0|^2 \varphi_{x_0, R}, \mu(dx, 0) \right\rangle = 0 \qquad (3.45)$$

is the case for some $x_0 \in \Omega$.

Besides Theorem 3.1 and Theorem 3.2, weak solution to (3.1) is provided with the *weak Liouville property*, valid to its full orbit on the whole plane (Lemma 3.4 in §3.8). The proof of this property is actually done by following that to classical solutions [Kurokiba–Ogawa (2003)]. More precisely, the *local second moment* and scaling invariance

$$u_\mu(x, t) = \mu^2 u(\mu x, \mu^2 t), \quad \mu > 0, \qquad (3.46)$$

of

$$u_t = \Delta u - \nabla \cdot u \nabla \Gamma * u, \quad u \geq 0 \quad \text{in } \mathbf{R}^2 \times (-\infty, +\infty), \qquad (3.47)$$

assures either $M = 0$ or $M = 8\pi$, for any weak solution

$$0 \leq a(dx, t) \in C_*(-\infty, +\infty; \mathcal{M}(\mathbf{R}^2))$$

to (3.47), defined similarly to the case of bounded domains, satisfying

$$a(\mathbf{R}^2, 0) = M < +\infty.$$

Recall that $\Gamma(x)$ is the fundamental solution to $-\Delta$ given by (3.41) and

$$(\Gamma * w)(x) = \int_{\mathbf{R}^2} \Gamma(x - y) w(y) \, dy.$$

We put, furthermore,

$$\mathcal{M}(\mathbf{R}^2) = C_\infty(\mathbf{R}^2)',$$

where $C_\infty(\mathbf{R}^2)$ denotes the set of continuous functions on $\mathbf{R}^2 \cup \{\infty\}$, the one point compactification of $\mathbf{R}^2$, taking the value zero at ∞. We have also the total mass conservation,

$$a(\mathbf{R}^2, t) = M, \quad -\infty < t < +\infty$$

for the weak solution $a = a(dy, t)$ to (3.47).

Turning back to the blowup in finite time of the classical solution to (3.1), we take $x_0 \in \mathcal{S}$ for the blowup set $\mathcal{S}$ defined by (3.29). Then we introduce the *backward self-similar transformation*

$$z(y, s) = (T - t)u(x, t)$$
$$y = (x - x_0)/(T - t)^{1/2}, \quad s = -\log(T - t), \qquad (3.48)$$

to obtain

$$z_s = \Delta z - \nabla \cdot z \nabla \left(w + \frac{|y|^2}{4} \right) \qquad \text{in } \bigcup_{s > -\log T} \Omega_s \times \{s\}$$

$$\frac{\partial z}{\partial \nu} - z \frac{\partial}{\partial \nu} \left(w + \frac{|y|^2}{4} \right) = 0 \qquad \text{on } \bigcup_{s > -\log T} \partial \Omega_s \times \{s\}, \quad (3.49)$$

where $\Omega_s = (T - t)^{-1/2}(\Omega - \{x_0\})$ and

$$w(y, s) = \int_{\Omega_s} G_s(y, y') z(y', s) \, dy', \quad G_s(y, y') = G(x, x').$$

Above described notion of the weak solution and the principle of its generation are extended and valid to (3.49). From the profile of Green's function $G(x, x')$ mentioned above in connection with (3.42) (§3.10), any $s_k \uparrow +\infty$ admits a subsequence, denoted by the same symbol, such that

$$z(y, s + s_k) dy$$

admits $*$-weak convergence to produce a weak solution to the full orbit on the whole and on a half space according to $x_0 \in \Omega$ and $x_0 \in \partial \Omega$, respectively (§3.10). Here and henceforth, $z(y, s)$ is set to be zero where it is not defined.

More precisely, if $x_0 \in \Omega$, we have

$$z(y, s + s_k) dy \rightharpoonup \zeta(dy, s) \quad \text{in } C_*(-\infty, +\infty; \mathcal{M}(\mathbf{R}^2)) \qquad (3.50)$$

as $k \to \infty$ up to a subsequence, where $\zeta = \zeta(dy, s)$ is a weak solution to

$$z_s = \Delta z - \nabla \cdot z \nabla (\Gamma * z + \frac{|y|^2}{4}) \quad \text{in } \mathbf{R}^2 \times (-\infty, +\infty) \qquad (3.51)$$

for $\Gamma = \Gamma(x)$ defined by (3.41).

If $x_0 \in \partial \Omega$, on the other hand, it holds that (3.50) with

$$\text{supp } \zeta(\cdot, s) \subset \overline{\mathbf{R}_+^2}, \quad \mathbf{R}_+^2 = \{(y_1, y_2) \mid y_2 > 0\}$$

under suitable rotation of the variable y. This $\zeta = \zeta(dy, s)$ is a weak solution to

$$z_s = \Delta z - \nabla \cdot z \nabla (E * z + \frac{|y|^2}{4}) \quad \text{in } \mathbf{R}_+^2 \times (-\infty, +\infty)$$

$$\frac{\partial z}{\partial \nu} - z \frac{\partial E * z}{\partial \nu} = 0 \qquad \text{on } \partial \mathbf{R}_+^2 \times (-\infty, +\infty), \qquad (3.52)$$

where

$$E(y, y') = \Gamma(y - y') - \Gamma(y - y'_*) \qquad (3.53)$$

and $y_* = (y_1, -y_2)$ for $y = (y_1, y_2)$.

Under this process of rescaling, some part of the total mass of $u(x,t)$ is lost as $t \uparrow T$, that is, the tail of the rescaled solution, as is usual in the blowup analysis developed for elliptic problem as in §2.6. Monotonicity formula (3.35) then ensures the following fundamental properties in this argument, called the *parabolic envelope* [Suzuki (2005); Senba (2007)]. It is indicated as (3.112)–(3.113) in §3.6, that is,

$$\zeta(\mathbf{R}^2, s) = m(x_0), \quad \langle |y|^2, \zeta(dy, s) \rangle \leq C, \quad -\infty < s < +\infty. \tag{3.54}$$

This property implies $\mathcal{S} \subset \Omega$, because there is no weak solution $\zeta = \zeta(dy, s)$ to (3.52) satisfying $\zeta(\mathbf{R}^2, 0) > 0$ ([Suzuki (2013)]). Hence we concentrate on the study of (3.51) with (3.54) to complete the proof of $m(x_0) \in 8\pi\mathbf{N}$ (§3.9).

Since ε-regularity is valid even to (3.49), the singular part of $\zeta(dy, s)$, denoted by $\zeta^s(dy, s)$, is composed of a finite sum of delta functions with uniformly bounded supports by the second inequality of (3.54):

$$\zeta^s(dy, s) = \sum_{j=1}^{m(s)} \tilde{m}_j(s) \delta_{y_j(s)}(dy)$$

$$m(s) \leq m(x_0)/\varepsilon_0, \quad |y_j(s)| \leq C, \quad \tilde{m}_j(s) \geq \varepsilon_0. \tag{3.55}$$

There is, furthermore, $R > 0$ such that $\zeta(dy, s)$ is composed of uniformly bounded regular parts $\zeta^{ac}(dy, s)$ in $B_R^c \times (-\infty, +\infty)$ by (3.54). Hence it holds that

$$\zeta(dy, s) = \zeta^{ac}(dy, s) \quad \text{in } (\mathbf{R}^2 \setminus B_R) \times (-\infty, +\infty),$$

and furthermore,

$$\|\zeta(\cdot, s)\|_{L^\infty(|y| \geq R)} \leq C, \quad -\infty < s < +\infty. \tag{3.56}$$

To establish this outer uniform estimate (3.56), we apply *scaling-back* to $\zeta(dy, s)$ used in [Suzuki (2005)],

$$\zeta(dy, s) = e^{-s} A(dy', s') \quad y' = e^{-s/2} y, \ s' = -e^{-s}, \tag{3.57}$$

to obtain a weak solution

$$A = A(dy', s') \in C_*(-\infty, 0; \mathcal{M}(\mathbf{R}^2))$$

to

$$A_{s'} = \Delta A - \nabla \cdot A\nabla\Gamma * A, \ A \geq 0 \quad \text{in } \mathbf{R}^2 \times (-\infty, 0). \tag{3.58}$$

Scaling invariance of (3.58) ensures a *scaling free ε-regularity* to the classical solution (Lemma 3.3 in §3.7). This property is also valid to the weak solution $A(dy', s')$ generated by a family of classical solutions.

Through (3.57) this ε-regularity produces a similar property to $\zeta(dy, s)$ indicated by (3.118) in §3.8, that is,

$$\zeta(B(y_0, 2r), s) < \varepsilon_0 \quad \Rightarrow \quad \|\zeta(\cdot, s)\|_{L^\infty(B(y_0, r))} \le Cr^{-2}$$

valid for any

$$y_0 \in \mathbf{R}^2, \ r > 0, \ -\infty < s < +\infty.$$

Then (3.56) is obtained for $R \gg 1$ again by the second inequality of (3.54). Inequality (3.56) and

$$\zeta(\mathbf{R}^2, s) = m(x_0)$$

now control the term $\Gamma * z$ in (3.51) for $|y| > R$, while the $|y|^2/4$ term there brings the mass of ζ far away as $s \uparrow +\infty$. This property, combined with the second inequality of (3.54) again, induces the *residual vanishing* indicated by

$$\zeta^{ac}(dy, s) = 0 \quad \text{in } (\mathbf{R}^2 \setminus B_R) \times (-\infty, +\infty),$$

and hence

$$\zeta(dy, s) = \zeta^s(dy, s) \quad \text{in } \mathbf{R}^2 \times (-\infty, +\infty)$$

by the strong maximum principle or unique continuation theorem for parabolic equations, because the density of $\zeta^{ac}(dy, s)$ is a classical solution to a parabolic equation in $\mathbf{R}^2 \times (-\infty, +\infty)$, except for the support of $\zeta^s(dy, s)$.

To establish this residual vanishing rigorously, the second moment in the outer region,

$$\left\langle \left(\frac{|y|^2}{R^2} - 1 \right)_+, \zeta(dy, s) \right\rangle, \quad R \gg 1$$

is efficient. This *outer second moment* is actually non-decreasing because of $|y|^2/4$ term ((3.125) in §3.9), and then there arises a contradiction to the second parabolic envelope as $s \uparrow +\infty$, if the residual vanishing is not achieved.

The weak solution to (3.58),

$$A = A(dy', s') \in C_*(-\infty, 0; \mathcal{M}(\mathbf{R}^2)),$$

is then composed of a finite sum of delta functions, denoted by

$$A(dy', s') = \sum_{j=1}^{m(s')} m'_j(s') \delta_{y'_j(s')}(dy'), \quad s' < 0,$$

satisfying

$$m'_j(s') > 0, \quad 1 \le j \le m(s'),$$

and

$$\sum_{j=1}^{m(s')} m'_j(s') = m(x_0).$$

Here we fix

$$s'_0 < 0, \quad 1 \le j \le m(s'_0),$$

and take $\beta > 0$ to define

$$\tilde{A}_\beta(dy', s') = \beta^2 A(dy, s), \quad y = \beta y' + y'_j(s'_0), \quad s = \beta^2 s' + s'_0.$$

Given $\beta_k \downarrow 0$, we have a subsequence, denoted by the same symbol, such that

$$\tilde{A}_{\beta_k}(dy', s') \rightharpoonup \tilde{A}(dy', s') \quad \text{in } C_*(-\infty, s'_0; \mathcal{M}(\mathbf{R}^2)),$$

and hence this limit

$$\tilde{A}(dy', s') = m'_j(s'_0)\delta_0(ds')$$

is a weak solution to

$$A_{s'} = \Delta A - \nabla \cdot A\nabla\Gamma * A, \ A \ge 0 \quad \text{in } \mathbf{R}^2 \times (-\infty, s'_0)).$$

We now take $s'_k \downarrow -\infty$, define

$$\hat{A}_k(dy', s') = \tilde{A}(dy', s' + s_k),$$

and send $k \to \infty$. The limit

$$\hat{A}(dy', s') = m'_j(s'_0)\delta_0(dy'), \quad -\infty < s' < +\infty,$$

therefore, is a weak solution to

$$\hat{A}_{s'} = \Delta\hat{A} - \nabla \cdot \hat{A}\nabla\Gamma * \hat{A}, \ \hat{A} \ge 0 \quad \text{in } \mathbf{R}^2 \times (-\infty, +\infty),$$

and it follows that

$$m'_j(s_0) = 8\pi$$

from the weak Liouville property, Lemma 3.6 in §3.8. We thus obtain

$$A(dy', s') = 8\pi \sum_{j=1}^m \delta_{y'_j(s')}(dy'), \ s' < 0, \quad m = m(x_0)/(8\pi). \tag{3.59}$$

Equality (3.59) means that the measure $A(dy', s')$ is composed of a finite number of sub-collapses,

$$8\pi\delta_{y'_j(s')}(dy'), \quad 1 \le j \le m,$$

which results in equality (3.30) in §3.3, that is,

$$m(x_0) = 8\pi m.$$

Then, *local second moment* applied to (3.58) assures the *sub-collapse dynamics*

$$\frac{dy'_j}{ds'} = 8\pi\nabla_{y_j}H^0_m(y'_1, \cdots, y'_m) \quad \text{a.e. } s' < 0$$

$$\lim_{s'\uparrow 0} y'_j(s') = 0, \ 1 \le j \le m, \tag{3.60}$$

with

$$H^0_m(y'_1, \cdots, y'_m) = \sum_{1 \le i < j \le m} \Gamma(y'_i - y'_j) \tag{3.61}$$

for $m \ge 2$, and

$$A(dy', s') = 8\pi\delta_0(dy'), \quad s' < 0 \tag{3.62}$$

for $m = 1$ (Remark 3.9 and Remark 3.10 in §3.9). The conclusion $f > 0$ on $\overline{\Omega} \setminus \mathcal{S}$ in (3.28), finally, is derived from the strong maximum principle or the unique continuation theorem for parabolic equations.

3.5   Blowup in Infinite Time

In this section, we describe the profile of the solution blowup in infinite time to (3.1), where $\Omega \subset \mathbf{R}^2$ is a bounded domain with smooth boundary $\partial\Omega$. This profile is proven in §3.12–§3.14.

First, assuming (3.31) in §5.13, we have

$$\exists t_k \uparrow +\infty, \quad \lim_{k\to\infty} \|u(\cdot, t_k)\|_\infty = +\infty. \tag{3.63}$$

Then there is a subsequence, denoted by the same symbol, such that

$$u(x, t + t_k)dx \rightharpoonup \mu(dx, t) \quad \text{in } C_*(-\infty, +\infty; \mathcal{M}(\overline{\Omega})), \tag{3.64}$$

where $\mu(dx, t)$ is a weak solution to (3.1). Second, the ε-regularity, Lemma 3.3 in §3.6 ensures $\sharp\mathcal{S}_t < +\infty$ and the singular part of this $\mu(dx, t)$ takes the form

$$\mu^s(dx, t) = \sum_{x_0 \in \mathcal{S}_t} m_t(x_0)\delta_{x_0}(dx), \quad m_t(x_0) \ge \varepsilon_0 \tag{3.65}$$

for each t, where
$$\mathcal{S}_t = \{x_0 \in \overline{\Omega} \mid \exists x_k \to x_0, \ u(x_k, t + t_k) \to +\infty\}.$$
It holds also that $\mathcal{S}_0 \neq \emptyset$ by (3.63).

To confirm $\mathcal{S}_T \subset \Omega$ for any $T \in \mathbf{R}$, we take $x_0 \in \mathcal{S}_T \cap \partial\Omega$ and the backward self-similar transformation
$$\zeta(dy, s) = (T - t)\mu(dx, t)$$
$$y = (x - x_0)/(T - t), \ s = -\log(T - t).$$
Each $s_k \uparrow +\infty$ takes a subsequence denoted by the same symbol such that the family of measures
$$\zeta(dy, s + s_k), \ k = 1, 2, \cdots,$$
generates a weak solution as $k \to \infty$,
$$\tilde{\zeta}(dy, s) \in C_*(-\infty, +\infty; \mathcal{M}(\overline{L})),$$
to (3.52), where $L \subset \mathbf{R}^2$ is a half-space. It satisfies
$$\tilde{\zeta}(\overline{L}, 0) = m_T(x_0) > 0,$$
which is a contradiction from the result stated in §3.4 to confirm the exclusion of boundary blowup in (3.1). Namely, equation (3.52) with (3.53) implies $m_T(x_0) = 0$ (Lemma 3.5 in §3.11). Thus we obtain $\mathcal{S}_t \subset \Omega$ for any $-\infty < t < +\infty$.

Given
$$-\infty < t_0 < +\infty, \quad x_0 \in \mathcal{S}_{t_0},$$
next, we take
$$\mu_\beta(dx, t) = \beta^2 \mu(dx', dt'), \quad x' = \beta x + x_0, \ t' = \beta^2 t + t_0$$
for $\beta > 0$. Any $\beta_k \downarrow 0$ admits a subsequence, denoted by the same symbol, such that
$$\mu_{\beta_k}(dx, t) \rightharpoonup \tilde{\mu}(dx, t) \quad \text{in } C_*(-\infty, +\infty; \mathcal{M}(\mathbf{R}^2)).$$
This $\tilde{\mu}(dx, t)$ is a weak solution to (3.47) satisfying
$$\tilde{\mu}(\mathbf{R}^2, 0) = m_{t_0}(x_0)$$
and hence it follows that $m_{t_0}(x_0) = 8\pi$ from the weak Liouville property (Lemma 3.6 in §3.13).

Equality (3.65) is now refined as
$$\mu^s(dx, t) = 8\pi \sum_{j=1}^{\ell} \delta_{x_j(t)}(dx), \quad -\infty < t < +\infty, \tag{3.66}$$

where $\ell \leq \lambda/(8\pi)$ is independent of t (Lemma 3.7 in §3.13). This profile is consistent to the instant blowup criterion, Theorem 3.2 in §3.4, and then, total mass quantization

$$\lambda = \|u_0\|_1 \in 8\pi\mathbf{N}$$

is a consequence of the residual vanishing, indicated by

$$\mu^{ac}(dx,t) = 0, \tag{3.67}$$

where $\mu^{ac}(dx,t)$ denotes the absolutely continuous part of $\mu(dx,t)$.

To establish this residual vanishing, first, we apply the arguments used for the exclusion of boundary blowup and the mass normalization of $\mu^s(dx,t)$ as in (3.66). Putting

$$\mathcal{O} = \{x(t) \mid -\infty < t < +\infty\}, \quad x(t) = (x_1(t), \cdots, x_\ell(t)), \tag{3.68}$$

we obtain $\overline{\mathcal{O}} \subset \Omega \setminus D$, where

$$D = \{(x_1, \cdots, x_\ell) \mid x_i = x_j, \exists i \neq j\}$$

(Lemma 3.8 in §3.13). Hence there is $0 < r \ll 1$ such that

$$B(x_i(t), r) \cap \mathcal{S}_t = \{x_i(t)\}, \quad -\infty < t < +\infty, \ 1 \leq i \leq \ell.$$

Extending the argument for radially symmetric solutions [Ohtsuka–Senba–Suzuki (2007); Suzuki (2015a)], second, we take the local second moment of the *defect mesure*. Namely, we derive (3.205) in §3.13 for

$$u_k(\cdot, t) = u(\cdot, t + t_k),$$

using the first volume derivative (Theorem 8.2 in §8.3), that is,

$$\frac{d}{dt}\int_{B(x_i(t),r)}(|x - x_i(t)|^2 - r^2)u_k$$

$$\leq 4\int_{B(x_i(t),r)}u_k - \frac{1}{2\pi}\left(\int_{B(x_i(t),r)}u_k\right)^2 + C\int_{B(x_i(t),r)}|x - x_i(t)|u_k.$$

Making $k \to \infty$, we arrive at (3.209) in §3.13, that is,

$$\frac{dI}{dt} \leq CI, \quad -\infty < t < +\infty$$

for $0 < r \ll 1$ in the sense of distributions in t, where

$$I(t) = \int_{B(x_i(t),r)}(|x - x_i(t)|^2 - r^2)f \leq 0$$

for $f = f(x,t) \geq 0$ defined by

$$f(x,t)dx = \mu^{ac}(dx,t).$$

Then it follows that

$$I(t) \not\equiv 0 \;\Rightarrow\; \lim_{t\uparrow+\infty} I(t) = -\infty,$$

a contradiction. We thus obtain $I(t) \equiv 0$, and hence (3.67) (Theorem 3.11 in §3.14).

We now use the local second moment to follow the collapse dynamics in (3.65). It is shown that $x_j = x_j(t)$ in (3.66),

$$\mu(dx,t) = 8\pi \sum_{j=1}^{\ell} \delta_{x_j(t)}(dx), \quad -\infty < t < +\infty,$$

satisfies

$$\frac{dx_j}{dt} = 8\pi \nabla_{x_j} H_\ell(x_1, \cdots, x_\ell), \quad -\infty < t < +\infty, \; 1 \le j \le \ell, \tag{3.69}$$

where

$$H_\ell = H_\ell(x), \quad x = (x_1, \cdots, x_\ell),$$

is the point vortex Hamiltonian defined by (3.33) (Lemma 3.12 in §3.14).

System (3.69) is an anti-gradient flow with analytic nonlinearity. There arises that

$$\frac{dH_\ell}{dt} = 8\pi |\nabla H_\ell|^2 \ge 0, \quad x = x(t),$$

and the theory of gradient inequality [Huang (2006)] assures the existence of critical points x_*, x^*, of H_ℓ, which realize $\mathcal{O}$ in (3.68) as a connecting orbit of them by the argument in §4.11:

$$\lim_{t\downarrow-\infty} x(t) = x_*, \quad \lim_{t\uparrow+\infty} x(t) = x^*.$$

In particular, the ℓ-th point vortex Hamiltonian, $H_\ell(x)$, must have at least one critical point in $\Omega^\ell \setminus D$ (§3.13). This property of the existence of the critical point of H_ℓ is sensitive to the shape of domain in accordance with the existence of the singular limit of the stationary states studied in the previous chapter, that is, the Boltzmann Poisson equation (Remark 3.16, Remark 3.17 in §3.12 and Remark 3.21 in §3.14).

3.6   Formation of Collapses

Description in §3.4 concerning the blowup in finite time on the Smoluchowski Poisson (3.1) equation becomes simpler if we take the system on the whole space $\mathbf{R}^2$,

$$u_t = \Delta u - \nabla \cdot (u \nabla \Gamma * u) \text{ in } \mathbf{R}^2 \times (0, T), \quad u|_{t=0} = u_0(x) \geq 0, \qquad (3.70)$$

where

$$(\Gamma * u)(x, t) = \int_{\mathbf{R}^2} \Gamma(x - x') u(x', t) \; dx'$$

for $\Gamma = \Gamma(x)$ defined by (3.41). In §3.6–§3.8, we show the quantized blowup mechanism realized in finite time to this (3.70).

First, the property

$$\|\nabla \Gamma * z\|_\infty \leq C(\|z\|_1 + \|z\|_\infty) \qquad (3.71)$$

guarantees that if

$$u_0 \in X \equiv L^1(\mathbf{R}^2) \cap L^\infty(\mathbf{R}^2) \qquad (3.72)$$

there is a unique semi-group solution local-in-time, $0 \leq u \in C([0, T), X)$:

$$u(t) = e^{t\Delta} u_0 + \int_0^t e^{(t-s)\Delta} \nabla \cdot (u \nabla \Gamma * u)(s) \; ds. \qquad (3.73)$$

It is smooth in (x, t) for $t > 0$, provided with the total mass conservation

$$\|u(\cdot, t)\|_1 = \lambda \equiv \|u_0\|_1. \qquad (3.74)$$

Second, if the maximal existence time denoted by $T = T_{\max}$ is finite, it follows that

$$\lim_{t \uparrow T} \|u(\cdot, t)\|_\infty = +\infty, \qquad (3.75)$$

because the existence time of the solution is estimated below by $\|u_0\|_\infty$.

Given $\varphi \in C_0^2(\mathbf{R}^2)$, we have the weak form of (3.70),

$$\frac{d}{dt} \int_{\mathbf{R}^2} u(x, t) \varphi(x) \; dx = \int_{\mathbf{R}^2} u(x, t) \Delta \varphi(x) \; dx$$

$$+ \frac{1}{2} \iint_{\mathbf{R}^2 \times \mathbf{R}^2} \rho_\varphi^0(x, x') u(x, t) u(x', t) \; dx dx' \qquad (3.76)$$

for

$$\rho_\varphi^0(x, x') = -\frac{x - x'}{2\pi |x - x'|^2} \cdot (\nabla \varphi(x) - \nabla \varphi(x')) \in L^\infty(\mathbf{R}^2 \times \mathbf{R}^2),$$

where $C_0^2(\mathbf{R}^2)$ denotes the set of C^2 functions in $\mathbf{R}^2$ with compact support. In fact, equality (3.76) follows from the method of symmetrization and

$$\nabla\Gamma(x) = -\frac{1}{2\pi}\frac{x}{|x|^2}.$$

Then, taking $\varphi = |x|^2$ is justified in (3.76) if we assume

$$I_0 \equiv \int_\Omega |x|^2 u_0(x)\ dx < +\infty, \tag{3.77}$$

which results in

$$\frac{d}{dt}\int_{\mathbf{R}^2} |x|^2 u(x,t)\ dx = 4\lambda - \frac{\lambda^2}{2\pi}. \tag{3.78}$$

In fact, first, the semi-group estimates

$$\|e^{t\Delta}z\|_{L^1(\mathbf{R}^2,(1+|x|^2)dx)} \leq C\|z\|_{L^1(\mathbf{R}^2,(1+|x|^2)dx)}$$

and

$$\|e^{t\Delta}z_{x_j}\|_{L^1(\mathbf{R}^2,(1+|x|^2)dx)} \leq Ct^{-1/2}\|z\|_{L^1(\mathbf{R}^2,(1+|x|^2)dx)},\ j = 1,2$$

arise from

$$(e^{t\Delta}z)(x) = \int_{\mathbf{R}^2} G(x-y,t)z(y)dy$$

for

$$G(x,t) = \left(\frac{1}{4\pi t}\right)^{n/2} e^{-|x|^2/4t},\ n = 2,$$

and $z_{x_j} = \partial z/\partial x_j$. These estimates guarantee the existence of the unique solution

$$u = u(\cdot,t) \in C([0,T),Y)$$

to (3.73) for $0 < T \ll 1$ under (3.72) and (3.77), where

$$Y = L^1(\mathbf{R}^2,(1+|x|^2)dx) \cap L^\infty(\mathbf{R}^2).$$

Second, this solution coincides with the one in X as far as it exists, because of the uniqueness of the latter in X. Then equality (3.78) is derived to this $u = u(\cdot,t)$ in the sense of distibutions in t by putting $\varphi(x) = |x|^2\psi(x/R)$ in (3.76) and making $R \uparrow +\infty$, where $0 \leq \psi = \psi(x) \in C_0^\infty(\mathbf{R}^2)$ is a cut-off function, equal to 1 on $B = B(0,1)$.

Finally, the solution $u = u(\cdot,t)$ exhibits (3.75) if its maximal existence time, denoted by T, is finite because of (3.74) and (3.78). In fact, equality (3.78) imlies

$$\sup_{0\leq t<T}\int_{\mathbf{R}^2} |x|^2 u(x,t)\ dx \leq T(4\lambda - \frac{\lambda^2}{2\pi})_+ + I_0 \tag{3.79}$$

and hence

$$\lim_{t \uparrow T} \int_{\mathbf{R}^2} |x|^2 u(x,t) dx < +\infty$$

besides (3.74).

Remark 3.1. Equality (3.78) implies also $T < +\infty$ if $\lambda > 8\pi$. This simple blowup criterion for (3.70), without concentration condition required as in (3.45) in §3.4 for bounded domains, is due to the scaling invariance of (3.70), indicated by (3.46). Local second moment, furthermore, reduces $u_0 \in Y$ to $u_0 \in L^1 \cap L^\infty(\mathbf{R}^2)$ ([Kurokiba–Ogawa (2003)]). By Theorem 3.3 in this section, on the other hand, $\lambda \leq 8\pi$ implies $T = +\infty$ in (3.70) if $u_0(x) \geq 0$ satisfies (3.72) and (3.77).

Remark 3.2. The above threshold $\lambda = 8\pi$ for the existence of the solution global in time to (3.70) is not valid to (3.1) on bounded domain. In fact, this case can be provided with stationary solutions, and in particular, there may arise $T = +\infty$ and $\|u(\cdot,t)\|_\infty \leq C$ even for $\lambda = \|u_0\|_1 > 8\pi$ (Remark 3.17 in §3.12 and Remark 3.21 in §3.14).

Remark 3.3. If the blowup in infinite time occures to (3.70), there arises a weak solution to (3.47) in §3.4 of full-orbit, $-\infty < t < +\infty$, similarly to (3.1) on bounded domain (Theorem 3.6 in §3.12). Since it contains a collapse with the normalized mass 8π, this property does not arise if $\lambda < 8\pi$. Hence it holds that $T = +\infty$ and $\|u(\cdot,t)\|_\infty \leq C$ if $\lambda < 8\pi$ occurs to (3.70).

Remark 3.4. The *free energy*

$$\mathcal{F}(u) = \int_{\mathbf{R}^2} u(\log u - 1) \, dx - \frac{1}{2} \int \int_{\mathbf{R}^2 \times \mathbf{R}^2} \Gamma(x - x')u \otimes u \, dx dx'$$

is defined for $u \otimes u = u(x,t)u(x',t)$, which satisfies

$$\frac{d}{dt}\mathcal{F}(u) = -\int_{\mathbf{R}^2} u|\nabla(\log u - v)|^2 \, dx \leq 0 \qquad (3.80)$$

formally. Then the dual Trudinger Moser inequality (3.12) on the whole space,

$$\inf\{\mathcal{F}(u) \mid u \geq 0, \ \|u\|_1 = 8\pi\} > -\infty \qquad (3.81)$$

will guarantee global-in-time existence of the uniformly bounded solution to (3.70), again for the case of $\lambda < 8\pi$. Equality (3.80), however, is not justified because we cannot make the right-hand side to be definite. Hence we do not use this free energy in the study of (3.70) on the whole space. Later in §3.11–§3.14, we clarify the role of free energy on bounded domains in accordance with the simplicity of collapses, that is, exclusion of the collision of sub-collapses.

Remark 3.5. If $\lambda < 8\pi$ occurs to (3.70) with $u_0 \in Y$, there is a uniformly bounded global-in-time solution as is confirmed in Remark 3.3. Then its ω-limit set, denoted by $\omega(u_0)$, is well-defined. It is non-empty, connected, uniformly bounded, and invariant under the semi-flow induced by (3.70) in X. If (3.80) is justified for this solution, the value $\mathcal{F}(u)$ is constant on $\omega(u_0)$ by the LaSalle principle. By the argument in §3.12 on bounded domains, based on the invariance of $\omega(u_0)$ under the semi-flow and the strong maximum principle, therefore, we obtain $0 < u_* = u_*(x) \in X$ satisfying

$$\log u_* - v_* = \text{constant}, \quad v_* = \Gamma * u_* \quad \text{in } \mathbf{R}^2$$

with

$$\int_{\mathbf{R}^2} u_* dx = \lambda$$

by (3.80), which implies

$$u_* = \frac{\lambda e^{v_*}}{\int_{\mathbf{R}^2} e^{v_*} dx} \text{ in } \mathbf{R}^2, \quad \int_{\mathbf{R}^2} e^{v_*} dx < +\infty, \tag{3.82}$$

and hence

$$-\Delta v = e^v \text{ in } \mathbf{R}^2, \quad \int_{\mathbf{R}^2} e^v dx < +\infty \tag{3.83}$$

for

$$v = v_* + \log \lambda - \log \left(\int_{\mathbf{R}^2} e^{v_*} dx \right). \tag{3.84}$$

Theorem 2.6 in §2.6, therefore, implies

$$\int_{\mathbf{R}^2} e^v dx = \lambda = 8\pi,$$

a contradiction. Hence equality (3.80) is generally invalid to (3.70) with $u_0 \in Y$.

Remark 3.6. Admitting (3.80) implies also a contradiction if there is a global-in-time solution, bounded in X with $\lambda = 8\pi$ and $u_0 \in Y$. In fact, even in this case there arises that $0 < u_* = u_*(x) \in X$ satisfying (3.82), similarly. It holds also that

$$\int_{\mathbf{R}^2} |x|^2 u_*(x) dx = \int_{\mathbf{R}^2} |x|^2 u_0(x) dx < +\infty \qquad (3.85)$$

by (3.78). We obtain, however, (3.83) for $v = v(x)$ defined by (3.84), and therefore,

$$\int_{\mathbf{R}^2} |x|^2 e^{v(x)} dx = +\infty$$

again by Theorem 2.6 in §2.6. Then it follows that

$$\int_{\mathbf{R}^2} |x|^2 u_*(x) dx = +\infty,$$

contradicting (3.85).

Remark 3.7. These global-in-time behaviors of the solution to (3.70) are described by the concentration compactness principle to the family of probability measures $\{\lambda^{-1} u(x,t) dx\}$ ([Lions (1984)]). First, if $\lambda < 8\pi$ and $u_0 \in Y$, there arises a *vanishing* to this family as $t \uparrow +\infty$. If only $u_0 \in L^1 \cap L^\infty(\mathbf{R}^2)$ is assumed, second, there can be a *dichotomy* to induce $T < +\infty$ for $\lambda > 8\pi$. Third, this alternative is prevented for $u_0 \in Y$, and we have *concentration* as $t \uparrow T < +\infty$ for $\lambda > 8\pi$, There may arise, finally, either concentration or vanishing as $t \uparrow +\infty$ for $\lambda = 8\pi$ and $u_0 \in Y$.

We are now concentrated on (3.70) with $u_0 \in Y$. To begin with, inequality (3.79) implies the boundedness of the blowup set defined by

$$\mathcal{S} = \{x_0 \in \mathbf{R}^2 \cup \{\infty\} \mid \exists x_k \to x_0, \ \exists t_k \uparrow T, \ \lim_{k \to \infty} u(x_k, t_k) = +\infty\}, \quad (3.86)$$

which ensures $\mathcal{S} \subset \mathbf{R}^2$. This fact is obtained by the following lemma, (3.79), and Chebyshev's inequality. It is actually the ε-regularity formulated in §3.4.

Lemma 3.1. *There is $\varepsilon_0 > 0$ such that*

$$\lim_{R \downarrow 0} \limsup_{t \uparrow T} \|u(\cdot, t)\|_{L^1(B(x_0, R))} < \varepsilon_0 \quad \Rightarrow \quad x_0 \notin \mathcal{S}.$$

Proof. The proof is divided into two parts. First, we show the existence of $\varepsilon_0 > 0$ such that

$$\limsup_{t \uparrow T} \|u(\cdot, t)\|_{L^1(\Omega \cap B(x_0, 4R))} < \varepsilon_0$$

$$\Rightarrow \quad \limsup_{t \uparrow T} \int_{\Omega \cap B(x_0, 2R)} [u(\log u - 1)](x, t) \, dx < +\infty \quad (3.87)$$

for $R > 0$. Second, it follows that

$$\limsup_{t \uparrow T} \int_{\Omega \cap B(x_0, 2R)} [u(\log u - 1)](x, t) \, dx < +\infty$$

$$\Rightarrow \quad \limsup_{t \uparrow T} \|u(\cdot, t)\|_{L^\infty(B(x_0, R))} < +\infty. \quad (3.88)$$

Here, a nice cut-off function is used for the proof. It is $\varphi = \varphi_{x_0, R}(x) \in C_0^2(\mathbf{R}^2)$ with the support radius $0 < R \le 1$ satisfying

$$0 \le \varphi = \varphi_{x_0, R}(x) = \begin{cases} 1, & x \in B(x_0, R/2) \\ 0, & x \in B(x_0, R)^c, \end{cases} \quad (3.89)$$

and

$$|\nabla \varphi_{x_0, R}| \le C R^{-1} \varphi^{5/6}, \quad |\nabla^2 \varphi_{x_0, R}| \le C R^{-2} \varphi^{2/3}. \quad (3.90)$$

This φ is obtained as $\varphi = \psi^6$ for $\psi = \psi_{x_0, R}$ satisfying (3.89).

Property (3.87) follows from the use of local entropy, while (3.88) is a consequence of standard argument of parabolic regularity; Gagliardo Nirenberg inequality and Moser's iteration (Chapter 11 of [Suzuki (2005)]). □

Here, we show the following theorem without using $\mathcal{F}(u)$, which assures that $T < +\infty$ implies $\lambda > 8\pi$ in (3.70) with (3.72) and (3.77).

Theorem 3.3. *Given*

$$0 \le u_0 \in L^1(\mathbf{R}^2) \cap L^\infty(\mathbf{R}^2), \quad |x|^2 u_0 \in L^1(\mathbf{R}^2), \quad (3.91)$$

let $u = u(\cdot, t)$ be the solution to (3.70). Assume blowup in finite time of u, that is, (3.75) with $T = T_{\max} < +\infty$. Then

$$u(x, t)dx = \mu(dx, t)$$

is continued up to $t = T$ as in

$$\mu(dx, t) \in C_*([0, T], \mathcal{M}(\mathbf{R}^2)) \quad (3.92)$$

and

$$\langle 1 + |x|^2, \mu(dx, t) \rangle \le C, \quad \mu(\mathbf{R}, t) = \lambda. \quad (3.93)$$

It holds that $\sharp \mathcal{S} < +\infty$, $\mathcal{S} \subset \mathbf{R}^2$, and

$$\mu(dx, T) = \sum_{x_0 \in \mathcal{S}} m(x_0)\delta_{x_0}(dx) + f(x)dx, \qquad (3.94)$$

for the blowup set $\mathcal{S}$ defined by (3.86), where

$$0 < f = f(x) \in L^1(\mathbf{R}^2) \cap C(\mathbf{R}^2 \setminus \mathcal{S}), \quad m(x_0) \in 8\pi\mathbf{N}.$$

Proof. We have readily shown $\mathcal{S} \subset \mathbf{R}^2$ by inequality (3.79) and Lemma 3.1. The weak form (3.76) implies the monotonicity formula,

$$\left| \frac{d}{dt} \int_{\mathbf{R}^2} u(x, t)\varphi(x)\ dx \right| \leq C\|\nabla\varphi\|_{C^1}(\lambda + \lambda^2), \qquad (3.95)$$

for each $\varphi \in C_0^2(\mathbf{R}^2)$. Since this inequality implies

$$\int_0^T \left| \frac{d}{dt} \int_{\mathbf{R}^2} u(x, t)\varphi(x)\ dx \right|\ dt < +\infty,$$

the continuation of $\mu(dx, t) = u(x, t)dx$ to

$$\mu(dx, t) \in C_*([0, T], \mathcal{M}(\mathbf{R}^2))$$

is achieved by (3.74) because $C_0^2(\mathbf{R}^2)$ is dense in $C_\infty(\mathbf{R}^2)$. Recall

$$\mathcal{M}(\mathbf{R}^2) = C_\infty(\mathbf{R}^2)'$$

and $C_\infty(\mathbf{R}^2)$ denotes the set of continuous functions on $\mathbf{R}^2 \cup \{\infty\}$ taking the value zero at ∞. Then (3.93) follows from (3.74) and (3.79).

Second, the ε-regularity, Lemma 3.1, implies

$$\mu(\{x_0\}, T) \geq \varepsilon_0, \quad \forall x_0 \in \mathcal{S}.$$

Then $\mathcal{S} < +\infty$ follows from the total mass conservation,

$$\mu(\mathbf{R}^2, T) = M.$$

It holds also that (3.94) with

$$m(x_0) \geq \varepsilon_0, \quad 0 \leq f = f(x) \in L^1(\mathbf{R}^2).$$

The property $f \in L^\infty_{loc}(\mathbf{R}^2 \setminus \mathcal{S})$ also follows from Lemma 3.1 and (3.92). Therefore, $f = f(x)$ is smooth in $\mathbf{R}^2 \setminus \mathcal{S}$ by the elliptic parabolic regularity of

$$u_t = \Delta u - \nabla \cdot (u\nabla v), \quad -\Delta v = u \quad \text{in } \mathbf{R}^2 \times (0, T)$$

derived from (3.70). Then the strong maximum principle guarantees

$$f = f(x) > 0, \quad \forall x \in \mathbf{R}^2 \setminus \mathcal{S}$$

unless $f \equiv 0$. This case of $f \equiv 0$ is not consistent with $T < +\infty$ because it implies $u_0 \equiv 0$ and hence $u \equiv 0$.

We have thus (3.92)–(3.94) with

$$m(x_0) \geq \varepsilon_0, \quad 0 < f = f(x) \in L^1(\mathbf{R}^2) \cap C(\mathbf{R}^2 \setminus \mathcal{S}),$$

and therefore, only *quantization of the collapse mass*,

$$m(x_0) \in 8\pi\mathbf{N}, \quad \forall x_0 \in \mathcal{S}$$

is left to complete the proof of Theorem 3.3. □

3.7 Improved ε-Regularity

In §3.7–§3.9 we show $m(x_0) \in 8\pi\mathbf{N}$ in (3.94) for any $x_0 \in \mathcal{S}$. The next lemma is a refinement of Lemma 3.1, based on the smoothing effect of several norms of the solution.

Lemma 3.2 ([Senba–Suzuki (2002b)]). *There are $\varepsilon_0 > 0$, $R > 0$, and $t_0 > 0$ such that if u is a classical solution to (3.70) and if*

$$\|u_0\|_{L^1(B(x_0,8R))} < \varepsilon_0, \quad x_0 \in \mathbf{R}^2, \ R > 0,$$

it follows that

$$\sup_{\tau \leq t < t_0} \|u(\cdot,t)\|_{L^\infty(B(x_0,R))} < +\infty \tag{3.96}$$

for any $\tau \in (0, t_0)$.

Proof. We note the following facts in advance.

(1) By (3.95) there is $t_1 \in (0, T)$ such that

$$\|u_0\|_{L^1(B(x_0,8R))} < \varepsilon_0/2 \Rightarrow \sup_{t \in (0,t_1)} \|u(\cdot,t)\|_{L^1(B(x_0,4R))} < \varepsilon_0. \tag{3.97}$$

(2) Property (3.87) relies on the inequality derived from (3.70),

$$\frac{d}{dt} \int_{\mathbf{R}^2} u(\log u - 1)\varphi \, dx + \frac{1}{4} \int_{\mathbf{R}^2} u^{-1}|\nabla u|^2 \varphi \, dx$$
$$\leq 2 \int_{\mathbf{R}^2} u^2 \varphi \, dx + C_\varphi \tag{3.98}$$

applied to $\varphi = \varphi_{x_0,4R}$ in (3.89) - (3.90) (Chapter 11 of [Suzuki (2005)]).

The first term on the right-hand side of (3.98) is absorbed into the second term on the left-hand side by the Gagliardo Nirenberg inequality, under the cost of smallness of the local L^1 norm of the solution as in the conclusion of (3.97), which results in

$$\frac{d}{dt}\int_{\mathbf{R}^2} u(\log u - 1)\varphi\,dx + \frac{1}{8}\int_{\mathbf{R}^2} u^{-1}|\nabla u|^2\varphi\,dx$$
$$\leq C_\varphi, \quad 0 \leq t < t_1. \tag{3.99}$$

At this stage, however, the estimate of

$$\int_{\mathbf{R}^2} u(\log u - 1)\varphi\,dx$$

at $t > 0$ involves that of $t = 0$. This term is a Zygmund norm concerning u_0 (§3.4 of [Suzuki (2015b)]), which is stronger than $\|u_0\|_{L^1(B(x_0, 2R))}$. Hence parabolic smoothing of this norm is necessary to get (3.96), differently from the proof of (3.87) in §3.6.

For this purpose, we derive

$$\frac{dJ}{dt} + 2\int_{\mathbf{R}^2} u^2\varphi\,dx \leq C_R$$

from (3.98) for

$$J = \int_{\mathbf{R}^2} (u\log u + e^{-1})\varphi\,dx,$$

recalling

$$s\log s + e^{-1} \geq 0, \quad s \geq 0.$$

Then we use

$$J = \int_{\mathbf{R}^2} (u\log u + e^{-1})\varphi dx \leq \left\{\int_{\mathbf{R}^2} [(u\log u + e^{-1})\varphi]^{3/2} dx\right\}^{2/3} |B_R|^{1/3},$$

to infer

$$3J^{3/2} \leq 3\int_{\mathbf{R}^2} (u\log u + e^{-1})^{3/2}\varphi^{3/2} dx \cdot |B_R|^{1/2}$$
$$\leq C_R \int_{\mathbf{R}^2} (u\log u + e^{-1})\varphi\,dx$$

and hence

$$\frac{dJ}{dt} + 3J^{3/2} \leq C_R + \int_{\mathbf{R}^2} [-2u^2 + C_R(u\log u + e^{-1})^{3/2}]\varphi\,dx.$$

An elementary inequality to the right-hand side then guarantees

$$\frac{dJ}{dt} + 3J^{3/2} \leq C_R'. \tag{3.100}$$

Here we use

$$\frac{d}{dt}t^{-2} + 3(t^{-2})^{3/2} = t^{-3}$$

to derive

$$J(t) \leq t^{-2}, \quad 0 < t \leq \min\{t_1, t_0\}, \tag{3.101}$$

where $t_0^{-3} = C'_R$. Inequality (3.101) now implies

$$J(s) + \int_s^t dt \int_\Omega u^{-1}|\nabla u|^2 \varphi \, dx \leq Cs^{-2}, \quad 0 < s \leq t \leq \min\{t_1, t_0\}$$

by (3.99). Then, modifying the iteration scheme used for the proof of (3.88), we obtain the conclusion (Chapter 12 of [Suzuki (2005)]). □

By Lemma 3.2 and the monotonicity formula (3.95) applied to direct and reverse directions of t, there exist $0 < \varepsilon_0 \ll 1$, $0 < \sigma_0 \ll 1$, and $C > 0$ such that

$$\|u_0\|_1 < \varepsilon_0 \quad \Rightarrow \quad \sup_{|t|<\sigma_0} \|u(\cdot, t)\|_\infty \leq C,$$

if $u = u(x, t)$ is a classical solution to

$$u_t = \Delta u - \nabla \cdot (u \nabla \Gamma * u) \text{ in } \mathbf{R}^2 \times (-T, T), \quad u_0 = u|_{t=0} \tag{3.102}$$

for $T > 0$. There is also a scaling invariance of (3.102) described by

$$u_\mu(x, t) = \mu^2 u(\mu x, \mu^2 t), \quad \mu > 0. \tag{3.103}$$

Thus if $u = u(x, t)$ is a solution then $u_\mu = u_\mu(x, t)$ also solves (3.102), which induces Lemma 3.2 to the following scaling invariant form.

Lemma 3.3. *There are positive constants ε_0, σ_0, and C such that*

$$\|u_0\|_{L^1(B(x_0, 2R))} < \varepsilon_0, \ u_0 = u|_{t=0}$$
$$\Rightarrow \sup_{t \in [-\sigma_0 R^2, \sigma_0 R^2] \cap (-T, T)} \|u(\cdot, t)\|_{L^\infty(B(x_0, R))} \leq CR^{-2} \tag{3.104}$$

for any $x_0 \in \mathbf{R}^2$ and $R > 0$, if $u = u(x, t) \geq 0$ is a classical solution to (3.102).

3.8   Scaling Limit

To prove $m(x_0) \in 8\pi \mathbf{N}$ in Theorem 3.3, we use the backward self-similar transformation

$$z(y,s) = (T-t)u(x,t)$$
$$y = (x-x_0)/(T-t)^{1/2}, \quad s = -\log(T-t), \tag{3.105}$$

to reach

$$z_s = \nabla \cdot (\nabla z - z\nabla(\Gamma * z + |y|^2/4)) \quad \text{in } \mathbf{R}^2 \times (-\log T, +\infty) \tag{3.106}$$

together with

$$z(y,s) \geq 0, \quad \|z(\cdot,s)\|_1 = \lambda.$$

Similarly to (3.95) it holds that

$$\left| \frac{d}{dt} \int_{\mathbf{R}^2} z(y,s)\varphi(y) \, dy \right| \leq C_\varphi, \quad \varphi \in C_0^2(\mathbf{R}^2). \tag{3.107}$$

This inequality is the monotonicity formula, which guarantees the generation of the weak scaling limit. Thus any $s_k \uparrow +\infty$ admits a subsequence, denoted by the same symbol, such that

$$z(y, s+s_k)dy \rightharpoonup \zeta(dy,s) \quad \text{in } C_*(-\infty, +\infty; \mathcal{M}(\mathbf{R}^2)), \tag{3.108}$$

similarly to Theorem 3.1 in §3.4. This $\zeta(dy,s) \geq 0$ is, furthermore, a weak solution $z = \zeta(dy,s)$ to

$$z_s = \nabla \cdot (\nabla z - z\nabla(\Gamma * z + |y|^2/4)) \quad \text{in } \mathbf{R}^2 \times (-\infty, +\infty). \tag{3.109}$$

To give its precise definition, let $\mathcal{E}$ be the closure in $L^\infty(\mathbf{R}^2 \times \mathbf{R}^2)$ of

$$\{\rho_\varphi + \psi \mid \varphi \in C_0^2(\mathbf{R}^2), \ \psi \in C_0(\mathbf{R}^2 \times \mathbf{R}^2)\}$$

for

$$\rho_\varphi^0(y,y') = -\frac{y-y'}{2\pi|y-y'|^2} \cdot (\nabla\varphi(y) - \nabla\varphi(y')) \in L^\infty(\mathbf{R}^2 \times \mathbf{R}^2),$$

where $C_0(\mathbf{R}^2 \times \mathbf{R}^2)$ denotes the set of continuous functions in $\mathbf{R}^2 \times \mathbf{R}^2$ with compact support. Although $\rho_\varphi(y,y')$ is not continuous at $y = y'$, this $\mathcal{E}$ is separable. Thanks to this property, there is a multiplicated operator

$$0 \leq \mathcal{K} \in L_*^\infty(0,T;\mathcal{E}')$$

in accordance with $\zeta(dy,s)$ generated by (3.108).

It satisfies

$$\|\mathcal{K}(\cdot,s)\|_{\mathcal{E}'} \leq M^2, \quad \mathcal{K}(\cdot,s)|_{C_0(\mathbf{R}^2 \times \mathbf{R}^2)} = \zeta(dy,s) \otimes \zeta(dy',s) \quad \text{a.e. } s,$$

and

$$\frac{d}{ds}\langle\varphi, \zeta(dy, s)\rangle = \langle\Delta\varphi + \frac{y}{2}\cdot\nabla\varphi, \zeta(dy, s)\rangle + \frac{1}{2}\langle\rho_\varphi^0, \mathcal{K}(\cdot, s)\rangle \qquad (3.110)$$

in the sense of distributions in $s \in (-\infty, +\infty)$ for each $\varphi \in C_0^2(\mathbf{R}^2)$, where $C_0(\mathbf{R}^2 \times \mathbf{R}^2)$ denotes the set of continuous functions in $\mathbf{R}^2 \times \mathbf{R}^2$ with compact support.

If the total variations in the sense of measures of the initial values of a family of weak solutions to (3.109) are uniformly bounded, and the associated multiplicated operators are also uniformly bounded in $L_*^\infty(0, T; \mathcal{E}')$, this family generates a weak solution up to a subsequence. These uniform conditions are always satisfied when we generate weak solutions sucessively, starting from the family of classical solutions $\{z(\cdot, t + t_k)\}$, although we do not mention explicitly.

Let $\varphi = \varphi_{x_0, R}(x) \in C_0^2(\mathbf{R}^2)$ be the cut-off function satisfying (3.89) and (3.90). Since

$$\|\nabla\varphi_{x_0, R}\|_{C^1} \leq CR^{-2}, \ 0 < R \leq 1$$

it holds that

$$\left|\int_{\mathbf{R}^2} \varphi_{x_0, R}(x)u(x, t) \ dx - \langle\varphi_{x_0, R}, \mu(dx, T)\rangle\right| \leq C_\lambda R^{-2}(T - t)$$

by (3.95) in §3.6:

$$\left|\frac{d}{dt}\int_{\mathbf{R}^2} u(x, t)\varphi_{x_0, R}(x)dx\right| \leq CR^{-2}, \ 0 < R \ll 1.$$

Any $b > 0$, therefore, admits

$$\left|\int_{\mathbf{R}^2} \varphi_{x_0, b(T-t)^{1/2}}u(x, t)dx - \langle\varphi_{x_0, b(T-t)^{1/2}}, \mu(dx, T)\rangle\right| \leq C_\lambda b^{-2},$$

which implies

$$\limsup_{t\uparrow T} \left|\langle\varphi_{x_0, b(T-t)^{1/2}}, \mu(dx, t)\rangle - m(x_0)\right| \leq C_\lambda b^{-2}, \qquad (3.111)$$

or

$$\limsup_{s\uparrow +\infty} \left|\int_{\mathbf{R}^2} \varphi_{x_0, b}(y)z(y, s)dy - m(x_0)\right| \leq C_\lambda b^{-2}.$$

We thus obtain

$$m(x_0) = \zeta(\mathbf{R}^2, s), \ -\infty < s < +\infty. \qquad (3.112)$$

Equality (3.112) is called the *first parabolic envelope* (Chapter 15 of [Suzuki (2005)]). In other words, the scaling limit $\zeta(dy, s)$ in (3.108) detects the prescaled collapse mass $m(x_0)$ in (3.94) on the whole space $\mathbf{R}^2$ of s.

Similarly, there is the *second parabolic envelope*

$$\langle |y|^2, \zeta(dy, s) \rangle \le C, \quad -\infty < s < +\infty \tag{3.113}$$

derived from

$$\left| \frac{d}{dt} \int_{\mathbf{R}^2} |x - x_0|^2 u(x,t) \varphi_{x_0,R}(x) dx \right| \le C, \ 0 < R \ll 1$$

([Senba (2007)]). This (3.113) implies the *tightness* of the family of measures $\{\zeta(dy, s)\}$ by Chebyshev's inequality. .

The scaling back $A(dy, s)$ of $\zeta(dy, s)$ is now defined by

$$\zeta(dy, s) = e^{-s} A(dy', s'), \ y' = e^{-s/2} y, \ s' = -e^{-s}. \tag{3.114}$$

It is a weak solution to

$$A_{s'} = \nabla \cdot (\nabla A - A \nabla \Gamma * A), \ A \ge 0 \quad \text{in } \mathbf{R}^2 \times (-\infty, 0) \tag{3.115}$$

defined similarly to the weak solution to (3.109), satisfying

$$A(\mathbf{R}^2, s') = m(x_0), \ -\infty < s' < 0.$$

The following lemma is concerned on the full orbit of (3.115), that is, the *weak Liouville property*.

Lemma 3.4. *If*

$$a = a(dy, s) \in C_*(-\infty, +\infty; \mathcal{M}(\mathbf{R}^2))$$

is a weak solution to

$$a_s = \nabla \cdot (\nabla a - a \nabla \Gamma * a), \ a \ge 0 \quad in \ \mathbf{R}^2 \times (-\infty, +\infty) \tag{3.116}$$

satisfying $a(\mathbf{R}^2, s) \le C$ *and* $a(\mathbf{R}^2, 0) = M$, *there arises that either* $M = 0$ *or* $M = 8\pi$.

Proof. The proof is similar to the case of classical solution in [Kurokiba–Ogawa (2003)]. First, we have the invariance of the total mass:

$$a(\mathbf{R}^2, s) = M, \quad -\infty < s < +\infty.$$

Second, we use the local second moment to infer that $M > 8\pi$ and $M < 8\pi$ contradict (3.116) for $s \ge 0$ and $s \le 0$, respectively, if the initial mass $a(dy, 0)$ is sufficiently concentrated at the origin.

Then, the scaling invariance of (3.116),

$$a^\mu(dy, s) = \mu^2 a(dy', s'), \ y' = \mu y, \ s' = \mu^2 s, \tag{3.117}$$

excludes this necessary condition of concentration. $\square$

Remark 3.8. This lemma does not require the uniform boundedness of the total second moment of $a(dy, s)$ as in

$$\langle |y|^2, a(dy, s) \rangle \leq C,$$

because the proof relies on the local second moment to $a(dy, s)$ and the scaling invariance of (3.116).

For (3.115) concerning the half orbit $-\infty < s < 0$, the above proof assures only $m(x_0) \geq 8\pi$. Here we confirm that the improved ε-regularity to the rescaled equation, Lemma 3.3, is valid for the weak solution $A(dy', s')$ to (3.115) generated by a family of classical solutions. Hence we have

$$A(B(y_0', 2R), s') < \varepsilon_0$$
$$\Rightarrow \|A(\cdot, s')\|_{L^\infty(B(y_0', R))} \leq CR^{-2}, \quad \forall s' < 0, \ y_0' \in \mathbf{R}^2, \ R > 0,$$

precisely.

Then we use the scaling back (3.114) to apply this result to (3.115):

$$\zeta(B(e^{s/2} y_0', 2Re^{s/2}), s) < \varepsilon_0$$
$$\Rightarrow \|\zeta(\cdot, s)\|_{L^\infty(B(e^{s/2} y_0, Re^{s/2})} \leq CR^{-2} e^{-s}, \quad \forall s \in \mathbf{R}, \ y_0' \in \mathbf{R}^2, \ R > 0,$$

which reads

$$\zeta(B(y_0, 2r), s) < \varepsilon_0$$
$$\Rightarrow \|\zeta(\cdot, s)\|_{L^\infty(B(y_0, r))} \leq Cr^{-2}, \quad \forall s \in \mathbf{R}, \ y_0 \in \mathbf{R}^2, \ r > 0. \quad (3.118)$$

This property, combined with the second parabolic envelope (3.113), assures

$$\|\zeta(dy, s)\|_{L^\infty(\mathbf{R}^2 \setminus B(0, R_0))} \leq C \quad -\infty < s < +\infty \qquad (3.119)$$

for $R_0 > 0$ sufficiently large. Writing

$$g(y, s) dy = \zeta^{ac}(dy, s),$$

we thus obtain

$$\|g(\cdot, s)\|_1 + \|g(\cdot, s)\|_{L^\infty(\mathbf{R}^2 \setminus B(0, R_0))} \leq C, \qquad (3.120)$$

which plays a fundamental role in controlling the term $\Gamma * z$ in (3.109).

It holds also that the singular part of $\zeta(dy, s)$, denoted by $\zeta^s(dy, s)$, is composed of a finite sum of delta functions of which supports are uniformly bounded in s, again by the above ε-regularity. The coefficients of these delta functions are quantized as 8π,

$$\zeta^s(dy, s) = 8\pi \sum_{y_0 \in \mathcal{S}(s)} \delta_{y_0}(dy), \quad \sharp \mathcal{S}(s) < +\infty, \quad \mathcal{S}(s) \subset B(0, 2R_0).$$

This result of [Senba–Suzuki (2003)] can be proven differently as in §3.4 using

$$A(dy', s')$$

defined by (3.114), where the weak Liouville property is applied to the scaling limit of $A(dy', s')$ by (3.115). We thus end up with

$$\zeta(dy, s) = 8\pi \sum_{j=1}^{m(s)} \delta_{y_j(s)}(dy) + g(y, s)dy \qquad (3.121)$$

in (3.108), where

$$m(s) \in \mathbf{N}, \ y_j(s) \in B(0, 2R_0),$$

and $g = g(y, s) \geq 0$ satisfying (3.120).

3.9   Residual Vanishing

Equality (3.121) implies (3.30), that is, the collapse mass quantization $m(x_0) \in 8\pi\mathbf{N}$, if the residual vanishing,

$$\zeta^{ac}(dy, s) = 0, \quad -\infty < s < +\infty, \qquad (3.122)$$

is proven. Inequality (3.122) actually holds by the following facts:

(1) The regular part of $\zeta(dy, s)$,

$$\zeta^{ac}(dy, s),$$

   is swept away to $y = \infty$ as $s \uparrow +\infty$ by the $|y|^2/4$ term in (3.109).
(2) By (3.120), a uniform L^p bound, $1 \leq p \leq \infty$, arises to the interaction term

$$z\nabla(\Gamma * z)$$

   of (3.109), in the outer region.
(3) The second parabolic envelope (3.113), on the other hand, implies that

$$\zeta(\mathbf{R}^2 \setminus B(0, R), s), \quad R \gg 1$$

   is uniformly small.

Proof of (3.30) in §3.3. We have $R_0 \gg 1$ satisfying

$$\zeta(\mathbf{R}^2 \setminus B_{R/8}, s) < \varepsilon_0/2, \quad -\infty < s < +\infty$$

for $R \geq 16R_0$ by (3.113), and hence

$$\|z_k(\cdot, s)\|_{L^1(\mathbf{R}^2 \setminus B_{R/4})} < \varepsilon_0, \quad -\tilde{s}_k < s < \tilde{s}_k$$

for

$$z_k(y, s) = z(y, s + s_k), \quad s_k \uparrow +\infty,$$

passing to a subsequence, by (3.108) in §3.8. Then we apply Lemma 3.3 for $u_k(x, t)$ associated with $z_k(y, s)$, to obtain

$$\|z_k(\cdot, s)\|_{L^\infty(\mathbf{R}^2 \setminus B_{R/2})} \leq C, \quad -\tilde{s}_k < s < \tilde{s}_k$$

similarly to (3.118). Consequently, it follows that

$$\|\nabla\Gamma * z_k(\cdot, s)\|_{L^\infty(\mathbf{R}^2 \setminus B_R)} \leq C, \quad -\tilde{s}_k < s < \tilde{s}_k$$

by

$$\|z_k(\cdot, s)\|_1 = M.$$

For

$$\varphi = \varphi(|y|) \in C^2(|y| \geq R)$$

satisfying

$$\varphi|_{r=R} = 0$$

and

$$0 \leq \varphi(r) \leq Cr^2, \quad \varphi_r \geq 0, \quad r = |y| > R,$$

it holds that

$$\frac{d}{ds} \int_{|y| \geq R} z_k \varphi \, dy \geq \int_{|y| > R} (\Delta z_k)\varphi + z_k \nabla(\Gamma * z_k + \frac{|y|^2}{4}) \cdot \nabla\varphi \, dy$$

$$\geq \int_{|y| \geq R} z_k \left(\Delta\varphi - C\varphi_r + \frac{r}{2}\varphi_r \right) dy, \quad -\tilde{s}_k < s < \tilde{s}_k.$$

Here we take the *outer second moment*

$$\varphi(r) = \xi(r/R), \quad \xi(r) = r^2 - 1$$

to achieve

$$\Delta\varphi + \frac{r}{2}\varphi_r \geq C\varphi_r, \quad r \geq R \tag{3.123}$$

with $R \gg 1$. In fact, inequality (3.123) is reduced to

$$\frac{1}{R^2}\Delta\xi + \left(\frac{r}{2} - \frac{C}{R} \right) \xi_r \geq 0, \quad r \geq 1. \tag{3.124}$$

Since $\Delta\xi = 4$ and $\xi_r = 2r$, inequality (3.124) is achieved for $R \gg 1$ as in

$$\frac{4}{R^2} + r^2 - \frac{C}{R}2r \geq 0, \quad r \geq 1.$$

Then there arises that

$$\frac{d}{ds}\int_{\mathbf{R}^2}\left(\frac{|y|^2}{R^2}-1\right)_+ z_k\,dy\geq 0,\quad -\tilde{s}_k<s<\tilde{s}_k.$$

Taking the limit, we obtain

$$\frac{d}{ds}\left\langle\left(\frac{|y|^2}{R^2}-1\right)_+,\zeta(dy,s)\right\rangle\geq 0,\quad -\infty<s<+\infty\qquad(3.125)$$

in the sense of distributions in s. If there is s_0 such that

$$\left\langle\left(\frac{|y|^2}{R^2}-1\right)_+,\zeta(dy,s_0)\right\rangle>0,$$

therefore, it follows that

$$\lim_{s\uparrow+\infty}\left\langle\left(\frac{|y|^2}{R^2}-1\right)_+,\zeta(dy,s)\right\rangle=+\infty,$$

contradicting to (3.113).

It thus holds that

$$\zeta^{ac}(dy,s)=0\quad\text{in }(\mathbf{R}^2\setminus B_R)\times(-\infty,+\infty).$$

Hence we obtain (3.122) by the strong maximum principle or the unique continuation theorem for parabolic equations to $g=g(y,s)$ because it satisfies

$$g_s=\Delta g-\nabla\cdot g\nabla(w+\frac{|y|^2}{4}),\quad -\Delta w=g\quad\text{in }H$$

for

$$H=\bigcup_{-\infty<s<+\infty}(\mathbf{R}^2\setminus\mathcal{S}_s)\times\{s\},\quad\mathcal{S}_s=\text{supp }\zeta^s(dy,s).$$

Since $g(y,s)\equiv 0$ in (3.121), we obtain

$$m(s)\cdot 8\pi=m(x_0)m\quad\forall s\in\mathbf{R}.$$

Hence $m(s)=m$ is independent of s and we end up with the proof of Theorem 3.3 by

$$\zeta(dy,s)=8\pi\sum_{j=1}^m\delta_{y_j(s)}(dy)\qquad(3.126)$$

and $m(x_0)=8\pi m.$ □

Remark 3.9. Putting $\varphi = |x|^2$ in (3.110) is justified by (3.112)–(3.113) as in (3.78) for (3.70) with (3.91) in §3.6, which results in

$$\frac{dI}{ds} = m(x_0) - \frac{m(x_0)^2}{2\pi} + I, \quad \text{a.e. } s \in \mathbf{R}$$

for the total second moment

$$I = \langle |y|^2, \zeta(dy, s) \rangle$$

([Senba (2007)]). If $m(x_0) = 8\pi$, therefore, it holds that

$$I(s) \equiv 0$$

by (3.113), and hence

$$\zeta(dy, s) = 8\pi \delta_0(dy), \quad -\infty < s < +\infty \tag{3.127}$$

([Suzuki (2013)]). We say that the blowup point $x_0 \in \mathcal{S}$ is *simple* if

$$m(x_0) = 8\pi.$$

Then it holds that

$$A(dy', s') = 8\pi \delta_0(dy'), \quad -\infty < s' < 0 \tag{3.128}$$

by (3.127). Equality (3.127) is valid, however, regardless of the choice of $s_k \uparrow +\infty$. Hence we obtain

$$z(y, s)dy \; \rightharpoonup \; 8\pi \delta_0(dy) \text{ in } \mathcal{M}(\mathbf{R}^2), \quad s \uparrow +\infty. \tag{3.129}$$

Remark 3.10. If $m(x_0) = 8\pi m$ for $m \geq 2$, on the contrary, we say that *collision of sub-collapses* occurs at $x_0 \in \mathcal{S}$. Equality (3.126) then implies

$$A(dy', s') = 8\pi \sum_{j=1}^{m} \delta_{y_j'(s')}(dy'), \quad -\infty < s' < 0. \tag{3.130}$$

In this case we obtain the sub-collapse dynamics,

$$\frac{dy_j'}{ds'} = 8\pi \nabla_{y_j'} H_\ell(y_1', \cdots, y_\ell'), \; 1 \leq j \leq m, \quad \text{a.e. } s' < 0, \tag{3.131}$$

where

$$H_\ell(y_1', \cdots, y_\ell') = \sum_{1 \leq i < j \leq m} \Gamma(y_i' - y_j'), \quad \Gamma(y') = \frac{1}{2\pi} \log \frac{1}{|y'|} \tag{3.132}$$

is the point vortex Hamiltonian on the whole space. Then the second equality in (3.60),

$$\lim_{s' \uparrow 0} y_j'(s') = 0, \; 1 \leq j \leq \ell$$

follows from

$$|y_j(s)| \leq C, \quad -\infty < s < +\infty \tag{3.133}$$

in (3.126).

Equation (3.131) with (3.132) is derived from the weak form of (3.115), that is,

$$\frac{d}{ds'}\langle \varphi, A(dy', s') \rangle = \langle \Delta\varphi, A(dy', s') \rangle + \frac{1}{2}\langle \rho_\varphi^0, \mathcal{K}(\cdot, s') \rangle, \quad \text{a.e. } s' < 0$$

$$\rho_\varphi^0(y', y'') = -\frac{y' - y''}{2\pi|y' - y''|^2} \cdot (\nabla\varphi(y') - \nabla\varphi(y'')). \tag{3.134}$$

Here, $\varphi \in C_0^2(\mathbf{R}^2)$,

$$0 \leq \mathcal{K} = \mathcal{K}(\cdot, s') \in L_*^\infty(-\infty, 0; \mathcal{E}')$$

is the associated multiplicated operator satisfying

$$\mathcal{K}(\cdot, s')|_{C_0(\mathbf{R}^2 \times \mathbf{R}^2)} = A(dy', s') \otimes A(dy'', s'),$$

and $\mathcal{E}$ is the closure of

$$\mathcal{E}_0 = \{\psi + \rho_\varphi^0 \mid \psi \in C_0(\mathbf{R}^2 \times \mathbf{R}^2), \ \varphi \in C_0^2(\mathbf{R}^2)\}$$

in $L^\infty(\mathbf{R}^2 \times \mathbf{R}^2)$. The local second moment applied to (3.134), thus, traces the movement of sub-collapses,

$$y_j'(s'), \quad 1 \leq j \leq m,$$

as in the proof of Lemma 3.12 in §3.14, which induces (3.131) and (3.132).

Remark 3.11. We have thus (3.126) with (3.133). Inequality (3.113), on the other hand, is uniform with respect to the selection of the sequence $t_k \uparrow T$, and hence so is R_0 in (3.119). We thus obtain the type II blowup rate,

$$\lim_{t \uparrow T}(T - t)\|u(\cdot, t)\|_{L^\infty(B(x_0, b(T-t)^{1/2}))} = +\infty, \quad \exists b > 0, \ \forall x_0 \in \mathcal{S}$$

in any case of $T < +\infty$. If $x_0 \in \mathcal{S}$ is simple, this equality is refined as

$$\lim_{t \uparrow T}(T - t)\|u(\cdot, t)\|_{L^\infty(B(x_0, b(T-t)^{1/2}))} = +\infty, \quad \forall b > 0 \tag{3.135}$$

by (3.127).

Remark 3.12. For radially symmetric solution

$$u = u(|x|, t),$$

it holds always that (3.128) and (3.135), while exact blowup rate is detected [Mizoguchi (2022)]. It coincides actually with the rate realized in the inner region constructed by the matched asymptotic expansion [Herrero–Velázquez (1996)],

$$L(t) = (T - t)^{1/2} e^{-\sqrt{|\log(T-t)|/2}}.$$

If $u = u(|x|, t)$ is a solution to (3.70) with $T < +\infty$, more precisely, it holds always that

$$u(x, t) = \frac{K}{L(t)^2} u_*(\sqrt{K} x / L(t))(1 + o(1)), \quad t \uparrow T$$

locally uniformly in $x \in \mathbf{R}^2$, where $K = 4^{-1} \exp(e^2 + \gamma)$, γ is the Euler constant, and

$$u_*(x) = \frac{8}{(1 + |x|^2)^2}$$

is a stationary solution associated with the Liouville equation (2.83) in §2.6,

$$-\Delta v = e^v, \ v \le v(0) = \log 8 \ \text{in } \mathbf{R}^2, \quad \int_{\mathbf{R}^2} e^v dx < +\infty,$$

by $u_* = e^v$, which results in $\|u_*\|_1 = 8\pi$.

Remark 3.13. There is a solution with $m(x_0) = 2$ to (3.70), constructed by the method of matched asymptotic expansion [Luckhaus–Sugiyama–Velázquez (2012)].

3.10 Bounded Domains

In §3.10 and §3.11, we study the solution to (3.1) blowup in finite time, where $\Omega \subset \mathbf{R}^2$ is a bounded domain with smooth boundary $\partial\Omega$. Differently from (3.70) with (3.91), there arises the case

$$T = +\infty, \quad \|u(\cdot, t)\|_\infty \le C$$

even for $\lambda = \|u_0\|_1 > 8\pi$ in accordance with the existence of stationary solutions.

First, the Poisson equation (3.5),

$$-\Delta v = u \ \text{in } \Omega, \quad v|_{\partial\Omega} = 0.$$

admits the Green function denoted by $G = G(x, x')$, and it holds that

$$v(x, t) = \int_\Omega G(x, x') u(x', t) \, dx'.$$

This $G = G(x, x')$ is smooth in $\overline{\Omega} \times \overline{\Omega} \setminus \overline{D}$ for $D = \{(x, x) \mid x \in \Omega\}$. The singularity on the diagonal D is distinguished according to the interior and the boundary. Second, interior singularity is described by

$$G(x, x') = \Gamma(x - x') + K(x, x'), \tag{3.136}$$

where

$$K = K(x, x') \in C^{1+\theta, \theta}(\Omega \times \overline{\Omega}) \cap C^{\theta, 1+\theta}(\overline{\Omega} \times \Omega) \tag{3.137}$$

for $0 < \theta < 1$. Third, boundary singularity around $x_0 \in \partial\Omega$ of $G(x, x')$ is described by the conformal diffeomorphism

$$X : \overline{\Omega \times B(x_0, 2R)} \to \overline{\mathbf{R}_+^2} = \{(X_1, X_2) \mid X_2 > 0\} \tag{3.138}$$

defined for $0 < R \ll 1$, satisfying

$$X(\partial\Omega \cap B(x_0, 2R)) \subset \partial\mathbf{R}_+^2. \tag{3.139}$$

Using the reflection

$$X = \begin{pmatrix} X_1 \\ X_2 \end{pmatrix} \mapsto X_* = \begin{pmatrix} X_1 \\ -X_2 \end{pmatrix}, \tag{3.140}$$

thus, a parametrix of the Poisson part of (3.1),

$$-\Delta v = u \text{ in } \Omega, \quad v|_{\partial\Omega} = 0,$$

near $\partial\Omega$, is given by

$$E(X, X') = \Gamma(X - X') - \Gamma(X - X'_*), \tag{3.141}$$

respectively. More precisely, we obtain

$$G(x, x') = E(X, X') + K(x, x') \tag{3.142}$$

with

$$K = K(x, x') \in (C^{1+\theta, \theta} \cap C^{\theta, 1+\theta})(\overline{\Omega \cap B(x_0, R)} \times \overline{\Omega \cap B(x_0, R)}) \tag{3.143}$$

([Suzuki (2013)]).

Using these properties of the Green function, we obtain the result stated in §5.13.

Theorem 3.4. *Let $u = u(x, t)$ be the classical solution to (3.1). Assume the blowup in finite time, $T < +\infty$, and define the blowup set by*

$$\mathcal{S} = \{x_0 \in \overline{\Omega} \mid \exists x_k \to x_0, \, \exists t_k \uparrow T, \, \lim_{k \to \infty} u(x_k, t_k) = +\infty\}. \tag{3.144}$$

Then it holds that

$$\sharp \mathcal{S} < +\infty, \quad \mathcal{S} \subset \Omega,$$

and

$$u(x,t)dx \rightharpoonup \sum_{x_0 \in \mathcal{S}} m(x_0)\delta_{x_0}(dx) + f(x)\, dx \quad in\ \mathcal{M}(\overline{\Omega}) \tag{3.145}$$

as $t \uparrow T$, where

$$0 < f = f(x) \in L^1(\Omega) \cap C(\overline{\Omega} \setminus \mathcal{S})$$

and

$$m(x_0) = 8\pi\ell,\ \ell \in \mathbf{N}. \tag{3.146}$$

There arises that (3.128) and (3.131) with (3.130) in §3.9 if $\ell = 1$ and $\ell \geq 2$, respectively, for $A(dy', s')$ generated by (3.105), (3.108), and (3.114) in §3.8.

The proof is the same as in Theorem 3.3 in §3.6 on the whole plane based on the weak form to (3.1),

$$\frac{d}{dt} \int_\Omega u\varphi\, dx = \int_\Omega u\Delta\varphi\, dx + \frac{1}{2} \iint_{\Omega \times \Omega} \rho_\varphi(x, x')u \otimes u\, dx dx'$$

valid to $\varphi \in \mathcal{X}$, where $u \otimes u = u(x,t)u(x',t)$,

$$\rho_\varphi(x, x') = \nabla\varphi(x) \cdot \nabla_x G(x, x') + \nabla\varphi(x') \cdot \nabla_{x'} G(x, x'),$$

and

$$\mathcal{X} = \left\{ \varphi \in C^2(\overline{\Omega}),\ \left.\frac{\partial\varphi}{\partial\nu}\right|_{\partial\Omega} = 0 \right\}. \tag{3.147}$$

In fact, this weak form induces the *monotonicity formula*

$$\left| \frac{d}{dt} \int_\Omega u\varphi\, dx \right| \leq C\|\nabla\varphi\|_{C^1}$$

and then (3.145) with $\sharp\mathcal{S} < +\infty$ follows from *ε-regularity* derived from the interior and boundary regularities of $G(x, x')$, indicated by (3.136) with (3.137) and (3.142) with (3.143), respectively.

By this weak formulation, the weak solution

$$\mu = \mu(dx, t) \in C_*([0, T], \mathcal{M}(\overline{\Omega})), \quad \mathcal{M}(\overline{\Omega}) = C(\overline{\Omega})'$$

to (3.1) is defined similarly as in §5.13. Then we obtain the principle of the generation of weak solution from bounded sequence of weak solutions with bounded associated multiplicated operators (Theorem 3.1 in §3.4). There arises also the instant blowup criterion, that is, over mass with concentration (Theorem 3.2 in §3.4).

Once the exclusion of boundary blowup, $\mathcal{S} \subset \Omega$, is proven in §3.11, there arises the collapse mass quantization, $m(x_0) \in 8\pi\mathbf{N}$, and the sub-collapse dynamics (3.128) and (3.131) by the method of weak scaling limit based on the weak Liouville property.

We conclude this section with a relation between the free energy and the simplicity of the blowup point, that is, $\ell = 1$ in (3.146) for any $x_0 \in \mathcal{S}$. In fact, free energy is easier to handle with for the case of bounded domains. For (3.1), this free energy is defined in accordance with the Green function $G = G(x, x')$, that is,

$$\mathcal{F}(u) = \int_\Omega u(\log u - 1) \, dx - \frac{1}{2}\iint_{\Omega \times \Omega} G(x, x')u \otimes u \, dxdx', \qquad (3.148)$$

Then it follows that

$$\frac{d}{dt}\mathcal{F}(u) = -\int_\Omega u|\nabla(\log u - v)|^2 \, dx \leq 0 \qquad (3.149)$$

for the classical solution $u = u(\cdot, t) > 0$ as is confirmed in §3.1. In fact, we have

$$u = u(x, t) > 0 \quad \text{in } \overline{\Omega} \times (0, T)$$

unless $u_0 \equiv 0$ by the strong maximum principle.

Theorem 3.5 ([Suzuki (2018)]). *Any $x_0 \in \mathcal{S}$ is simple if*

$$T < +\infty, \quad \lim_{t \uparrow T} \mathcal{F}(u(\cdot, t)) > -\infty \qquad (3.150)$$

in (3.1).

Proof. By (3.148) and (3.150) it holds that

$$\int_0^T \int_\Omega u|\nabla(\log u - v)|^2 \, dxdt < +\infty. \qquad (3.151)$$

Writing (3.224) as

$$u_t = \nabla \cdot u\nabla(\log u - v), \quad u\frac{\partial}{\partial\nu}(\log u - v)\bigg|_{\partial\Omega} = 0,$$

we obtain

$$\left|\frac{d}{dt}\int_\Omega \varphi u \, dx\right| \leq \left|\int_\Omega u\nabla(\log u - v) \cdot \nabla\varphi \, dx\right|$$

$$\leq \|\nabla\varphi\|_\infty \lambda^{1/2}\|u^{1/2}\nabla(\log u - v)\|_2$$

for any $\varphi \in \mathcal{X}$ in (3.147), where $\lambda = \|u_0\|_1$.

With $\varphi = \varphi_{x_0,R}$ satisfying (3.89)–(3.90) in §3.6, therefore, it follows that

$$|\langle \varphi_{x_0,R}, \mu(dx,t)\rangle - \langle \varphi_{x_0,R}, \mu(dx,T)\rangle|$$

$$\leq \int_t^T \left|\frac{d}{dt}\int_\Omega \varphi_{x_0,R}\cdot u\,dx\right|dt$$

$$\leq \|\nabla\varphi_{x_0,R}\|_\infty \lambda^{1/2}\int_t^T \|u^{1/2}\nabla(\log u - v)\|_2\,dt$$

$$\leq C\lambda^{1/2}R^{-1}(T-t)^{1/2}\left\{\int_t^T\int_\Omega u|\nabla(\log u - v)|^2\,dxdt\right\}^{1/2}$$

for $x_0 \in \mathcal{S}$. Putting $R = b(T-t)^{1/2}$, we thus obtain

$$|\langle \varphi_{x_0,b(T-t)^{1/2}}, \mu(dx,t)\rangle - \langle \varphi_{x_0,b(T-t)^{1/2}}, \mu(dx,T)\rangle|$$

$$\leq C\lambda^{1/2}b^{-1}\left\{\int_t^T\int_\Omega u|\nabla(\log u - v)|^2\,dxdt\right\}^{1/2}$$

for any $b > 0$. The equality

$$\lim_{b\uparrow+\infty}\limsup_{t\uparrow T}|\langle \varphi_{x_0,b(T-t)^{1/2}}, \mu(dx,t)\rangle - m(x_0)| = 0,$$

derived from (3.111), is thus improved as

$$\limsup_{t\uparrow T}|\langle \varphi_{x_0,b(T-t)^{1/2}}, \mu(dx,t)\rangle - m(x_0)| = 0, \quad \forall b > 0 \qquad (3.152)$$

by (3.151), which implies

$$\zeta(B(0,b),s) = m(x_0), \quad \forall b > 0.$$

Then it follows that

$$\zeta(dy,s) = m(x_0)\delta_0(dy), \quad -\infty < s < +\infty$$

in (3.108), and hence

$$A(dy',s') = m(x_0)\delta_0(dy'), \quad s' < 0. \qquad (3.153)$$

The case $\ell \geq 2$ in (3.130) thus does not occur by Theorem 3.4, and therefore, it holds that $m(x_0) = 8\pi$. $\qquad\square$

Remark 3.14. If $x_0 \in \mathcal{S}$ is simple there arises also the *free energy transmission*

$$\lim_{t\uparrow T}\mathcal{F}_{x_0,b(T-t)^{1/2}}(u(\cdot,t)) = +\infty, \quad \forall b > 0, \qquad (3.154)$$

where

$$\mathcal{F}_{x_0,R}(u) = \int_{\Omega\cap B(x_0,R))} u(\log u - 1)\,dx$$

$$- \frac{1}{2}\iint_{(\Omega\cap B(x_0,R))\times(\Omega\cap B(x_0,R)))} u\otimes u\,dxdx'$$

stands for the local free energy (Chaper 16 of [Suzuki (2005)]).

3.11   Exclusion of Boundary Blowup Points

Here we show $\mathcal{S} \subset \Omega$ in Theorem 3.4, following [Suzuki (2013)]. Let $u = u(x,t) > 0$ be the classical solution to (3.1) satisfying $T < +\infty$, and define the blowup set $\mathcal{S}$ by (3.144).

Given $x_0 \in \mathcal{S}$, we define $z(y,s)$ by (3.105), and get the weak scaling limit. Hence if $x_0 \in \Omega$, any $s_k \uparrow +\infty$ admits a subsequence, denoted by the same symbol, satisfying (3.108). This $\zeta(dy,s)$ is a weak solution to (3.109) satisfying (3.112)–(3.113),

$$\zeta(\mathbf{R}^2, s) = m(x_0), \quad \langle |y|^2, \zeta(dy,s)\rangle \leq C, \quad -\infty < s < +\infty, \qquad (3.155)$$

and then $m(x_0) \in 8\pi\mathbf{N}$ follows with the sub-collapse dynamics stated in Theorem 3.4 in §3.10.

If $x_0 \in \partial\Omega$ were true, this $\zeta(dy,s)$ is supported on the closure of a half space L. Hence it holds that

$$\zeta = \zeta(dy,s) \in C_*(-\infty,+\infty; \mathcal{M}(\mathbf{R}^2)), \quad \mathrm{supp}\,\zeta(dy,s) \subset \overline{L}$$

and

$$\zeta(\overline{L}, s) = m(x_0) > 0, \quad \langle |y|^2, \zeta(dy,s) \leq C, \quad -\infty < s < +\infty.$$

It is also a weak solution to

$$z_s = \nabla \cdot (\nabla z - z\nabla(E * z + \frac{|y|^2}{4}) \quad \text{in } L \times (-\infty, +\infty)$$

with

$$\frac{\partial z}{\partial \nu} - z\frac{\partial}{\partial \nu}(E * z + \frac{|y|^2}{4})\Big|_{\partial L} = 0.$$

We assume $L = \mathbf{R}_+^2$ without loss of generality, to obtain (3.52)–(3.53) in §3.4. Then we get a contradiction by the following lemma, and hence $\partial\Omega \cap \mathcal{S} = \emptyset$.

Lemma 3.5 ([Suzuki (2013)]). *If*

$$\zeta = \zeta(dy,s) \in C_*(-\infty,+\infty; \mathcal{M}(\mathbf{R}^2))$$

is a weak solution to (3.52)–(3.53) satisfying (3.155), then it follows that $m(x_0) = 0$.

Proof. Let

$$\mathcal{Y}_0 = \{\varphi \in C_0^2(\overline{\mathbf{R}}^2) \mid \frac{\partial\varphi}{\partial\nu}\Big|_{\partial\mathbf{R}_+^2} = 0\}$$

$$\rho_\varphi^0(y,y') = \nabla\varphi(y) \cdot \nabla_y E(y,y') + \nabla\varphi(y') \cdot \nabla_{y'} E(y,y'),$$

for $E = E(y, y')$ defined by (3.53) in §3.4, and $\mathcal{E}$ be the closure of

$$\mathcal{E}_0 = \{\psi + \rho_\varphi^0 \mid \psi \in C_0^2(\overline{\mathbf{R}_+^2} \times \overline{\mathbf{R}_+^2}), \ \varphi \in \mathcal{Y}_0\}$$

in $L^\infty(\Omega \times \Omega)$. Then it holds that

$$\frac{d}{ds}\langle \varphi, \zeta(dy, s)\rangle = \left\langle \Delta\varphi + \frac{y}{2} \cdot \nabla\varphi, \zeta(dy, s)\right\rangle + \frac{1}{2}\left\langle \rho_\varphi^0, \mathcal{K}(\cdot, s)\right\rangle_{\mathcal{E},\mathcal{E}'}, \quad \text{a.e. } s \in \mathbf{R}$$

for any $\varphi \in \mathcal{Y}_0$, where

$$0 \le \mathcal{K} = \mathcal{K}(\cdot, s) \in L_*^\infty(-\infty, +\infty; \mathcal{E}')$$
$$\mathcal{K}(\cdot, s)|_{C_0(\mathbf{R}^2 \times \mathbf{R}^2)} = \zeta(dy, s) \otimes \zeta(dy', s). \tag{3.156}$$

Take $\xi = \xi(|y|) \in C_0^2(\mathbf{R}^2)$ in $0 \le \xi \le 1$, $\xi = 1$ on $B(0, 1)$, and $\xi = 0$ on $B(0, 2)^c$, and put $\varphi = |y|^2 \xi_R$ for $\xi_R(y) = \xi(y/R)$. It holds that $\varphi \in \mathcal{Y}_0$, and therefore,

$$\int_{s_1}^{s_2}\left\langle \Delta\varphi + \frac{y}{2} \cdot \nabla\varphi, \zeta(dy, s)\right\rangle + \frac{1}{2}\left\langle \rho_\varphi^0, \mathcal{K}(\cdot, s)\right\rangle_{\mathcal{E},\mathcal{E}'} \ ds$$
$$= [\langle \varphi, \zeta(dy, s)\rangle]_{s=s_1}^{s=s_2}, \quad -\infty < s_1 < s_2 < +\infty. \tag{3.157}$$

Letting $R \uparrow +\infty$ in (3.157), we apply the dominated convergence theorem to the left-hand side of (3.157), regarding (3.155). Since

$$\Delta|y|^2 = 4, \quad y \cdot \nabla|y|^2 = 2|y|^2$$

we obtain

$$\lim_{R\uparrow+\infty} \int_{s_1}^{s_2}\langle \Delta\varphi + \frac{y}{2} \cdot \nabla\varphi, \zeta(dy, s)\rangle ds$$
$$= 4(s_2 - s_1)m(x_0) + \int_{s_1}^{s_2} I(s) \ ds, \tag{3.158}$$

where

$$I(s) = \langle |y|^2, \zeta(dy, s)\rangle.$$

The second term on the left-hand side of (3.157) is treated by (3.156). Since

$$(y - y_*') \cdot y + (y' - y_*) \cdot y' = |y|^2 - 2y \cdot y_*' + |y_*|^2$$
$$= |y - y_*'|^2 = |y' - y_*|^2$$

we obtain

$$\nabla\Gamma(y - y') \cdot (\nabla|y|^2 - \nabla|y'|^2) = \frac{1}{\pi}$$
$$\nabla\Gamma(y - y_*') \cdot \nabla|y|^2 + \nabla\Gamma(y' - y_*) \cdot \nabla|y'|^2 = \frac{1}{\pi},$$

and therefore,

$$\rho^0_{|y|^2}(y, y') = 0. \tag{3.159}$$

Then it holds that

$$\begin{aligned}
\rho^0_\varphi(y, y') &= \nabla(|y|^2 \xi_R(y)) \cdot \nabla_y E(y, y') + \nabla(|y'|^2 \xi_R(y')) \cdot \nabla_{y'} E(y, y') \\
&= \{|y|^2 \nabla \xi_R(y) \cdot \nabla_y E(y, y') + |y'|^2 \nabla \xi_R(y') \cdot \nabla_{y'} E(y, y')\} \\
&\quad + \{\xi_R(y) \nabla |y|^2 \cdot \nabla_y E(y, y') + \xi_R(y') \nabla |y'|^2 \cdot \nabla_{y'} E(y, y')\} \\
&= I + II
\end{aligned} \tag{3.160}$$

for $\varphi = |y|^2 \xi_R$.

For the term II on the right-hand side of (3.160), we divide $\mathbf{R}^2_+ \times \mathbf{R}^2_+$ in the yy' plane into three parts. First, for $|y| < 2R$, $y \in \mathbf{R}^2_+$ we have

$$II = (\xi_R(y) - \xi_R(y'))\nabla |y|^2 \cdot \nabla_y E(y, y') \tag{3.161}$$

by (3.159). Then we obtain

$$\begin{aligned}
|II| &\leq \|\nabla \xi\|_\infty \cdot \frac{|y - y'|}{R} \cdot 2|y| \cdot \frac{1}{\pi |y - y'|} \\
&\leq \frac{2}{\pi} \|\nabla \xi\|_\infty \varphi_{0,4R}(y) \cdot \frac{|y|}{R}
\end{aligned}$$

by

$$\frac{|y - y'|}{|y - y'_*|} \leq 1, \quad y, y' \in \mathbf{R}^2_+.$$

A similar estimate is valid for $|y'| < 2R$ by the symmetry, while there arises

$$|y| \geq 2R, \ |y'| \geq 2R \quad \Rightarrow \quad II = 0.$$

We thus obtain

$$|II| \leq C(\varphi_{0,4R}(y)\frac{|y|}{R} + \varphi_{0,4R}(y')\frac{|y'|}{R}), \quad y, y' \in \mathbf{R}^2_+. \tag{3.162}$$

Inequality (3.162) implies

$$\begin{aligned}
&|\langle II, \mathcal{K}(\cdot, s)\rangle_{\mathcal{E}, \mathcal{E}'}| \\
&\leq \langle C(\varphi_{0,4R}(y)\frac{|y|}{R} + \varphi_{0,4R}(y')\frac{|y'|}{R}), \mathcal{K}(\cdot, s)\rangle_{\mathcal{E}, \mathcal{E}'} \\
&= C\left\langle \varphi_{0,4R}(y)\frac{|y|}{R} + \varphi_{0,4R}(y')\frac{|y'|}{R}, \zeta(dy, s) \otimes \zeta(dy', s) \right\rangle_{\mathcal{E}, \mathcal{E}'}
\end{aligned}$$

by (3.156), while the right-hand side of (3.162) is uniformly bounded. The dominated convergence theorem again, therefore, assures

$$\lim_{R \uparrow + \infty} \int_{s_1}^{s_2} \langle II, \mathcal{K}(\cdot, s)\rangle_{\mathcal{E}, \mathcal{E}'} ds = 0. \tag{3.163}$$

For the first term on the right-hand side of (3.160), denoted by I, we use

$$I = (|y|^2 - |y'|^2)\nabla\xi_R(y) \cdot \nabla_y E(y, y')$$
$$+ |y'|^2(\nabla\xi_R(y) - \nabla\xi_R(y')) \cdot \nabla_{y'} E(y, y')$$
$$= III + IV$$

and divide $\mathbf{R}_+^2 \times \mathbf{R}_+^2$ into four parts. First, for $|y| < 4R$ and $|y'| < 4R$ it holds that

$$|III| \leq C\left(\varphi_{0,8R}(y)|y| + \varphi_{0,8R}(y')|y'|\right) \cdot \frac{|y|}{R}$$
$$|IV| \leq C\left(\varphi_{0,8R}(y') + \varphi_{0,8R}(y)\right) \cdot \frac{|y'|^2}{R^2}.$$

Second, there arises

$$|y| \geq 2R, \quad |y'| \geq 2R \quad \Rightarrow \quad III = IV = 0.$$

Third, for $|y| < 2R$ and $|y'| \geq 4R$, we return to

$$III + IV = I = |y|^2\nabla\xi_R(y) \cdot \nabla_y E(y, y') + |y'|^2\nabla\xi_R(y') \cdot \nabla_{y'} E(y, y')$$

and use $|y - y'| \geq 2R$, to obtain

$$\left||y|^2\nabla\xi_R(y) \cdot \nabla_y E(y, y')\right| \leq C\varphi_{0,4R}(y)\frac{|y|^2}{R^2}$$
$$|y'|^2\nabla\xi_R(y') \cdot \nabla_{y'} E(y, y') = 0.$$

For $|y| \geq 4R$ and $|y'| < 2R$, finally, there arises

$$|y|^2\nabla\xi_R(y) \cdot \nabla_y E(y, y') = 0$$
$$\left||y'|^2\nabla\xi_R(y') \cdot \nabla_{y'} E(y, y')\right| \leq C\varphi_{0,4R}(y')\frac{|y'|^2}{R^2}$$

similarly.

We end up with

$$|I| \leq C\left(\varphi_{0,8R}(y)(1 + |y|) + \varphi_{0,8R}(y')(1 + |y'|)\right)$$
$$\cdot\left(\frac{|y|}{R} + \frac{|y|^2}{R^2} + \frac{|y'|}{R} + \frac{|y'|^2}{R^2}\right), \quad y, y' \in \mathbf{R}_+^2$$

and then the dominated convergence theorem implies

$$\lim_{R\uparrow+\infty} \int_{s_1}^{s_2} \langle I, \mathcal{K}(\cdot, s)\rangle_{\varepsilon,\varepsilon'} ds = 0 \tag{3.164}$$

by (3.112)–(3.113).

From (3.163)–(3.164) it follows that

$$\lim_{R\uparrow+\infty}\int_{s_1}^{s_2}\langle\rho_\varphi^0,\mathcal{K}(\cdot,s)\rangle_{\mathcal{E},\mathcal{E}'}ds = 0, \tag{3.165}$$

and hence we obtain

$$I(s_2) - I(s_1) = \int_{s_1}^{s_2} 4m(x_0) + I(s)\ ds, \quad -\infty < s_1 < s_2 < \infty$$

by (3.157), (3.158), and (3.165). The mapping

$$s \in (-\infty,+\infty) \quad \mapsto \quad I(s) = \langle|y|^2,\zeta(dy,s)\rangle \tag{3.166}$$

is locally absolutely continuous, and there arises

$$\frac{dI}{ds} = 4m(x_0) + I \quad \text{a.e. } s. \tag{3.167}$$

Then it follows that

$$\lim_{s\uparrow+\infty} I(s) = +\infty$$

from $m(x_0) > 0$, which contradicts (3.113). Thus we obtain $m(x_0) = 0$. $\quad\square$

The proof of Theorem 3.4 in §3.10 is thus complete.

3.12   Global-in-Time Solution

In §3.12–§3.14, we show the quantized blowup mechanism to (3.1) for the solution blowup in infinite time. Here, blowup in infinite time indicates

$$T = +\infty, \quad \limsup_{t\uparrow+\infty}\|u(\cdot,t)\|_\infty = +\infty. \tag{3.168}$$

In the other case of

$$T = +\infty, \quad \sup_{t\geq 0}\|u(\cdot,t)\|_\infty < +\infty, \tag{3.169}$$

the orbit $\mathcal{O} = \{u(\cdot,t)\}$ is pre-compact in $C^2(\overline{\Omega})$, and then, the ω-limit set defined by

$$\omega(u_0) = \{u_\infty \in C^2(\overline{\Omega}) \mid \exists t_k \uparrow +\infty,\ \lim_{k\to\infty}\|u(\cdot,t_k) - u_\infty\|_{C^2} = 0\} \tag{3.170}$$

is non-empty, compact, and connected [Henry (1981)]. In accordance with the free energy, acting as a Lyapunov function to (3.1), this $\omega(u_0)$ is contained in the set of stationary solutions.

In fact, first, this $\omega(u_0)$ is invariant under the flow, which means that the solution $\tilde{u} = \tilde{u}(\cdot,t)$ to (3.224) and (3.225), or (3.224) and (3.227), taking initial value $u_\infty \in \omega(u_0)$, remains $\tilde{u}(\cdot, t) \in \omega(u_0)$, as far as it exists: $0 < t \leq t_0$ with some $t_0 > 0$.

Since

$$\|u_\infty\|_1 = \|\tilde{u}(\cdot, t)\|_1 = \|u_0\|_1 = \lambda > 0, \tag{3.171}$$

second, we obtain $u_\infty \not\equiv 0$, and hence $\tilde{u}(x, t) > 0$ on $\overline{\Omega}$ for $0 < t \leq t_0$ by the strong maximum principle. Then $\mathcal{F}(\tilde{u}(\cdot, t))$ is well-defined by (3.148), and is independent of t from the LaSalle principle (§3.3 of [Suzuki (2015b)]. It thus follows that

$$\frac{d}{dt}\mathcal{F}(\tilde{u}(\cdot, t)) = 0, \quad t > 0 \tag{3.172}$$

in (3.149), and equalities (3.171) and (3.172) imply

$$\log \tilde{u}(\cdot, t) - \tilde{v}(\cdot, t) = \text{constant}, \quad \tilde{v}(\cdot, t) = (-\Delta)^{-1}\tilde{u}(\cdot, t)$$

for $(-\Delta)^{-1}$ defined by

$$((-\Delta)^{-1}u)(x) = \int_\Omega G(x, x')u(x')\, dx'.$$

Then we obtain

$$\tilde{u}(\cdot, t) = \frac{\lambda e^{\tilde{v}(\cdot, t)}}{\int_\Omega e^{\tilde{v}(x, t)}dx}$$

by $\|\tilde{u}(\cdot, t)\|_1 = \lambda$. Sending $t \downarrow 0$, we get

$$\tilde{v}(\cdot, t) \to v_\infty \equiv (-\Delta)^{-1}u_\infty \quad \text{uniformly on } \overline{\Omega}$$

as well as

$$-\Delta v_\infty = \frac{\lambda e^{v_\infty}}{\int_\Omega e^{v_\infty}dx}, \quad v_\infty|_{\partial\Omega} = 0 \tag{3.173}$$

and hence $v_\infty \in E_\lambda$, where E_λ denotes the set of solutions to the Boltzmann Poisson equation (3.11) in §3.1.

Remark 3.15. Theory of gradient inequality in Chapter 4, furthermore, guarantees $\sharp\omega(u_0) = 1$, and therefore, $u(\cdot, t)$ converges in C^∞ topology to a stationary state u_∞ defined by

$$u_\infty = \frac{\lambda e^{v_\infty}}{\int_\Omega e^{v_\infty}dx}$$

for $v_\infty = v_\infty(x)$ satisfying (3.173).

By Theorem 3.4 in §3.10, on the other hand, blowup in finite time $T < +\infty$ does not arise to (3.1) in the case of $\lambda = 8\pi$. Therefore, if there is no classical stationary solution for this value of λ, the orbit $\mathcal{O}$ cannot be bounded. Blowup in infinite time, (3.168), thus occurs to (3.1) and $\lambda = 8\pi$ if Ω is close to a disc, because $E_{8\pi} = \emptyset$ in this case (Remark 3.21 in §3.14).

Here we recall the argument in §3.5 for the blowup in infinite time, (3.168). In fact, there arises that (3.63), that is,

$$\exists t_k \uparrow +\infty, \quad \lim_{k \to \infty} \|u(\cdot, t_k)\|_\infty = +\infty. \tag{3.174}$$

Principle of the generation of weak solution, Theorem 3.1 in §3.4, then, works to

$$u_k(\cdot, t) = u_k(\cdot, t + t_k), \quad k = 1, 2, \cdots,$$

and passing to a subsequence we obtain

$$u(x, t + t_k)dx \rightharpoonup \mu(dx, t) \quad \text{in } C_*(-\infty, +\infty; \mathcal{M}(\overline{\Omega})), \tag{3.175}$$

where $\mu = \mu(dx, t)$ is a weak solution to (3.1). We have now the following theorem concerning the singular part of $\mu(dx, t)$.

Theorem 3.6. *It holds that*

$$\mu^s(dx, t) = 8\pi \sum_{x_0 \in \mathcal{S}_t} \delta_{x_0}(dx), \quad -\infty < t < +\infty \tag{3.176}$$

in (3.175), where $\mathcal{S}_t$ is the blowup set defined by

$$\mathcal{S}_t = \{x_0 \in \overline{\Omega} \mid \exists x_k \to x_0, \ \lim_{k \to \infty} u(x_k, t + t_k) = +\infty\}.$$

It holds also that $\mathcal{S}_t \subset \Omega$ and $\sharp\mathcal{S}_t = \ell$ is independent of t.

The regular part of $\mu(dx, t)$, second, vanishes as follows.

Theorem 3.7. *It holds that*

$$\mu^{ac}(dx, t) = 0, \quad -\infty < t < +\infty,$$

and therefore, blowup in infinite time, (3.168), does not occur to (3.1) if

$$\lambda = \|u_0\|_1 \notin 8\pi\mathbf{N}.$$

Writing

$$\mu(dx, t) = 8\pi \sum_{j=1}^{\ell} \delta_{x_j(t)}(dx), \quad -\infty < t < +\infty, \qquad (3.177)$$

furthermore, we obtain the collapse dynamics as follows.

Theorem 3.8. *It holds that*

$$\frac{dx_j}{dt} = 8\pi \nabla_{x_j} H_\ell(x_1, \cdots, x_\ell), \quad 1 \leq j \leq \ell, \ -\infty < t < +\infty, \qquad (3.178)$$

where

$$H_\ell = H_\ell(x_1, \cdots, x_\ell)$$

is the ℓ-th point vortex Hamiltonian defined by (3.33) in §5.13. This solution $x = (x_j(t))$ to (3.178), furthermore, makes a pre-compact orbit

$$\{x(t) \mid -\infty < t < +\infty\}$$

in $\Omega^\ell \setminus D$ with its closure contained there, where

$$D = \{x = (x_1, \cdots, x_\ell) \in \Omega^\ell \mid \exists i \neq j, \ x_i = x_j\} \qquad (3.179)$$

is the diagonal set. Consequently, there arises a critical point of

$$H_\ell = H_\ell(x_1, \cdots, x_\ell),$$

and therefore, blowup in infinite time deos not occur even to $\lambda = 8\pi\ell$, $\ell \in \mathbf{N}$, if there is no critical point of H_ℓ.

These theorems illustrate rather similar features of the blowup solution in infinite time to those of the family of stationary solutions, that is,

(1) quantized blowup mechanism without collision.
(2) Hamiltonian control of the location of blowup points.

This Hamiltonian control, however, is kinetic as in (3.175), (3.177), and (3.178).

Remark 3.16. If Ω is convex, there is no critical point of $H_\ell = H_\ell(x_1, \cdots, x_\ell)$ for $\ell \geq 2$, and $H_1 = R(x)$ takes a unique critical point [Grossi–Takahashi (2010)]. Therefore, blowup in infinite time occurs only to $\lambda = 8\pi$ for such domains, and furthermore,

$$u(x, t)dx \ \rightharpoonup \ 8\pi\delta_{x_0}(dx) \quad \text{in } \mathcal{M}(\overline{\Omega})$$

occurs as $t \uparrow +\infty$ in this case, where $x_0 \in \Omega$ is the critical point of $R = R(x)$.

Remark 3.17. A class of convex domains admits stationary solutions even for $\lambda \geq 8\pi$ ([Chang, Cheng, and Lin (2003)], §4.3 of [Suzuki (2020)]). If Ω is a ball and the initial value is radially symmetric, however, $\lambda > 8\pi$ implies blowup in finite time of the solution, similarly to (3.70) with (3.74) and (3.77) in §3.6.

3.13   Initial Mass Quantization

Here we show Theorem 3.6 and Theorem 3.7 in §3.12. The proof is divided into several steps.

In the first step, we notice the ε–regularity, Lemma 3.2 in §3.7, which implies that the singular part of $\mu(dx,t)$ in (3.175), denoted by $\mu^s(dx,t)$, is a finite sum of delta functions:

$$\mu^s(dx,t) = \sum_{x_0 \in \mathcal{S}_t} m_t(x_0)\delta_{x_0}(dx), \quad m_t(x_0) \geq \varepsilon_0. \qquad (3.180)$$

Then we show the exclusion of boundary blowup and collapse mass normalization, which means $\mathcal{S} \subset \Omega$ and $m_t(x_0) = 8\pi$ for any t and $x_0 \in \mathcal{S}_t$, respectively. Here we provide a proof of both of these properties using the scaling limit [Suzuki (2018)].

Lemma 3.6. *It holds that $\mathcal{S}_t \subset \Omega$ and $m_t(x_0) = 8\pi$ in (3.180).*

Proof. The fact $\partial\Omega \cap \mathcal{S}_t = \emptyset$ for (3.224) with (3.227) is due to the parametrix $E(x,x')$ in (3.141), and the proof is similar to the case of blowup in finite time as is described in §3.5.

In fact, assuming $x_0 \in \partial\Omega \cap \mathcal{S}_T$ in (3.180), we take

$$\zeta(dy,s) = (T-t)\mu(dx,t)$$
$$y = (x-x_0)/(T-t)^{1/2}, \quad s = -\log(T-t).$$

This $\zeta(dy,s)$ is put it to be zero where it is not defined. Then there arises the generation of weak solution, and hence any $s_k \uparrow +\infty$ admits a subsequence denoted by the same symbol and the limit measure $\tilde{\zeta}(dy,s)$ such that

$$\zeta(dy,s+s_k) \rightharpoonup \tilde{\zeta}(dy,s) \quad \text{in } C_*(-\infty,+\infty;\mathcal{M}(\mathbf{R}^2)).$$

This $\tilde{\zeta}(dy,s)$ has the support included in a closed half space independent of s, denoted by $\overline{L}$, and is a weak solution to (3.52) in §3.4:

$$z_s = \Delta z - \nabla \cdot z\nabla\left(E*z + \frac{|y|^2}{4}\right) \quad \text{in } L \times (-\infty,+\infty)$$

$$\left.\frac{\partial z}{\partial\nu} - z\frac{\partial}{\partial\nu}\left(E*z + \frac{|y|^2}{4}\right)\right|_{\partial L} = 0.$$

It holds also that $\tilde{\zeta}(\overline{L},s) = m_T(x_0) > 0$, and then we get a contradiction by Lemma 3.5 in §3.11.

Given $x_0 \in \Omega \cap \mathcal{S}_0$, we show $m_0(x_0) = 8\pi$ in (3.180). Then we get the desired conclusion by translating the time variable t. For this purpose we assume $x_0 = 0 \in \Omega$ without loss of generality, and take

$$\mu_\beta(dx,t) = \beta^2\mu(dx',dt'), \quad x' = \beta x, \quad t' = \beta^2 t$$

for $\beta > 0$. It is a weak solution in $(\beta^{-1}\Omega) \times (-\infty, +\infty)$, and each $\beta_k \downarrow 0$ admits a subsequence, denoted by the same symbol, and the limit measure $\tilde{\mu}(dx, t)$ such that

$$\mu_{\beta_k}(dx, t) \rightharpoonup \tilde{\mu}(dx, t) \quad \text{in } C_*(-\infty, +\infty; \mathcal{M}(\mathbf{R}^2)).$$

This $\tilde{\mu}(dx, t)$ is a weak solution to (3.116) in §3.8. Then we obtain

$$\tilde{\mu}(\mathbf{R}^2, t) \leq \lambda, \quad \tilde{\mu}(\mathbf{R}^2, 0) = m_0(x_0),$$

which implies $m_0(x_0) = 8\pi$ by Lemma 3.4. $\qquad \square$

By Lemma 3.6 it holds that

$$\mu(dx, t) = \sum_{x_0 \in \mathcal{S}_t} 8\pi \delta_{x_0}(dx) + f(x, t)dx \tag{3.181}$$

in (3.175), with $\mathcal{S}_t \subset \Omega$ and $0 \leq f = f(\cdot, t) \in L^1(\Omega)$. Since

$$\mu(dx, t) \in C_*(-\infty, +\infty; \mathcal{M}(\overline{\Omega})), \tag{3.182}$$

the set

$$Q = \bigcup_{-\infty < t < +\infty} (\Omega \setminus \mathcal{S}_t) \times \{t\}$$

is open in $\Omega \times (-\infty, +\infty)$. From the elliptic parabolic regularity, furthermore, this $f = f(x, t) \geq 0$ is smooth in Q, and satisfies

$$f_t = \nabla \cdot (\nabla f - f\nabla v), \quad -\Delta v = f \quad \text{in } Q$$

with smooth v. Hence it is positive everywhere unless identically zero, because $u(\cdot, t) > 0$ in Ω for $0 < t \ll 1$.

The second step for the proof of initial mass quantization is the following lemma.

Lemma 3.7. *The number $\sharp \mathcal{S}_t = \ell(t)$ in (3.181) is independent of t, denoted by $\ell(t) = \ell$.*

Proof. Since $\mu(dx, t) \in C_*(-\infty, +\infty; \mathcal{M}(\overline{\Omega}))$ in (3.175) is a weak solution to (3.1), there is $0 \leq \nu \in L_*^\infty(-\infty, +\infty; \mathcal{E}')$ satisfying

$$\nu|_{C(\overline{\Omega} \times \overline{\Omega})} = \mu \otimes \mu$$

and

$$\frac{d}{dt} \langle \xi, \mu(dx, t) \rangle = \langle \Delta \xi, \mu(dx, t) \rangle + \frac{1}{2} \langle \rho_\xi, \nu(t) \rangle \tag{3.183}$$

in the sense of distributions in t for each $\xi \in \mathcal{X}$.

Here, we put

$$\mathcal{X} = \{\xi \in C^2(\overline{\Omega}) \mid \frac{\partial \xi}{\partial \nu}\Big|_{\partial \Omega} = 0\}$$

and

$$\rho_\xi(x, x') = \nabla\xi(x) \cdot \nabla_x G(x, x') + \nabla\xi(x') \cdot \nabla_{x'} G(x, x'),$$

and furthermore, $\mathcal{E}$ is the closure of

$$\mathcal{E}_0 = \{\rho_\xi + \psi \mid \xi \in \mathcal{X}, \ \psi \in C(\overline{\Omega} \times \overline{\Omega})\}$$

in $L^\infty(\Omega \times \Omega)$. Thus we obtain

$$-\int_{\mathbf{R}} \eta'(t) \langle \xi, \mu(dx, t) \rangle \ dt = \int_{\mathbf{R}} \eta(t) [\langle \Delta\xi, \mu(dx, t) \rangle + \frac{1}{2} \langle \rho_\xi, \nu(t) \rangle] \ dt \quad (3.184)$$

for any $\eta = \eta(t) \in C_0^1(\mathbf{R})$.

Let

$$\mathcal{S}_t = \{x_i(t) \mid 1 \leq i \leq \ell(t)\} \quad (3.185)$$

in (3.181), recalling $\ell(t) = \sharp \mathcal{S}_t$. Take

$$-\infty < t_0 < +\infty, \ 1 \leq i \leq \ell(t_0),$$

and put $x_0 = x_i(t_0)$. Then we have $0 < \delta \ll 1$ and $0 < r \ll 1$ such that

$$\sharp(\mathcal{S}_t \cap B(x_0, 4r)) \leq 1, \quad |t - t_0| < \delta \quad (3.186)$$

by (3.182).

We may assume $x_0 = 0$ without loss of generality. Let

$$\tilde{m}(t) = \begin{cases} 8\pi, \ \sharp(\mathcal{S}_t \cap B(0, r)) = 1 \\ 0, \quad \text{otherwise,} \end{cases} \quad (3.187)$$

and note that (3.182) implies

$$\limsup_{t \to t_0} \tilde{m}(t) \leq \tilde{m}(t_0) = 8\pi. \quad (3.188)$$

Let $\{x_i(t)\} = \mathcal{S}_t \cap B(0, r)$ for $|t - t_0| < \delta$ if $\tilde{m}(t) = 8\pi$ and put

$$\tilde{x}_i(t) = \begin{cases} x_i(t), \ \tilde{m}(t) = 8\pi \\ 0, \quad \tilde{m}(t) = 0. \end{cases} \quad (3.189)$$

Let, furthermore,

$$\xi = |x|^2 \varphi(x)$$

for $\varphi = \varphi_r(x)$, where $\varphi_r = \varphi_r(x)$, $0 < r \ll 1$, is the cut-off function,

$$0 \le \varphi_r = \varphi_r(x) \in C_0^\infty(\Omega),$$

satisfying

$$\varphi_r(x) = \begin{cases} 1, \ x \in B(0, r/2) \\ 0, \ x \in \mathbf{R}^2 \setminus B(0, r). \end{cases}$$

We obtain

$$\|\nabla \xi\|_{C^1} \le C$$

by

$$\frac{\partial^2 \xi}{\partial x_i \partial x_j} = 2\delta_{ij}\varphi + 2x_j \frac{\partial \varphi}{\partial x_j} + 2x_j \frac{\partial \varphi}{\partial x_i} + |x|^2 \frac{\partial^2 \varphi}{\partial x_i \partial x_j},$$

and hence

$$|\langle \Delta \xi, \mu(dx, t) \rangle| + \frac{1}{2} |\langle \rho_\xi, \nu(t) \rangle| \le C_\lambda \equiv C(\lambda + \lambda^2), \ t \in \mathbf{R}.$$

There arises, therefore, that

$$\left| \frac{d}{dt} \langle \xi, \mu(dx, t) \rangle \right| \le C_\lambda$$

in the sense of distributions in t by (3.184), which implies

$$|\langle \xi, \mu(dx, t_2) \rangle| - \langle \xi, \mu(dx, t_1) \rangle| \le C_\lambda |t_2 - t_1|, \ t, t' \in \mathbf{R}. \tag{3.190}$$

Since

$$\lim_{r \downarrow 0} \int_\Omega |x|^2 \varphi_r(x) f(x, t) dx = 0, \quad t \in \mathbf{R}$$

for

$$f(\cdot, t) \in L^1(\Omega)$$

in (3.181), inequality (3.190) implies

$$\left| \tilde{m}(t_2) |\tilde{x}_i(t_2)|^2 - \tilde{m}(t_1) |\tilde{x}_i(t_1)|^2 \right| \le C_\lambda |t_2 - t_1|$$
$$\forall t_1, t_2 \in (t_0 - \delta, t_0 + \delta)$$

for $\tilde{m}(t)$ and $\tilde{x}_i(t)$ defined by (3.187) and (3.189), respectively. We thus obtain

$$\tilde{m}(t) |\tilde{x}_i(t)|^2 \in W^{1,\infty}(t_0 - \delta, t_0 + \delta). \tag{3.191}$$

Given $a \in \mathbf{R}^2$ in $|a| \ll 1$, we apply the above argument, replacing

$$B(0, r), \quad |x|^2 \varphi_r(x)$$

by

$$B(a, r), \quad |x - a|^2 \varphi_r(x - a),$$

respectively. Then we obtain, as in (3.191),

$$\tilde{m}(t)|\tilde{x}_i(t) - a|^2 \in W^{1,\infty}(t_0 - \delta, t_0 + \delta). \tag{3.192}$$

Assume

$$\exists t_k \to t_0, \quad \tilde{m}(t_k) = 0.$$

Then it holds that

$$0 = \lim_{k \to \infty} \tilde{m}(t_k)|\tilde{x}_i(t_k) - a|^2 = \tilde{m}(t_0)|\tilde{x}_i(t_0) - a|^2 = 8\pi|a|^2$$

by (3.192), a contradiction for $0 < |a| \ll 1$. We thus obtain

$$\tilde{m}(t) = 8\pi, \quad |t - t_0| < \delta', \tag{3.193}$$

for some $0 < \delta' < \delta$, and therefore,

$$\tilde{x}_i(t) = x_i(t), \quad |t - t_0| < \delta'$$

in (3.189). Hence it holds that

$$|x_i(t) - a|^2 = |x_i(t)|^2 - 2(x_i(t), a) + |a|^2$$
$$\in W^{1,\infty}(t_0 - \delta', t_0 + \delta')$$

for any $|a| \ll 1$. Putting $a = 0$, we obtain

$$|x_i(t)|^2 \in W^{1,\infty}(t_0 - \delta', t_0 + \delta'),$$

and therefore,

$$(x_i(t), a) \in W^{1,\infty}(t_0 - \delta', t_0 + \delta')$$

for any $0 < |a| \ll 1$. We thus end up with

$$x_i = x_i(t) \in W^{1,\infty}(t_0 - \delta', t_0 + \delta'), \quad |\dot{x}_i| \leq C. \tag{3.194}$$

Since the properties (3.193) and (3.194) are valid for any $t_0 \in \mathbf{R}$ and $1 \leq i \leq \ell(t)$, the number $\ell(t) \in \mathbf{N}$ is locally constant. Hence it is independet of t, denoted by ℓ. □

Remark 3.18. The bound C of $|\dot{x}|$ in (3.194) is estimated by r in (3.186) form the above proof.

We have readily shown (3.175) in §3.12 with

$$\mu(dx, t) = 8\pi \sum_{j=1}^{\ell} \delta_{x_j(t)}(dx) + f(x, t)dx, \qquad (3.195)$$

and

$$0 \le f(\cdot, t) \in L^1(\Omega), \quad f \in C\left(\bigcup_{-\infty < t < +\infty} (\overline{\Omega} \setminus \mathcal{S}_t) \times \{t\} \right), \qquad (3.196)$$

where

$$\mathcal{S}_t = \{x_j(t) \mid 1 \le j \le \ell\}.$$

We put $\mathcal{O} = \{x(t)\}$ for $x(t) = (x_1(t), \cdots, x_\ell(t))$ and recall

$$D = \{(x_1, \cdots, x_\ell) \mid x_i = x_j, \ \exists i \ne j\}.$$

The third step for the proof of initial mass quantization is the analysis of $\mathcal{O}$.

Lemma 3.8. *The closure of* $\{x(t)\}$ *is contained in* $\Omega^\ell \setminus D$.

Proof. We apply the scaling argument used for the proof of $\partial\Omega \cap \mathcal{S}_t = \emptyset$ in Lemma 3.6 in §3.13. Then it follows that

$$\liminf_{t \to \pm\infty} \operatorname{dist}(x_j(t), \partial\Omega) > 0, \quad 1 \le j \le \ell. \qquad (3.197)$$

Similarly, assuming the existence $t_k \to \pm\infty$ and $i \ne j$ such that

$$\lim_{k \to \infty} |x_i(t_k) - x_j(t_k)| = 0,$$

we take a subsequence satisfying

$$\mu(dx, t + t_k) \ \rightharpoonup \ \tilde{\mu}(dx, t) \quad \text{in } C_*(-\infty, +\infty; \mathcal{M}(\overline{\Omega}))$$

and

$$\lim_{k \to \infty} x_i(t_k) = \lim_{k \to \infty} x_j(t_k) = x_0 \in \Omega.$$

This $\tilde{\mu}(dx, t)$, however, is a weak solution to (3.224) with (3.227) satisfying

$$\tilde{\mu}(\{x_0\}, 0) \ge 16\pi.$$

This property induces a contradiction by the proof of Lemma 3.6, using scaling limit and the weak Liouville peperty. □

Remark 3.19. Writing

$$\mu^s(dx,t) = 8\pi \sum_{j=1}^{\ell} \delta_{x_j(t)}(dx), \tag{3.198}$$

we obtain

$$x_j \in W^{1,\infty}(-\infty,+\infty;\mathbf{R}^2), \ 1 \le j \le \ell. \tag{3.199}$$

by Lemma 3.8 and Remark 3.18.

We have thus reached the fourth step for the proof of initial mass quantization, Theorem 3.7 of vanishing of the regular part of $\mu(dx,t)$, denoted by $\mu^{ac}(dx,t)$. This vanishing, called the residual vanishing, is proven for radially symmetric solutions through the use of defect measures [Ohtsuka–Senba–Suzuki (2007)]. Here we apply the argument for the general case [Suzuki (2015a)], using Liouville's first volume derivative (Theorem 8.2 in §8.3).

Take $1 \le i \le \ell$, and put

$$x_i = x_i(t), \ u_k(x,t) = u(x,t+t_k), \ v_k(x,t) = v(x,t+t_k). \tag{3.200}$$

Then we obtain $r_1 > 0$ such that

$$\mathcal{S}_t \cap B(x_i(t),2r_1) = \{x_i(t)\}, \quad -\infty < t < +\infty$$

by Lemma 3.8.

Lemma 3.9. *It holds that*

$$\frac{d}{dt}\int_{B(x_i,r)}(|x-x_i|^2 - r^2)u_k dx$$

$$\le 4\int_{B(x_i,r)} u_k \, dx + 2\int_{B(x_i,r)}(x-x_i)\cdot u_k\nabla v_k dx$$

$$- 2\int_{B(x_i,r)}(x-x_i)\cdot\dot{x}_i u_k \, dx \tag{3.201}$$

for $0 < r \le r_1$.

Proof. Theorem 8.2 in §8.3 implies

$$\frac{d}{dt}\int_{B(x_i,r)}|x-x_i|^2 u_k \, dx$$

$$= \int_{B(x_i,r)}\frac{\partial}{\partial t}(|x-x_i|^2 u_k) + \dot{x}_i\cdot\nabla(|x-x_i|^2 u_k) \, dx$$

$$= \int_{B(x_i,r)}|x-x_i|^2 u_{kt} - 2(x-x_i)\cdot\dot{x}_i u_k + 2\dot{x}_i\cdot(x-x_i)u_k$$

$$+\dot{x}_i|x-x_i|^2\cdot\nabla u_k dx$$

$$= \int_{B(x_i,r)}|x-x_i|^2 u_{kt} + \dot{x}_i\cdot|x-x_i|^2\nabla u_k \, dx. \tag{3.202}$$

Here, it holds that

$$\int_{B(x_i,r)} |x - x_i|^2 u_{kt} \, dx = \int_{B(x_i,r)} |x - x_i|^2 \nabla \cdot (\nabla u_k - u_k \nabla v_k) \, dx$$

$$= \int_{\partial B(x_i,r)} |x - x_i|^2 \left(\frac{\partial u_k}{\partial \nu} - u_k \frac{\partial v_k}{\partial \nu} \right) dS$$

$$- \int_{B(x_i,r)} 2(x - x_i) \cdot (\nabla u_k - u_k \nabla v_k) \, dx$$

$$= r^2 \int_{\partial B(x_i,r)} \frac{\partial u_k}{\partial \nu} - u_k \frac{\partial v_k}{\partial \nu} \, dS - 2 \int_{\partial B(x_i,r)} (x - x_i) \cdot \nu \, u_k \, dS$$

$$+ 4 \int_{B(x_i,r)} u_k \, dx + 2 \int_{B(x_i,r)} (x - x_i) \cdot u_k \nabla v_k \, dx$$

and hence the first term of the right-hand side of (3.202) is estimated as

$$\int_{B(x_i,r)} |x - x_i|^2 u_{kt} \, dx \le r^2 \int_{B(x_i,r)} \nabla \cdot (\nabla u_k - u_k \nabla v_k) \, dx$$

$$+ 4 \int_{B(x_i,r)} u_k \, dx + \int_{B(x_i,r)} 2(x - x_i) \cdot u_k \nabla v_k \, dx$$

$$= r^2 \int_{B(x_i,r)} u_{kt} \, dx + 4 \int_{B(x_i,r)} u_k \, dx$$

$$+ 2 \int_{B(x_i,r)} (x - x_i) \cdot u_k \nabla v_k \, dx. \tag{3.203}$$

For the second term of the right-hand side of (3.202), we note

$$\int_{B(x_i,r)} \dot{x}_i \cdot |x - x_i|^2 \nabla u_k \, dx$$

$$= \int_{\partial B(x_i,r)} (\dot{x}_i \cdot \nu)|x - x_i|^2 u_k \, dS - \int_{B(x_i,r)} 2(x - x_i) \cdot \dot{x}_i u_k \, dx$$

$$= r^2 \int_{\partial B(x_i,r)} (\dot{x}_i \cdot \nu) u_k \, dx - \int_{B(x_i,r)} 2(x - x_i) \cdot \dot{x}_i u_k \, dx$$

$$= r^2 \int_{B(x_i,r)} \dot{x}_i \cdot \nabla u_k \, dx - \int_{B(x_i,r)} 2(x - x_i) \cdot \dot{x}_i u_k \, dx. \tag{3.204}$$

By (3.202), (3.203), and (3.204), there arises that (3.201) because

$$\frac{d}{dt} \int_{B(x_i,r)} u_k \, dx = \int_{B(x_i,r)} u_{kt} + \dot{x}_i \cdot \nabla u_k \, dx$$

follows from Liouville's first volume derivative, Theorem 8.2 in §8.3 again.

□

Lemma 3.10. *There are r_2, t_0, and k_0 such that*

$$\frac{d}{dt}\int_{B(x_i,r)}(|x-x_i|^2-r^2)u_k dx \le 4\int_{B(x_i,r)}u_k\ dx$$

$$-\frac{1}{2\pi}(\int_{B(x_i,r)}u_k\ dx)^2 + C\int_{B(x_i,r)}|x-x_i|u_k\ dx \qquad (3.205)$$

for

$$0 < r \le r_2, \quad t \ge t_0, \quad k \ge k_0.$$

Proof. The last term on the right-hand side of (3.201) we use Lemma 3.7 to deduce

$$|\dot{x}_i| \le C.$$

For the second term, we divide v_k as

$$v_k(x,t) = \int_{B(x_i,r)}\Gamma(x-x')u_k(x',t)\ dx' + \int_{B(x_i,r)}K(x,x')u_k(x',t)\ dx'$$

$$+ \int_{\Omega\setminus B(x_i,r)}G(x,x')u_k(x',t)\ dx' \equiv v_k^0(x,t) + v_k^1(x,t) + v_k^2(x,t).$$

We use the method of symmetrization for v_k^0, to obtain

$$2\int_{B(x_i,r)}(x-x_i)\cdot u_k\nabla v_k^0\ dx = -\frac{1}{2\pi}(\int_{B(x_i,r)}u_k\ dx)^2. \qquad (3.206)$$

As for v_k^1, we have

$$\|\nabla v_k^1(\cdot,t)\|_{L^\infty(B(x_i,r))} \le C,$$

because $K = K(x,x')$ is smooth in $\Omega \times \Omega$.

We have also

$$\|\nabla v_k^2(\cdot,t)\|_{L^\infty(B(x_i,r))} \le C, \quad k \gg 1. \qquad (3.207)$$

In fact, $u_k(\cdot,t)$ is locally uniformly bounded in $\Omega\setminus\mathcal{S}_t$ with respect to $k \gg 1$, while

$$B(x_i(t),4r)\cap\mathcal{S}_t = \{x_i(t)\}, \quad -\infty < t < +\infty \qquad (3.208)$$

holds for $0 < r \ll 1$. Now we divide v_k^2 as in

$$v_k^2(x,t) = \int_{\Omega\setminus\mathcal{S}_t^{2r}}G(x,x')u_k(x',t)dx' + \int_{\mathcal{S}_t^{2r}\setminus B(x_i,r)}G(x,x')u_k(x',t)dx'.$$

For the first term on the right-hand side of this inequality, the result follows from

$$\|u_k(\cdot,t)\|_1 = \lambda$$

and the smoothness of $G = G(x, x')$ in $\Omega \times \Omega \setminus D$. For the second term, we use

$$\sup_x \int_\Omega |\nabla_x G(x, x')| \, dx' < +\infty$$

with (3.196), to reach (3.207). We thus end up with (3.205) for some r_2, t_0, and k_0. □

We are ready to execute the final step, completion of the proof of Theorem 3.7 in §3.12.

Lemma 3.11. *It holds that $f \equiv 0$ in (3.181), and therefore, (3.31) implies (3.32) in §5.13.*

Proof. We continue to write $x_i = x_i(t)$. Making $k \to \infty$ in (3.205), we obtain

$$\frac{d}{dt} \int_{B(x_i, r)} (|x - x_i|^2 - r^2) f \, dx \leq 4 \left(8\pi + \int_{B(x_i, r)} f \, dx \right)$$

$$- \frac{1}{2\pi} \left(8\pi + \int_{B(x_i, r)} f \, dx \right)^2 + C \int_{B(x_i, r)} |x - x_i| f \, dx$$

$$\leq -4 \int_{B(x_i, r)} f \, dx + C \int_{B(x_i, r)} |x - x_i| f \, dx$$

in the sense of distributions in t by (3.175), (3.195), and (3.208), and hence

$$\frac{d}{dt} \int_{B(x_i, r)} (|x - x_i|^2 - r^2) f \, dx$$

$$\leq -2 \int_{B(x_i, r)} f \, dx + C \int_{B(x_i, r)} |x - x_i|^2 f \, dx$$

$$= C \int_{B(x_i, r)} (|x - x_i|^2 - r^2) f \, dx + (Cr^2 - 2) \int_{B(x_i, r)} f \, dx$$

$$\leq \left(C + \frac{2 - Cr^2}{r^2} \right) \int_{B(x_i, r)} (|x - x_i|^2 - r^2) f \, dx$$

$$= C(r) \int_{B(x_i, r)} (|x - x_i|^2 - r^2) f \, dx \tag{3.209}$$

with $C(r) > 0$ for $0 < r \ll 1$ and $t \geq t_0$.

Let

$$I(t) = \int_{B(x_i, r)} (|x - x_i|^2 - r^2) f \, dx \leq 0.$$

By (3.209) if there is $t_1 \geq t_0$ such that $I(t_1) < 0$ then it holds that

$$\lim_{t\uparrow+\infty} I(t) = -\infty,$$

a contradiction. Hence we have $I(t) \equiv 0$, which implies

$$f(\cdot, t) = 0 \quad \text{in } B(x_i, r) = B(x_i(t), r).$$

Then we obtain $f \equiv 0$ from the strong maximum principle or the unique continuation theorem for parabolic equations. $\qquad\square$

3.14 Collapse Dynamics

Here we show Theorem 3.8 in §3.12. We have readily proven

$$\mu(dx, t) = 8\pi \sum_{j=1}^{\ell} \delta_{x_j(t)}(dx)$$

with (3.199), and hence this theorem is reduced to the following lemma.

Lemma 3.12. *It holds that (3.178) in §3.12.*

Proof. We have

$$\nu(t)|_{C(\overline{\Omega}\times\overline{\Omega})} = 64\pi^2 \sum_{j,k=1}^{\ell} \delta_{x_j(t)}(dx) \otimes \delta_{x_k(t)}(dx') \tag{3.210}$$

in (3.183) with

$$|x_j(t) - x_i(t)| \geq c_0, \quad j \neq i, \ |t - t_0| < \delta \tag{3.211}$$

for $0 < c_0 \ll 1$.

Here we use $\xi = |x|^2 \varphi$ with $\varphi = \varphi_r(x)$, to obtain

$$\rho_\xi(x, x') = -\frac{1}{\pi} + 2x \cdot \nabla_x K(x, x') + 2x' \cdot \nabla_{x'} K(x, x')$$

for

$$(x, x') \in B(0, r/2) \times B(0, r/2),$$

with $K = K(x, x')$ defined by (3.136). Therefore, it follows that

$$\begin{aligned}
\langle \Delta\xi, &\mu(dx, t)\rangle + \frac{1}{2}\langle \rho_\xi, \nu(t)\rangle \\
&= 64\pi^2 x_i(t) \cdot (\nabla_x K(x_i(t), x_i(t)) + \nabla_{x'} K(x_i(t), x_i(t)) \\
&\quad + 64\pi^2 x_i(t) \cdot \sum_{j\neq i}(\nabla_x G(x_i(t), x_j(t)) + \nabla_{x'} G(x_i(t), x_j(t)) \\
&= 128\pi^2 x_i(t) \cdot \nabla_{x_i} H_\ell(x_1(t), \cdots, x_\ell(t)). \tag{3.212}
\end{aligned}$$

By (3.184), (3.210), and (3.212), we arrive at

$$\frac{1}{2}\frac{d}{dt}|x_i|^2 = 8\pi x_i \cdot \nabla_{x_i} H_\ell(x_1, \cdots, x_\ell). \tag{3.213}$$

We obtain, similarly,

$$\frac{1}{2}\frac{d}{dt}|x_i - a|^2 = 8\pi(x_i - a) \cdot \nabla_{x_i} H_\ell(x_1, \cdots, x_\ell)$$

for $|a| \ll 1$, and hence

$$(x_i - a) \cdot \frac{dx_i}{dt} = 8\pi(x_i - a) \cdot \nabla_{x_i} H_\ell(x_1, \cdots, x_\ell).$$

From the arbitrariness of a in $|a| \ll 1$, it follows that (3.178). □

The proof of Theorem 3.6, Theorem 3.7, Theorem 3.8 in §3.12 is complete.

Remark 3.20. Since closure of the orbit $\{x(t)\}$ in $\Omega^\ell \setminus D$ is compact by Lemma 3.8, the ω-limit and α-limit sets of this orbit are contained in the critical point of H_ℓ from the LaSalle principle applied to the anti-gradient flow (4.63). Theory of gradient inequality in §4.11 clarifies the profile of $\mathcal{O} = \{x(t)\}$ in more details, as is described in §3.5. This $\mathcal{O}$ is thus a connecting orbit of two critical points of the real-analytic function $H_\ell = H_\ell(x)$ which may be the same.

A result analogous to Theorem 3.5 in §3.10 is the following theorem. It is applicable if $\lambda = 8\pi$ and Ω is close to a disc, in which case $\mathcal{F}(u)$ is bounded by the Trudinger Moser inequality and there is no classical solution to the Boltzmann Poisson equation (3.11) (Remark 3.21). Recall (3.148) and (3.149) in §3.10:

$$\mathcal{F}(u) = \int_\Omega u(\log u - 1)\, dx - \frac{1}{2}\iint_{\Omega\times\Omega} G(x,x')u \otimes u\, dxdx'$$

$$\frac{d}{dt}\mathcal{F}(u) = -\int_\Omega u|\nabla(\log u - v)|^2 dx. \tag{3.214}$$

Theorem 3.9 ([Senba–Suzuki (2002b)]). *If*

$$T = +\infty, \quad \lim_{t\uparrow+\infty}\|u(\cdot,t)\|_\infty = +\infty, \quad \lim_{t\uparrow+\infty}\mathcal{F}(u(\cdot,t)) > -\infty, \tag{3.215}$$

then $\mu(dx,t)$ generated in (3.175) is independent of t and takes the form

$$\mu(dx,t) = 8\pi\sum_{j=1}^{\ell}\delta_{x_j^*}(dx), \tag{3.216}$$

with $x_j^* \in \Omega$, $1 \leq j \leq \ell$, *satisfying*

$$\nabla_{x_j} H_\ell(x_1^*, \cdots, x_\ell^*) = 0, \quad 1 \leq j \leq \ell. \tag{3.217}$$

Proof. We have readily shown (3.175) and

$$\mu(dx, t) = 8\pi \sum_{i=1}^{\ell} \delta_{x_i(t)}(dx), \quad x_i(t) \neq x_j(t), \ i \neq j \tag{3.218}$$

with $x_i = x_i(t) \in \Omega$ satisfying (3.178) in §3.12, where

$$\lambda = \|u_0\|_1 = 8\pi\ell.$$

First, since

$$\|\nabla u - u\nabla v\|_1^2 \leq \lambda \int_\Omega u^{-1}|\nabla u - u\nabla v|^2 dx = \lambda \int_\Omega u|\nabla(\log u - v)|^2 dx$$

it holds that

$$\int_0^\infty \|\nabla u - u\nabla v\|_1^2 \, dt < +\infty \tag{3.219}$$

by (3.214) and (3.215). Second, the equation

$$-\Delta v = u,$$

is equivalent to

$$v_{z\bar{z}} = -\frac{u}{4}$$

for $z = x_1 + \sqrt{-1}x_2$ and $\bar{z} = x_1 - \sqrt{-1}x_2$, where $x = (x_1, x_2)$. Then it holds that

$$s_{\bar{z}} = v_{zz\bar{z}} - v_z v_{z\bar{z}} = -\frac{1}{4}u_z + \frac{1}{4}uv_z = -\frac{1}{4}(u_z - uv_z), \tag{3.220}$$

where

$$s = v_{zz} - \frac{1}{2}v_z^2.$$

We thus obtain

$$\int_0^\infty \|s_{\bar{z}}\|_1^2 \, dt < +\infty \tag{3.221}$$

by (3.219). Elliptic L^1 estimate [Brezis–Strauss (1973)], third, implies

$$v_k(\cdot, t) = v(\cdot, t + t_k) \rightharpoonup \tilde{v}(\cdot, t) \quad \text{in } W^{1,q}(\Omega), \quad 1 < q < 2$$

for any $t \in (-\infty, +\infty)$ in accordance with (3.175), where

$$\tilde{v}(\cdot, t) = 8\pi \sum_{i=1}^{\ell} G(\cdot, x_i(t)) \qquad (3.222)$$

by (3.218).

Given $t_k \uparrow +\infty$, we have a subsequence denoted by the same symbol such that $t_{k+1} > t_k + 2$. Then it follows that

$$\lim_{k \to \infty} \int_{t_k - 1}^{t_k + 1} \|s_{\bar{z}}\|_1^2 \, dt = 0,$$

from (3.221) and

$$\sum_k \int_{t_k - 1}^{t_k + 1} \|s_{\bar{z}}\|_1^2 \, dt \le \int_0^\infty \|s_{\bar{z}}\|_1^2 \, dt < +\infty,$$

which impliles

$$\tilde{s}(\cdot, t)_{\bar{z}} = 0 \text{ in } \overline{\Omega} \setminus \mathcal{S}_t, \quad -1 < t < 1$$

for $\tilde{s} = \tilde{s}(\cdot, t)$ defined by

$$\tilde{s} = \tilde{v}_{zz} - \frac{1}{2}\tilde{v}_z^2.$$

Then the complex analysis developed for the stationary solutions in §2.5 guarantees

$$\nabla_{x_j} H_\ell(x_1(t), \cdot, x_\ell(t)) = 0, \ 1 \le j \le \ell, \quad -1 < t < 1. \qquad (3.223)$$

Hence $x(t) = (x_1(t), \cdots, x_\ell(t))$ is a stationary solution to (4.63), denoted by

$$(x_1(t), \cdots, x_\ell(t) = (x_1^*, \cdots, x_\ell^*), \quad -\infty < t < +\infty,$$

and consequently, (3.223) is valid for any $-\infty < t < +\infty$. We thus obtain (3.216)–(3.217).

It holds also that

$$v_k \to v_* \quad \text{locally uniformly in } (\overline{\Omega} \setminus \mathcal{S}) \times (-\infty, +\infty)$$

for $v_k = v(\cdot, t + t_k)$, where

$$v_*(x) = 8\pi \sum_{j=1}^{\ell} G(x, x_j^*), \quad \mathcal{S} = \{x_j^* \mid 1 \le j \le \ell\}.$$

□

Remark 3.21. Theorem 3.5 and Theorem 3.7 in §3.12, and Theorem 3.9 in this section control the dynamics of (3.1) as follows. First, if $0 < \lambda < 8\pi$, there is a uniformly bounded global-in-time solution, which converges to a stationary solution. If $\lambda = 8\pi$, second, it holds always that $T = +\infty$. If blowup in infinite time occurs in this case of $\lambda = 8\pi$, the limit measure $\mu(dx, t)$ generated by (3.175) takes the form $\mu(dx, t) = 8\pi\delta_{x_0}(dx)$ with $x_0 \in \Omega$ satisfying $\nabla R(x_0) = 0$. This case actually arises either if Ω is close to a disc. If u_0 is so concentrated as in $\mathcal{F}(u_0) \ll -1$ and if $\lambda \notin 8\pi\mathbf{N}$, third, it holds that $T < +\infty$. In fact, the set of solutions to the Boltzmann Poisson equation, (2.58) in §2.5, is compact in $C^2(\overline{\Omega})$ for each $\lambda \in (0, +\infty) \setminus 8\pi\mathbf{N}$, and therefore, there arises the non-existence of the stationary solution u_* satisfying $\mathcal{F}(u_*) \ll -1$ besides that of the singular limits of the solution to the Boltzmann Poisson equation for $\lambda \notin 8\pi\mathbf{N}$. If Ω is simply connected and $\lambda \gg 1$, fourth, it holds always that $T < +\infty$, because there is no stationary solutions nor their singular limits in this case, similarly ([Chang, Cheng, and Lin (2003)], §7.5.4 of [Suzuki (2020)]). If Ω is convex and $\lambda > 8\pi$, fifth, blowup in infinite time does not occur because there is no critical points of H_ℓ for $\ell \geq 2$ ([Grossi–Takahashi (2010)]). Hence there arises either blowup in finite time or uniformly bounded global-in-time solution if Ω is convex and $\lambda > 8\pi$, and the latter case is excluded if $\lambda \gg 1$. There can, however, exist stationary solutions even if Ω is convex and $\lambda > 8\pi$ (§4.3.7 of [Suzuki (2020)]). If Ω is convex and $\lambda = 8\pi$, finally, it holds that

$$u(x, t)dx \rightharpoonup 8\pi\delta_{x_0}(dx), \quad \nabla R(x_0) = 0$$

as $t \uparrow +\infty$, and such $x_0 \in \Omega$ is unique [Grossi–Takahashi (2010)].

Remark 3.22. An analogous object in higher space dimensions is a *degenerate parabolic equation* studied by [Suzuki–Takahashi (2009a,b); Suzuki–Tasaki (2010); Suzuki–Takahashi (2012)]. There are critical mass for the blowup in finitite time, ε-regularity, and finiteness of the type II blowup points, in spite of the lack of the monotonicity formula.

3.15   Simplified System of Chemotaxis

A variant of the Smoluchowski Poisson equation arises in mathematical biology in the context of a simplified system of chemotaxis [Jäger–Luckhaus (1992)]. It is composed of the Smoluchowski part

$$u_t = \nabla \cdot (\nabla u - u\nabla v) \text{ in } \Omega \times (0, T), \quad \left.\frac{\partial u}{\partial \nu} - u\frac{\partial v}{\partial \nu}\right|_{\partial\Omega} = 0 \qquad (3.224)$$

and the Poisson part

$$-\Delta v = u - \frac{1}{|\Omega|}\int_\Omega u\ dx, \quad \frac{\partial v}{\partial \nu}\Big|_{\partial\Omega} = 0, \quad \int_\Omega v\ dx = 0, \tag{3.225}$$

where $\Omega \subset \mathbf{R}^2$ is a bounded domain with smooth boundary. The initial condition is provided with

$$u|_{t=0} = u_0(x) > 0. \tag{3.226}$$

From the structure of the set of stationary solutions [Senba–Suzuki (2000)], its quantized blowup mechanism was suspected. Under this concept of *potential of self-organization*, realized in the stationary state, system of equations (3.225)–(3.226) has been studied for a long time [Suzuki (2005, 2015b, 2018)].

Here, only Green's function $G(x, x')$ to the Poisson part (3.225) is different from that of (3.1) formulated by

$$-\Delta v = u, \quad v|_{\partial\Omega} = 0. \tag{3.227}$$

Particularly, its boundary behavior different from that to (3.227) in §3.10 ensures actual boundary blowup points to (3.224)–(3.225) in §3.10, where the collapse masses are reduced to a half of those of interior blowup points. Thus, using the conformal diffeomorphism X in (3.138)–(3.139) and the reflection X_* in (3.140), the parametrix of $G(x, x')$ to (3.225) near $\partial\Omega$ is given by

$$E(X, X') = \Gamma(X - X') + \Gamma(X - X'_*) \tag{3.228}$$

(Chapter 5 of [Suzuki (2005)]).

First, the stationary problem of (3.224)–(3.225) is formulated as

$$-\Delta v = \lambda\left(\frac{e^v}{\int_\Omega e^v dx} - \frac{1}{|\Omega|}\right), \quad \frac{\partial v}{\partial \nu}\Big|_{\partial\Omega} = 0, \quad \int_\Omega v\ dx = 0. \tag{3.229}$$

It is a sort of the Boltzmann Poisson equation, and actually, analogous result of Theorem 2.1 on the quantized blowup mechanism arises, provided with the control of the point vortex Hamiltonian of the location to blowup points of the family of solutions. The Hamiltonian for this problem takes the same form as in (3.227),

$$H_\ell(x_1, \cdots, x_\ell) = \frac{1}{2}\sum_{j=1}^{\ell} R(x_j) + \sum_{1\le i<j\le \ell} G(x_i, x_j). \tag{3.230}$$

The Robin function $R(x) = K(x, x)$ is defined by $K = K(x, x')$, which takes, however, a different form from that to (3.227), that is,

$$K(x, x') = \begin{cases} G(x, x') + \frac{1}{2\pi}\log|x - x'|, & x' \in \Omega \\ G(x, x') + \frac{1}{\pi}\log|x - x'|, & x' \in \partial\Omega. \end{cases}$$

Thus, if

$$\lim_{k \to \infty} \lambda_k = \lambda_0, \quad \lim_{k \to \infty} \|v_k\|_\infty = +\infty, \tag{3.231}$$

arises to a sequence of classical solutions

$$(\lambda_k, v_k), \ k = 1, 2 \cdots$$

to (3.229), it holds that $\lambda_0 \in 4\pi \mathbf{N}$. Passing to a subsequnce, we obtain

$$v_k \to v_0 \quad \text{locally uniformly in } \overline{\Omega} \setminus \mathcal{S}$$

for

$$v_0(x) = \sum_{x_0 \in \mathcal{S}} m_*(x_0) G(x, x_0), \ m_*(x_0) = \begin{cases} 8\pi, \ x_0 \in \Omega \\ 4\pi, \ x_0 \in \partial\Omega, \end{cases}$$

where

$$\mathcal{S} = \{x_0 \in \overline{\Omega} \mid \exists x_k \to x_0, \ v_k(x_k) \to +\infty\}$$

stands for the blowup set of $\{v_k\}$. It holds that $\sharp \mathcal{S} < +\infty$ and

$$\ell = 2 \cdot \sharp(\partial\Omega \cap \mathcal{S}) + \sharp(\Omega \cap \mathcal{S}) \tag{3.232}$$

with $\lambda_0 = 4\pi\ell$. There is, furthermore, a Hamiltonian control of the location of the blowup points. Hence $x_* = (x_1^*, \cdots, x_\ell^*)$ is a critical point of $H_\ell(x)$, $x = (x_1, \cdots, x_\ell)$, in (3.230), if

$$\mathcal{S} = \{x_j^* \mid 1 \le j \le \ell\}.$$

The structure of the set of stationary solutions to (3.229), however, is rather different from that to the Boltzmann Poisson equation (3.11) in §3.1. First of all, since problem (3.229) admits the trivial solution $v = 0$, there is a compact orbit $\mathcal{O}$ to (3.224) and (3.225) global in time for any Ω and λ. Second, several results on the set of stationary solutions studied in [Senba–Suzuki (2000)] for the slightly perturbed problem,

$$-\Delta v + av = u \text{ in } \Omega, \quad \left. \frac{\partial v}{\partial \nu} \right|_{\partial\Omega} = 0$$

for $a > 0$, are valid to (3.225).

Concerning the blowup in infinite time,

$$\exists t_k \uparrow +\infty, \quad \lim_{k \to \infty} \|u(\cdot, t_k)\|_\infty = +\infty,$$

second, we still have (3.175), passing to a subsequence, and there arises that

$$\mu(dx, t) = \sum_{x_0 \in \mathcal{S}_t} m_*(x_0) \delta_{x_0}(dx),$$

where $\mathcal{S}_t$ is the blowup set of

$$u_k(\cdot, t) = u(\cdot, t + t_k), \quad k \to \infty.$$

The numbers

$$\sharp\left(\partial\Omega \bigcap \mathcal{S}_t\right), \quad \sharp\left(\Omega \bigcap \mathcal{S}_t\right)$$

are independent of t. Letting

$$\partial\Omega \cap \mathcal{S}_t = \{x_1(t), \cdots, x_{\ell_1}(t)\},$$
$$\Omega \cap \mathcal{S}_t = \{x_{\ell_1+1}(t), \cdots, x_{\ell_1+\ell_2}(t)\},$$

we obtain

$$\tau \cdot \frac{dx_i}{dt} = \tau \cdot \nabla_{x_i} H_\ell(x_1, \cdots, x_\ell), \quad 1 \le i \le \ell_1$$

$$\frac{dx_i}{dt} = \nabla_{x_i} H_\ell(x_1, \cdots, x_\ell), \quad \ell_1 + 1 \le i \le \ell = \ell_1 + \ell_2, \quad (3.233)$$

where τ is the unit tangential vector. Hence the boundary and interior collapses are subject to their own dynamics, and both trajectories

$$\mathcal{O}_1 = \{(x_1(t), \cdots, x_{\ell_1}(t)\} \subset (\partial\Omega)^{\ell_1} \setminus D_1,$$
$$\mathcal{O}_2 = \{(x_{\ell_1+1}(t), \cdots, x_{\ell_1+\ell_2}(t))\} \subset \Omega^{\ell_2} \setminus D_2,$$

are pre-compact, where

$$D_1 = \{x = (x_j) \in (\partial\Omega)^{\ell_1} \mid \exists i \ne j, \ x_i = x_j\},$$
$$D_2 = \{x = (x_j) \in \Omega^{\ell_2} \mid \exists i \ne j, \ x_i = x_j\}.$$

Blowup in infinte time does not arise, in particular, if

$$\lambda = \|u_0\|_1 \notin 4\pi\mathbf{N}.$$

It does not occur even for the quantized initial mass

$$\lambda \in 4\pi\mathbf{N}$$

if $H_\ell(x_1, \cdots, x_\ell)$ does not take the critical point

$$x_* = (x_1^*, \cdots, x_{\ell_1}^*, x_{\ell_1+1}^*, \cdots, x_{\ell_1+\ell_2}^*) \in (\partial\Omega)^{\ell_1} \times \Omega^{\ell_2}$$

for any $\ell_1, \ell_2 \in \mathbf{N} \cup \{0\}$ satisfying

$$\ell = \ell_1 + \ell_2, \ 4\pi\ell_1 + 8\pi\ell_2 = \lambda.$$

If the blowup in finite time, $T < +\infty$, occurs to (3.224)–(3.225), finally, there arises that (3.28) with

$$m(x_0) \in m_*(x_0)\mathbf{N}, \quad 0 < f = f(x) \in L^1(\Omega) \cap C(\overline{\Omega} \setminus \mathcal{S}),$$

where $\mathcal{S}$ is the blowup set defined by (3.29) in §5.13, which implies

$$\sharp \mathcal{S} < +\infty.$$

If $z(y, s)$ is defined by (3.48) in §3.4 for $x_0 \in \mathcal{S}$, any $s_k \uparrow +\infty$ admits a subsequence, denoted by the same symbol, satisfying (3.108) in §3.8 for $x_0 \in \Omega$ and

$$z(y, s + s_k)dy \rightharpoonup \zeta(dy, s) \quad \in C_*(-\infty, +\infty; \mathcal{M}(\overline{L}))$$

for $x_0 \in \partial\Omega$, respectively, where L is a half space containing the origin on ∂L, which is assumed to be $\mathbf{R}_+^2$ without loss of generality.

Its scaling back

$$A(dy', s')$$

defined by (3.114) in §3.8 takes the form

$$A(dy', s') = m_*(x_0) \sum_{j=1}^{m} \delta_{y_j'(s)}(dy'), \quad s' < 0$$

with $m = m(x_0)/m_*(x_0)$. If $m = 1$, it holds that

$$A(dy', s') = m_*(x_0)\delta_0(dy').$$

If $m \geq 2$ and $x_0 \in \Omega$, there arises that (3.60)–(3.61) in §3.4. If

$$m \geq 2, \quad x_0 \in \partial\Omega,$$

we obtain

$$\frac{dy_j'}{ds} = 4\pi\nabla_{y_j'}\tilde{H}_m^0(y_1', \cdots, y_m') \text{ a.e. } s' < 0$$

$$\lim_{s'\uparrow 0} y_j'(s') = 0, \ 1 \leq j \leq m$$

for

$$\tilde{H}_m^0(y_1', \cdots, y_m') = \sum_{1 \leq i < j \leq m} E(y_i' - y_j')$$

and $E(X, X')$ defined by (3.228), and consequently,

$$s' \in (-\infty, 0] \mapsto y_j'(s') \in \partial\Omega, \ 1 \leq j \leq m$$

is absolutely continuous.

Chapter 4

Diffusion Geometry

Several problems in differential geometry have been solved by significant use of the theory of partial differential equations. Among them is the proof of the Poincaré conjecture completed by [Perelman (2002, 2003)]. This approach, called the method of normalized Ricci flow (NRF), is due to [Hamilton (1982)]. This chapter is devoted to the case of two space dimension, 2D-NRF, resolved by R. Hamilton himself [Hamilton (1988)]. First, we formulate 2D–NRF (§4.1). Then this NRF is reduced to a parabolic equation with logarigthmic diffusion, which takes the form of model (B) equation associated with Helmholtz's free energy (§4.2–§4.4). In spite of this common thermo-dynamical structure with the Smoluchowski Poisson equation described in §3.4, 2D–NRF does not exihibit any blowup of the solution at the crtical mass 8π because of the lack of scaling invariance [Hamilton (1988)]. We try to recover the result by pure analytic argument, using Trudinger Moser inequality, Benilan Crandall inequality, principle of concentration compactness, and the theory gradient inequality (§4.5–§4.11). Inefficiency of the theory of center manifold is also noticed in accordance with the rate of convergence to the stationary state (§4.12–§4.14).

4.1   2D Normalized Ricci Flow

Normalized Ricci flow in two space dimension describes the time evolution of the metric $g = g(t)$ on the compact Riemann surface Ω without boundary. It is given by

$$\frac{\partial g}{\partial t} = (r - R)\, g, \tag{4.1}$$

where $R = R(t)$ is the scalar (Gauss) curvature of $(\Omega, g(t))$, $r = r(t)$ is the volume mean

$$r = \frac{\int_\Omega R(t)\,d\mu_t}{\int_\Omega d\mu_t},\tag{4.2}$$

and $\mu = \mu_t$ is its area element. The profile of the solution to this equation, global-in-time existence and convergence as $t \uparrow +\infty$ to the metric with constant mean curvature, has opened the resolution of the Poincaré conjecture in three space dimension [Perelman (2002, 2003)].

We have

$$\frac{\partial R}{\partial t} = \Delta_g R + R(R - r)\tag{4.3}$$

from (4.1), where Δ_g denotes the Laplace Beltrami operator associated with $g = g_t$. If $R|_{t=0}$ has a definite sign on Ω, therefore, so does $R = R(t)$ for $t \geq 0$ in (4.1) by the maximum principle for parabolic equations. In this case, R. Hamilton showed

$$g(\cdot, t) \to g_\infty, \ t \to \infty \quad \text{exponentially in } C^\infty \text{ topology},\tag{4.4}$$

where g_∞ is a metric on Ω with constant scalar curvature [Hamilton (1988)].

The significant case in this proof is $R = R_g > 0$ everywhere rather than $R_g \leq 0$ everywhere. In the former case, Gauss–Bonnet's theorem,

$$\int_\Omega R_g d\mu_g = 4\pi\chi(\Omega), \ \chi(\Omega) = 2 - 2g(\Omega),\tag{4.5}$$

ensures $g(\Omega) = 0$, where $\chi(\Omega)$ and $g(\Omega)$ denote the Euler characteristics and the genus of Ω, respectively. Hence by the uniformization theorem, there arises that $\Omega = S^2$. Chow showed, conversely, that if $\Omega = S^2$ then R becomes positive definite on Ω in finite time unless $R_0 \leq 0$ everywhere [Chow (1991)].

In the following, we assume $R > 0$ on Ω. We thus obtain $g(\Omega) = 0$, and therefore, again the uniformization theorem ensures that the metric g on Ω is conformal to the one with constant scalar curvature, denoted by g_0. Hence there arises that

$$\Omega = S^2, \ g(t) = e^{w(\cdot, t)} g_0\tag{4.6}$$

with smooth $w = w(\cdot, t)$. Then we obtain

$$R(t) = e^{-w}(-\Delta w + R_0), \ d\mu_g = e^w dx\tag{4.7}$$

and

$$\lambda \equiv \int_\Omega R_g d\mu_g = \int_\Omega R_0 dx = |\Omega| R_0 = 4\pi(2 - 2g(\Omega)) = 8\pi,\tag{4.8}$$

where $dx = d\mu_0$, $\Delta = \Delta_0$, R_0, and $|\Omega|$ denote the volume element, the Laplace Bertrami operator, the scalar curvature, and the volume of Ω with respect to g_0, respectively. This $\lambda = 8\pi$, again, is the critical mass in the generalized 2D-NFR, (4.18) in §4.2.

It thus holds that

$$r = \frac{8\pi}{\int_\Omega e^w dx},\qquad(4.9)$$

and hence problem (4.1)–(4.2) is reduced to

$$\frac{\partial e^w}{\partial t} = \Delta w + 8\pi\left(\frac{e^w}{\int_\Omega e^w dx} - \frac{1}{|\Omega|}\right)\quad\text{in }\Omega\times(0,T)\qquad(4.10)$$

with

$$w\big|_{t=0} = w_0(x).\qquad(4.11)$$

The result [Hamilton (1988)] reads, in the context of (4.10), that the solution $v = v(\cdot,t)$ exists global in time and satisfies

$$w(\cdot,t)\to w_\infty\quad\text{exponentially in }C^\infty\text{ topology,}\qquad(4.12)$$

where v_∞ is a stationary solution:

$$-\Delta w_\infty = 8\pi\left(\frac{e^{w_\infty}}{\int_\Omega e^{w_\infty} dx} - \frac{1}{|\Omega|}\right).\qquad(4.13)$$

Analytic method was tried to simplify this proof ([Bartz–Struwe–Ye (1994)]). First, the a priori estimate

$$|\nabla_{S^2} w| \le C\qquad(4.14)$$

is derived via the moving plane on the sphere, called the *moving sphere method*. Second, inequality (4.14) induces global-in-time existence of the solution and the pre-compactness of the orbit via the Harnack inequality derived from (4.14). Finally, the normalized stationary solution,

$$-\Delta v = 8\pi\left(\frac{e^v}{\int_\Omega e^v dx} - \frac{1}{|\Omega|}\right),\qquad \int_\Omega v\, dx = 0,\qquad(4.15)$$

is unique for $\Omega = S^2$, which assues that the ω-limit set to (4.10)–(4.11) defined by

$$\omega(v_0) = \{w_\infty \mid \exists t_k \to +\infty,\ \lim_{k\to\infty}\|w(\cdot,t_k) - w_\infty\|_\infty = 0\}$$

is a singleton. Geometrically, this uniqueness means that the metric on the round sphere with constant Gaussian curvature must be the standard one.

Analytic proof of this property, on the other hand, is in [Chanillo–Kiessling (1994); Chen–Lin (1997); Lin (2000)].

To examine the structure of the ω-limit set, we use the functional

$$J_{8\pi}(w) = \frac{1}{2}\|\nabla w\|_2^2 - 8\pi \log \left(\int_\Omega e^{w - \overline{w}} dx \right), \quad \overline{w} = \frac{1}{|\Omega|} \int_\Omega w \, dx.$$

This $J_{8\pi}(w)$ takes the role of the Lyapunov function:

$$\frac{d}{dt} J_{8\pi}(w) = \int_\Omega \nabla w \cdot \nabla w_t dx - 8\pi \int_\Omega \left(\frac{e^w}{\int_\Omega e^w dx} - \frac{1}{|\Omega|} \right) w_t$$

$$= - \int_\Omega e^w w_t^2 dx \le 0.$$

Hence the ω-limit set of the pre-compact, global-in-time orbit is contained in the set of stationary solutions, $w_t = 0$ in (4.10) by LaSalle's principle as in §3.12.

If $w = w_\infty$ belongs to the ω-limit set of (4.10), therefore, there arises that

$$-\Delta w_\infty = 8\pi \left(\frac{e^{w_\infty}}{\int_\Omega e^{w_\infty} dx} - \frac{1}{|\Omega|} \right) \quad \text{in } \Omega,$$

and hence (4.15) for

$$v = w_\infty - \overline{w}_\infty, \quad \overline{w}_\infty = \frac{1}{|\Omega|} \int_\Omega w_\infty \, dx. \tag{4.16}$$

It then follows that $v = 0$ from the above result on $\Omega = S^2$, and therefore, this $w_\infty = \overline{w}_\infty$ is a constant.

It holds, on the other hand, that

$$\frac{d}{dt} \int_\Omega e^w dx = 0$$

in (4.10), and hence

$$\int_\Omega e^{w_\infty} dx = \int_\Omega e^{w_0} dx. \tag{4.17}$$

We thus obtain

$$w_\infty = \log \left(\frac{1}{|\Omega|} \int_\Omega e^{w_0} dx \right).$$

In particular, the ω-limit set to (4.13) is a singleton, and there arises (4.12) in C^∞ topology. The exponential rate of convergence is also provided by the analytic method, using conformal transformation on the two-dimensional sphere S^2 ([Struwe (2002)]).

4.2 Analytic Approach

Here we take a general form of (4.10),

$$\frac{\partial e^w}{\partial t} = \Delta w + \lambda \left(\frac{e^w}{\int_\Omega e^w dx} - \frac{1}{|\Omega|} \right) \quad \text{in } \Omega \times (0, T), \quad w|_{t=0} = w_0(x) \quad (4.18)$$

where Ω is a compact Riemann surface without boundary and $\lambda > 0$ is a constant. The analytic method described in §4.1 still uses the geometric structure of 2D–NRF, such as the method of moving sphere or conformal transformation, valid to

$$\Omega = S^2, \quad \lambda = 8\pi. \quad (4.19)$$

If the surface Ω has a positive genus $g(\Omega) \geq 1$ and λ is determined by (4.8), then $\lambda \leq 0$. Hence the geometric problem (4.1)–(4.2) is reduced to (4.18) either for (4.19) or $\lambda \leq 0$. The latter case is easier to treat in the theory of partial differential equations, provided with uniform a priori estimates and uniqueness of the steady state. Hence we assume $\lambda > 0$ in (4.18).

Problem (4.18) is thus no longer 2D–NRF in the other case of Ω and λ than (4.19). Similar conclusion, however, is derived for any Ω and

$$0 < \lambda \leq 8\pi.$$

Hence we provide an analytic proof of the above result on (4.1), differently from the other work [Bartz–Struwe–Ye (1994); Struwe (2002)] applied to (4.19), developing a more universal perspective for investing the dynamical behavior of (4.18).

Even in the other case of (4.19) it holds that

$$\frac{d}{dt} \int_\Omega e^w dx = 0 \quad (4.20)$$

and

$$\frac{d}{dt} J_\lambda(w) = - \int_\Omega e^w w_t^2 dx \quad (4.21)$$

in (4.18) for

$$J_\lambda(w) = \frac{1}{2} \|\nabla w\|_2^2 - \lambda \log \left(\int_\Omega e^{w-\overline{w}} dx \right), \quad \overline{w} = \frac{1}{|\Omega|} \int_\Omega w.$$

If w_∞ is in its ω limit set, therefore, there arises that

$$-\Delta w_\infty = \lambda \left(\frac{e^{w_\infty}}{\int_\Omega e^{w_\infty} dx} - \frac{1}{|\Omega|} \right), \quad \int_\Omega e^{w_\infty} dx = \int_\Omega e^{w_0} dx. \quad (4.22)$$

Then we obtain

$$-\Delta v = \lambda \left(\frac{e^v}{\int_\Omega e^v dx} - \frac{1}{|\Omega|} \right) \text{ in } \Omega, \qquad \int_\Omega v \, dx = 0 \qquad (4.23)$$

for v defined by (4.16) in §4.1.

Equation (4.23) is the stationary state of the simplified system of chemotaxis studied in §3.15, defined on the compact Riemann surface Ω without boundary,

$$u_t = \nabla \cdot (\nabla u - u \nabla v), \quad -\Delta v = u - \frac{1}{|\Omega|} \int_\Omega u \, dx \text{ in } \Omega \times (0, T)$$

$$\int_\Omega v \, dx = 0, \quad u|_{t=0} = u_0(x) > 0 \qquad (4.24)$$

with $\lambda = \|u_0\|_1$. We have quantized blowup mechanism and Hamiltonian control even to (4.24) in three levels; stationary state, blowup in finite time, and blowup in infinite time as in Chapter 2 and Chapter 3.

Given the family $(\lambda, v) = (\lambda_k, v_k)$, $k = 1, 2, \cdots$, of solutions to (4.23) satisfying (3.231),

$$\lim_{k \to \infty} \lambda_k = \lambda_0, \quad \lim_{k \to \infty} \|v_k\|_\infty = +\infty,$$

in particular, it holds that $\lambda_0 \in 8\pi\mathbf{N}$ as in Theorem 2.1 in §2.5. Blowup in infinite time, furthermore, can occur to (4.24) for $\lambda = \|u_0\|_1 = 8\pi$, under the presence of a singular limit of the classical solution to (4.23) as in Theorems 3.6, 3.7 and 3.8 in §3.12.

As for (4.18) with $\lambda = 8\pi$, however, there exists always global-in-time solution, converging to a steady state, regardless of the existene of the singular limits of (4.23) and of multiple existence of the steady states, for any compact Riemann surface Ω without boundary.

To approach (4.18), first, we recall (4.21) to infer that

$$r = \frac{\lambda}{\int_\Omega e^w dx}$$

is a constant. Second, there aries

$$\frac{\partial e^w}{\partial t} \leq \Delta w + re^w, \quad w|_{t=0} = w_0(x),$$

and therefore, the comparison theorem applied to the ODE part,

$$\frac{de^W}{dt} = re^W, \quad W|_{t=0} = \|W_0\|_\infty,$$

ensures

$$e^{w(x,t)} \leq e^{\|w_0\|_\infty} \cdot e^{rt} \equiv W(t). \qquad (4.25)$$

Lower estimate of $w = w(\cdot, t)$, however, is not easy with the lack of geometric structure of (4.18).

4.3 Geometric Argument

It may be worth noting the geometric argument to derive the estimate below of the solution to (4.18) with (4.19) in §4.2 ([Hamilton (1988)]. In fact, here, we rewrite (4.18) as

$$w_t = r - R, \tag{4.26}$$

using

$$R = e^{-w}\left(-\Delta w + \frac{\lambda}{|\Omega|}\right).$$

There arises that

$$\int_\Omega w_t \, d\mu_t = \int_\Omega w_t e^w dx = \frac{d}{dt}\int_\Omega e^w dx = 0,$$

and then the curvature potential $f = f(\cdot, t)$ is detemined by

$$\Delta_t f = R - r, \tag{4.27}$$

where Δ_t is the Laplace Beltrami operator associated with the metric $g(t) = e^{w(\cdot,t)}g_0$. Then we obtain

$$\Delta_t\left(\frac{\partial f}{\partial t}\right) = \Delta_t(\Delta_t f + rf),$$

to normalize this f as

$$\frac{\partial f}{\partial t} = \Delta_t f + rf$$

(Chapter 5 of [Chow–Knopf (2004)]).

If (4.18) is the 2D–NRF as in (4.19), the *Bochner Weitzenböck formula* induces

$$\frac{\partial H}{\partial t} = \Delta_t H - 2\,|M|^2 + rH, \tag{4.28}$$

where $H = R - r + |\nabla^t f|^2$ and

$$M_{ij} = \nabla_i^t \nabla_j^t f - \frac{1}{2}\Delta_t f \cdot g_{ij} \tag{4.29}$$

with ∇_i^t, $i = 1, 2$, being co-variant derivatives associated with $g(t) = e^{w(\cdot,t)}g_0$, that is,

$$\nabla_i^t = d\xi^i + \sum_{k=1}^{2} \xi^k \omega_k^i$$

of (1.33) in §1.7.

We obtain

$$H \leq \|H_0\|_\infty e^{rt}$$

and hence

$$R \leq r + \|H_0\|_\infty e^{rt} \tag{4.30}$$

by (4.28)–(4.29), and then

$$w_t \geq - \|H_0\|_\infty e^{rt}$$

and hence

$$w \geq w_0 - r^{-1} \|H_0\|_\infty e^{rt} \tag{4.31}$$

follows from (4.26) and (4.30). This estimate of $w = w(\cdot, t)$ from below now ensures $T = +\infty$.

Inequalities (4.25) and (4.31) are not sufficient to recover all the results of [Hamilton (1988)]. In fact, we use the modified Ricci flow, increasing of the surface entropy, and Harnack inequality of Li Yau type. Above argument, however, suggests the role of the estimate of w below in (4.18). The analytic approach developed so far actually provides an alternative proof of this fact in the case of (4.19) ([Bartz–Struwe–Ye (1994); Struwe (2002)]). Estimate of w below is thus essential even to general Ω and λ as we are confirming.

Here we describe the argument [Hamilton (1988)] to derive (4.4) in §4.1, where $g = g(\cdot, t)$ is a solution to (4.1) with

$$R|_{t=0} > 0 \text{ on } \Omega$$

and R_{g_∞} is a constant on Ω. In fact, we use the *curvature potential* f in (4.27) and the *trace free part* $M_g = (M_{ij})$ of Hess(f) in (4.29).

It follows that

$$2M_g = (r - R)g + \mathcal{L}_{\nabla f} g, \tag{4.32}$$

where $\mathcal{L}_{\nabla f}$ is the *Lie derivative* associated with the vector field ∇f. Equality (4.32) then induces the *modified Ricci flow*,

$$\frac{\partial \tilde{g}}{\partial t} = 2M_{\tilde{g}} \tag{4.33}$$

for $\tilde{g} = \tilde{g}(t) = T_t^* g(t)$, where $\{T_t\}$ denotes the one-parameter family of diffeomorphisms generated by ∇f. The quantity

$$|M_g|^2 = |\nabla\nabla f|^2 - \frac{1}{2}(\Delta f)^2, \tag{4.34}$$

furthermore, is invariant under $\{T_t\}$:

$$|M_g|^2 = |M_{\tilde{g}}|^2. \tag{4.35}$$

Then we notice the non-decreasing of the *surface entropy*,

$$\frac{d}{dt}\int_\Omega R\log R\, d\mu \le 0,$$

which implies

$$\sup_t \|R(\cdot,t)\|_\infty < +\infty$$

by the parabolic regularity applied to (4.3) as in Lemma 3.1 for the Smolu-chowski Poisson equation in §3.6 in the Euler coordinate. Then the Harnack inequality of Li Yau type guarantees the reverse inequality,

$$\inf_t \min_\Omega R(\cdot,t) > 0.$$

It follows, therefore,

$$|M_{g(t)}|^2 \le Ce^{-\gamma t} \tag{4.36}$$

from the comparison theorem to

$$\frac{\partial}{\partial t}|M|^2 = \Delta|M|^2 - 2|\nabla M|^2 - 2R|M|^2,$$

where $\gamma > 0$ and $M = M_{g(t)} = M_g$.

We thus end up with

$$|M_{\tilde{g}(t)}| \le Ce^{-\gamma t}$$

by (4.35)–(4.36), which results in

$$\tilde{g}(t) = T_t^*g(t) \to \tilde{g}_\infty \quad \text{exponentially in } C^\infty \text{ topology}$$

with $\tilde{g}_\infty$ satisfying $M_{\tilde{g}_\infty} = 0$ by (4.33), which implies (4.4) with $R_{g_\infty} = 0$.

The above geometric result ensures the following property to (4.18) with (4.19) from the viewpoint of dynamical systems:

(1) global-in-time existence of the solution.
(2) pre-compactness of the orbit.
(3) uniqueness of the ω-limit set.
(4) exponential rate of convergence to the limit state in C^∞ topology as $t \uparrow +\infty$.

Any quantity used in the above geometric argument, however, is associated with the metric $g = g(t)$, and is invalid to (4.18) other than (4.19). Here we show that the first three properties hold to (4.18) for any Ω and $\lambda \le 8\pi$, and that the last one is valid if the stationary state is non-degenerate. We thus take pure analytic approach for the proof of these properties.

4.4   Logarithmic Diffusion

Coming back to (4.18), we put $u = re^w$, $t = r^{-1}\tau$, to obtain

$$u_\tau = \Delta \log u + u - \frac{1}{|\Omega|} \int_\Omega u \, dx$$

$$u|_{\tau=0} = u_0(x) > 0, \quad \int_\Omega u_0 \, dx = \lambda \qquad (4.37)$$

Analogous equation with logarithmic diffusion is

$$u_\tau = \Delta \log u \ \text{ in } \Omega \times (0, T), \quad u|_{t=0} = u_0(x) > 0, \qquad (4.38)$$

which is known to represent several physical models. If

$$\Omega = \mathbf{R}^2, \quad u_0 \in L^1(\mathbf{R}^2)$$

the total mass of $u \in C([0, T), L^1(\Omega))$ decreases as

$$\int_{\mathbf{R}^2} u(x, \tau) dx = \int_{\mathbf{R}^2} u_0 \, dx - 4\pi\tau,$$

and

$$T = \int_{\mathbf{R}^2} u_0 \, dx/(4\pi)$$

is the maximal existence time of the solution (Chapter 8 of [Vázques (2006)]). We have $u \equiv 0$ at $t = T$, and hence quenching of the solution.

Equation (4.37) takes the form

$$u_\tau = \Delta(\log u - v), \quad -\Delta v = u - \frac{1}{|\Omega|} \int_\Omega u \, dx, \quad \int_\Omega v \, dx = 0,$$

which is comparable to the simlified system of chemotaxis (4.24) in §4.2,

$$u_\tau = \nabla \cdot u\nabla(\log u - v), \quad -\Delta v = u - \frac{1}{|\Omega|} \int_\Omega u \, dx, \quad \int_\Omega v \, dx = 0. \qquad (4.39)$$

Henceforth, we rewrite t for τ in (4.37), to study

$$u_t = \Delta \log u + u - \frac{1}{|\Omega|} \int_\Omega u \, dx, \quad u|_{t=0} = u_0(x) > 0. \qquad (4.40)$$

We write, furthermore, $(-\Delta)^{-1}u = v$ for

$$-\Delta v = u - \frac{1}{|\Omega|} \int_\Omega u \, dx, \quad \int_\Omega v \, dx = 0. \qquad (4.41)$$

Hence it holds that

$$u_t = \Delta(\log u - v), \quad v = (-\Delta)^{-1}u \ \text{ in } \Omega \times (0, T) \qquad (4.42)$$

by (4.40).

As in (4.24), equation (4.42) is associated with Helmholtz's free energy,

$$\mathcal{F}(u) = \int_{\Omega} u(\log u - 1)dx - \frac{1}{2}\left\langle(-\Delta)^{-1}u, u\right\rangle,$$

where $\langle\cdot,\cdot\rangle$ is the paring between the field and particle distribution identified with the L^2-inner product [Suzuki (2015b)]. Defining

$$\delta\mathcal{F}(u)$$

by

$$\frac{d}{ds}\mathcal{F}(u + sz)\Big|_{s=0} = \langle z, \delta\mathcal{F}(u)\rangle,$$

we obtain

$$\delta\mathcal{F}(u) = \log u - (-\Delta)^{-1}u,$$

and hence (4.42) is equivalent to

$$u_t = \Delta\mathcal{F}(u) \text{ in } \Omega \times (0, T), \quad u|_{t=0} = u_0(x) > 0. \tag{4.43}$$

Similarly to (3.7), equation (4.43) is the model (B) equation associated with the free energy $\mathcal{F}(u)$, but with the lack of scaling invariance (3.2) in §3.1.

In fact we have

$$\frac{d}{dt}\int_{\Omega} u \, dx = \int_{\Omega} u_t \, dx = \int_{\Omega} \Delta\mathcal{F}(u) \, dx = 0$$

and

$$\frac{d}{dt}\mathcal{F}(u) = \langle u_t, \mathcal{F}(u)\rangle = -\int_{\Omega} |\nabla\mathcal{F}(u)|^2 dx$$

$$= -\int_{\Omega} |\nabla(\log u - v)|^2 \, dx \le 0. \tag{4.44}$$

For

$$\int_{\Omega} u(x, t)dx = \int_{\Omega} u_0 dx \equiv \lambda, \tag{4.45}$$

equation (4.40) is reduced to

$$u_t = \Delta\log u + u - \frac{\lambda}{|\Omega|} \text{ in } \Omega \times (0, T), \quad u|_{t=0} = u_0(x) > 0. \tag{4.46}$$

Furthermore,

$$\overline{u}(t) = \|u_0\|_{\infty} e^t$$

is a super solution to (4.46), and it holds that

$$0 < u(x, t) \le \|u_0\|_{\infty} e^t \text{ in } \Omega \times [0, T).$$

4.5   Benilan's Inequality

We begin with the global in time existence of the solution to (4.46) for $0 < \lambda \le 8\pi$. Here we derive an a priori lower estimate of u. Equation (4.46) actually arises as a limit case of the porous media equation, which admits the Benilan Crandall inequality with prescribed $h(t)$,

$$\frac{u_t(x,t)}{u(x,t)} \le h(t),$$

relative to the Aronson Benilan inequality (Chapter 1 of [Vázques (2006)]).

Lemma 4.1. *The solution $u = u(\cdot, t)$ to (4.46) satisfies*

$$\frac{u_t(x,t)}{u(x,t)} \le \frac{e^t}{e^t - 1} \quad \text{in } \Omega \times (0,T). \tag{4.47}$$

Proof. Equation (4.46) implies

$$u_{tt} = \Delta\left(\frac{u_t}{u}\right) + u_t$$

and hence

$$\frac{u_{tt}u - u_t^2}{u^2} = \frac{1}{u}\Delta\left(\frac{u_t}{u}\right) + \frac{u_t}{u} - \left(\frac{u_t}{u}\right)^2.$$

We thus obtain

$$p_t = u^{-1}\Delta p + p - p^2 \quad \text{in } \Omega \times (0,T) \tag{4.48}$$

for $p = u_t/u$, which admits a spatially homogeneous solution

$$\bar{p} = \bar{p}(t) = \frac{e^t}{e^t - 1} = \frac{1}{1 - e^{-t}}$$

satisfying

$$\bar{p}_t = \bar{p} - \bar{p}^2, \quad \lim_{t\downarrow 0}\bar{p}(t) = +\infty.$$

Hence it follows that $p \le \bar{p}$, or, (4.47). $\qquad\square$

Inequality (4.47) implies

$$\frac{\partial}{\partial t}\left(\frac{u}{e^t - 1}\right) = \frac{u_t - u \cdot e^t/(e^t - 1)}{e^t - 1} \le 0, \tag{4.49}$$

and therefore, in case $T = T_{\max} < +\infty$ there exists

$$\lim_{t\uparrow T} u(x,t) = u(x,T) \in [0,\infty), \quad \forall x \in \Omega. \tag{4.50}$$

Here we take

$$w = \log u, \tag{4.51}$$

to reduce (4.46) to a similar form of (4.18), that is,

$$\frac{\partial e^w}{\partial t} = \Delta w + e^w - \frac{\lambda}{|\Omega|}, \quad \int_\Omega e^w dx = \lambda. \tag{4.52}$$

Inequality (4.47) then means

$$w_t(x,t) \leq \frac{e^t}{e^t - 1} \quad \text{in } \Omega \times (0, T). \tag{4.53}$$

Put

$$\overline{w} = \frac{1}{|\Omega|} \int_\Omega w \, dx$$

and define the functional

$$
\begin{aligned}
J_\lambda(w) &= \frac{1}{2}\|\nabla w\|_2^2 - \lambda \left\{ \log \left(\int_\Omega e^w dx \right) - \overline{w} \right\} \\
&= \frac{1}{2}\|\nabla(w - \overline{w})\|_2^2 - \lambda \log \left(\int_\Omega e^{w - \overline{w}} dx \right).
\end{aligned} \tag{4.54}
$$

For the solution $w = w(\cdot, t)$ to (4.52), we obtain

$$
\begin{aligned}
\frac{d}{dt} J_\lambda(w) &= \int_\Omega \nabla w \cdot \nabla w_t - \lambda \left(\frac{e^w}{\int_\Omega e^w} - \frac{1}{|\Omega|} \right) w_t \, dx \\
&= - \int_\Omega e^w w_t^2 dx
\end{aligned} \tag{4.55}
$$

in accordance with (4.44) in §4.4. As is stated in §3.2 the Trudinger Moser inequality on compact Riemann surface without boundary arises as Fontana's inequality [Fontana (1993)] in the form of

$$J_{8\pi}(w)) \geq -C, \quad w \in H^1(\Omega), \quad \int_\Omega w = 0. \tag{4.56}$$

Here is a key lemma for the estimate of w from below, derived by [Kavallaris–Suzuki (2010)] and [Kavallaris–Suzuki (2015)] for sub-critical case $0 < \lambda < 8\pi$ and critical case $\lambda = 8\pi$, respectively.

Lemma 4.2. *Let $w = w(\cdot, t)$ be the solution to (4.52) for $0 < \lambda \leq 8\pi$. Then it holds that*

$$\liminf_{t \uparrow T} \overline{w}(t) > -\infty. \tag{4.57}$$

To show this lemma, particularly for the critical case $\lambda = 8\pi$, we use the following fact proven in §4.6. The case of bounded domain $\Omega \subset \mathbf{R}^2$ with $-\Delta$ under the Dirichlet condition is treated in [Ohtsuka (2005)]. An analogous result is Lemma 6.1 of [Caglioti–Lions–Marchioro–Pulvirenti (1995)].

Lemma 4.3. *Let Ω be a compact Riemann manifold without boundary, embedded into $\mathbf{R}^N$ isometrically:*

$$\Omega \hookrightarrow \mathbf{R}^N.$$

Let $\{u_k\}$ be a family of positive measurable functions on Ω satisfying

$$\|u_k\|_1 = 8\pi, \quad \mathcal{F}(u_k) \leq C, \quad \lim_{k\to\infty} \langle (-\Delta)^{-1} u_k, u_k \rangle = +\infty$$

and

$$\lim_{k\to\infty} \int_\Omega x u_k dx = 8\pi x_\infty \in \mathbf{R}^N.$$

Then, it holds that $x_\infty \in \Omega$ and

$$u_k(x) dx \rightharpoonup 8\pi \delta_{x_\infty}(dx) \quad in \ \mathcal{M}(\Omega).$$

Proof of Lemma 4.2. In (4.54) we have

$$J_\lambda(w) = (8\pi - \lambda) \log \left(\int_\Omega e^{w - \overline{w}} dx \right) + J_{8\pi}(w)$$
$$= (8\pi - \lambda)(\log \lambda - \overline{w}) + J_{8\pi}(w),$$

and therefore, this lemma is obvious for $0 < \lambda < 8\pi$ by (4.55)–(4.56).

Assume $\lambda = 8\pi$. Since

$$\int_\Omega e^w dx = 8\pi$$

it holds that

$$J_{8\pi}(w) = \frac{1}{2}\|\nabla w\|_2^2 - 8\pi \log \left(\int_\Omega e^{w - \overline{w}} dx \right)$$
$$= \frac{1}{2}\|\nabla w\|_2^2 + 8\pi \overline{w} - 8\pi \log(8\pi)$$

and hence

$$-C \leq \frac{1}{2}\|\nabla w\|_2^2 + 8\pi \overline{w} \leq C \qquad (4.58)$$

by (4.55)–(4.56). Inequality (4.56) implies also

$$\frac{1}{2} \int_\Omega |\nabla w|^2 \cdot \frac{1}{4} dx = \frac{1}{2} \int_\Omega |\nabla(w/2)|^2 dx$$
$$= 8\pi \log \left(\int_\Omega e^{w/2} dx \right) - \frac{8\pi}{|\Omega|} \int_\Omega (w/2) \, dx - C,$$

and hence

$$\frac{1}{2}\int_\Omega |\nabla w|^2 dx \geq 4 \cdot 8\pi \log\left(\int_\Omega e^{w/2}dx\right) - \frac{8\pi}{|\Omega|} \cdot 2 \int_\Omega w \, dx - C. \qquad (4.59)$$

Then it follows that

$$\overline{w} \geq 4\log\left(\int_\Omega e^{w/2}dx\right) - C \qquad (4.60)$$

from (4.58)–(4.59).

First, we take the case $T = T_{\max} < +\infty$, which assures the pointwise convergence (4.50). Inequality (4.60) now implies

$$\liminf_{t\uparrow T} \overline{w} = -\infty \quad \Rightarrow \quad \liminf_{t\uparrow T}\int_\Omega e^{w/2}dx = \liminf_{t\uparrow T}\int_\Omega u(x,t)^{1/2}dx = 0,$$

which implies

$$\int_\Omega u(x,T)^{1/2}dx = 0$$

by the dominated convergence theorem. We thus obtain

$$u(x,T) = 0 \quad \text{a.e. } x \in \Omega,$$

and therefore,

$$8\pi = \lim_{t\uparrow T}\int_\Omega u(x,t) \, dx = \int_\Omega u(x,T) \, dx = 0$$

again by the dominated convergene theorem, a contradiction. Hence (4.57) holds in this case.

Letting $T = +\infty$, second, we show (4.57). In fact, since

$$e^w w \geq -e, \quad w \in \mathbf{R}$$

it holds that

$$H(t) \equiv \int_\Omega e^w w \, dx \geq -e\,|\Omega|,$$

and furthermore, (4.52) and (4.56) imply

$$\frac{dH}{dt} = \int_\Omega e^w(w_t + ww_t)dx = \frac{d}{dt}\int_\Omega e^w dx + \int_\Omega \frac{\partial e^w}{\partial t} \cdot w \, dx$$

$$= \int_\Omega \left[\Delta w + \left(e^w - \frac{8\pi}{|\Omega|}\right)\right] w \, dx$$

$$= -\|\nabla w\|_2^2 + \int_\Omega e^w w \, dx - 8\pi\overline{w}$$

$$\leq \int_\Omega e^w w \, dx + 8\pi\overline{w} + C = H + 8\pi\overline{w} + C.$$

Assume the contrary to (4.57),

$$\liminf_{t\uparrow+\infty} \overline{w}(t) = -\infty.$$

Then inequality (4.53) assures the existence of $t_k \uparrow +\infty$ and $\delta > 0$ such that

$$t_{k+1} > t_k + \delta$$

and

$$8\pi\overline{w}(t) + C \le -k, \quad t_k - \delta < t < t_k$$

for $k = 1, 2, \cdots$. Then it holds that

$$\frac{d}{dt}\left(e^{-t}H\right) \le -ke^{-t}, \quad t_k - \delta < t < t_k.$$

Take $t \in (t_k - \delta, t_k - \delta/2)$, operate $\int_t^{t_k} \cdot \, dt$ to this inequality, and obtain

$$e^{-t_k}H(t_k) \le e^{-t}H(t) + k(e^{-t_k} - e^{-t}),$$

which means

$$\begin{aligned}
H(t) &\ge e^{t-t_k}H(t_k) + k(1 - e^{t-t_k}) \\
&\ge -e^{-\delta+1}|\Omega| + k(1 - e^{-\delta/2}), \quad t_k - \delta < t < t_k - \delta/2.
\end{aligned}$$

Hence there arises that

$$\lim_{k\to\infty} \inf_{t\in(t_k-\delta, t_k-\delta/2)} \int_\Omega (e^w w)(x,t)dx = +\infty \tag{4.61}$$

We have, on the other hand,

$$\sum_{k=1}^\infty \int_{t_k-\delta}^{t_k-\delta/2} dt \int_\Omega e^w w_t^2 dx \le \int_0^\infty dt \int_\Omega e^w w_t^2 dx < +\infty \tag{4.62}$$

by (4.55), together with

$$\|e^w w_t\|_1^2 \le \int_\Omega e^w dx \cdot \int_\Omega e^w w_t^2 dx = 8\pi \int_\Omega e^w w_t^2 dx. \tag{4.63}$$

Inequalities (4.62)–(4.63) then imply

$$\lim_{k\to\infty} \int_{t_k-\delta}^{t_k-\delta/2} \|e^w w_t(\cdot, t)\|_1^2 \, dt = 0,$$

and therefore, equality (4.61) assures the existence of

$$t_k' \in (t_k - \delta, t_k - \delta/2)$$

such that

$$\lim_{k\to\infty} \int_\Omega (e^w w)(x, t_k') \, dx = +\infty, \quad \lim_{k\to\infty} \left\| \frac{\partial e^w}{\partial t}(\cdot, t_k') \right\|_1 = 0. \qquad (4.64)$$

Here we use $u = u(x, t)$ in (4.51). We confirm the assumptions of Lemma 4.3 for $u_k = u(\cdot, t_k')$. In fact, we obtain

$$\lim_{k\to\infty} \int_\Omega u_k \log u_k dx = +\infty \qquad (4.65)$$

by the first equality of (4.64), and also

$$\mathcal{F}(u_k) \le \mathcal{F}(u_0) \qquad (4.66)$$

by (4.44). These (4.65)–(4.66) imply

$$\lim_{k\to\infty} \langle (-\Delta)^{-1} u_k, u_k \rangle = +\infty.$$

It is obvious that $\|u_k\|_1 = 8\pi$ by (4.45) for $\lambda = 8\pi$, and finally, passing to a subsequence, we obtain $x_\infty \in \mathbf{R}^N$ such that

$$\lim_{k\to\infty} \int_\Omega x u_k dx = x_\infty \in \mathbf{R}^N.$$

Lemma 4.3, then, assures $x_\infty \in \Omega$, and passing to a subsequence we obtain

$$e^{w(x, t_k')} dx = u_k dx \rightharpoonup 8\pi \delta_{x_\infty} \quad \text{in } \mathcal{M}(\Omega). \qquad (4.67)$$

We now apply the second equality of (4.64) and (4.67) to (4.52). From the elliptic L^1 estimate [Brezis–Strauss (1973)] and the second equality of (4.64), therefore, it follows that

$$w(\cdot, t_k') \rightharpoonup 8\pi G(\cdot, x_\infty) \text{ in } W^{1,q}(\Omega), \quad 1 < q < 2, \qquad (4.68)$$

where $G = G(x, x')$ denotes the Green function to (4.41). This (4.68) implies

$$\lim_{k\to\infty} \int_\Omega e^{w(\cdot, t_k')} dx = +\infty, \qquad (4.69)$$

as in [Brezis–Merle (1991)] by Fatou's lemma and the estimate

$$\left| G(x, x') + \frac{1}{2\pi} \log \text{dist}(x, x') \right| \le C, \quad x, x' \in \Omega.$$

Equality (4.69) contradicts

$$\int_\Omega e^w dx = \lambda = 8\pi$$

and the proof is complete. $\qquad\qquad\qquad\qquad\qquad\qquad\qquad\qquad\qquad$ □

4.6   Concentration of Probability Measures

This section is devoted to the proof of Lemma 4.3, which provides a criterion of the concentration of absolutely continuous probability measures. We put

$$P(\Omega) = \{\rho \in L^1(\Omega) \mid \rho \geq 0, \ \|\rho\|_1 = 1\}$$

and

$$\mathcal{I}(\rho) = \frac{1}{2} \iint_{\Omega \times \Omega} G(x, x')\rho \otimes \rho \ dx dx' - \frac{1}{8\pi} \int_\Omega \rho \log \rho \ dx, \ \ \rho \in P(\Omega),$$

where $\rho \otimes \rho = \rho(x)\rho(x')$.

Given

$$0 \leq u \in L^1(\Omega), \ \|u\|_1 = 8\pi,$$

we have $\rho = u/8\pi \in P(\Omega)$ and

$$\mathcal{I}(\rho) = -\frac{1}{64\pi^2} \left\{ \int_\Omega u(\log u - 1) \ dx - \frac{1}{2} \iint_{\Omega \times \Omega} G(x, x')u \otimes u \ dx dx' \right\}$$

$$- \frac{1}{64\pi^2} \{1 - \log(8\pi)\}$$

$$= -\frac{1}{64\pi^2} \mathcal{F}(u) + \text{constant}. \tag{4.70}$$

By the dual form of Fontana' inequality, (3.24), we obtain

$$\sup \{\mathcal{I}(\rho) \mid \rho \in P(\Omega)\} < +\infty.$$

Here we put

$$\mathcal{K}(\rho) = \frac{1}{2} \iint_{\Omega \times \Omega} G(x, x')\rho \otimes \rho \ dx dx'$$

$$\mathcal{E}(\rho) = - \int_\Omega \rho(\log \rho - 1) \ dx.$$

It holds that

$$\mathcal{I} = \mathcal{K} + \mathcal{E}/(8\pi).$$

Lemma 4.3 is reduced to the following lemma by (4.70).

Lemma 4.4. *If the sequence $\rho_k \in P(\Omega)$, $k = 1, 2, \cdots$, satisfies*

$$\lim_{k \to \infty} \mathcal{K}(\rho_k) = +\infty, \ \lim_{k \to \infty} \mathcal{I}(\rho_k) = I_\infty > -\infty, \ \lim_{k \to \infty} \int_\Omega x\rho_k dx = x_\infty$$

for $x_\infty \in \mathbf{R}^N$, it holds that $x_\infty \in \Omega$ and

$$\rho_k(x)dx \rightharpoonup \delta_{x_\infty}(dx) \quad in \ \mathcal{M}(\Omega). \tag{4.71}$$

For the proof, we use the following lemma, where

$$\mathcal{I}_\beta = \mathcal{K} + \mathcal{E}/\beta.$$

Lemma 4.5. *Each $d > 0$ and $m > 0$ admit $C > 0$ and $\beta > 8\pi$ such that if $A_i \subset \Omega$, $i = 1, 2$, are measurable sets satisfying*

$$\text{dist}(A_1, A_2) \geq d, \quad \int_{A_i} \rho \, dx \geq m, \quad i = 1, 2 \tag{4.72}$$

for $\rho \in P(\Omega)$, then it holds that $\mathcal{I}_\beta(\rho) \leq C$.

Proof. Let $A_3 = \Omega \setminus (A_1 \bigcup A_2)$, and put

$$\chi_i = \chi_{A_i} \rho, \quad 1 \leq i \leq 3.$$

It holds that

$$\mathcal{K}(\rho) = \frac{1}{2} \iint G(x, x') \rho \otimes \rho \, dx dx'$$
$$= E_{11} + E_{22} + E_{33} + 2E_{12} + 2E_{23} + 2E_{31},$$

where $G = G(x, x')$ is the Green function to (4.41) and

$$E_{ij} = \frac{1}{2} \iint_{\Omega \times \Omega} G(x, x') \rho_i \otimes \rho_j dx dx'.$$

Since $G(x, x')$ is smooth in $x \neq x'$, it holds that $E_{12} \leq C = C(d, m)$ by (4.72). Thus we obtain

$$\mathcal{K}(\rho) \leq a^2 + b^2 + c^2 + 2bc + 2ca + C \tag{4.73}$$

for

$$E_{11} = a^2, \quad E_{22} = b^2, \quad E_{33} = c^2,$$

because the Schwarz inequality implies

$$E_{ij}^2 \leq E_{ii} E_{jj}.$$

Inequality (4.70), on the other hand, implies

$$S_i \leq \frac{-8\pi E_{ii} + C}{M_i}, \quad 1 \leq i \leq 3$$

for

$$S_i = -\int_{A_i} \rho \log \rho, \quad M_i = \int_{A_i} \rho,$$

and therefore,

$$\mathcal{E}(\rho) = -\int_\Omega \rho \log \rho \leq -8\pi \left(\frac{a^2}{M_1} + \frac{b^2}{M_2} + \frac{c^2}{M_3} \right) + C \tag{4.74}$$

with

$$M_1, M_2 \geq m, \quad M_1 + M_2 + M_3 = 1. \tag{4.75}$$

Inequalities (4.73), (4.74), and (4.75), then imply the existence of

$$\delta = \delta(m) > 0$$

such that

$$\mathcal{E}(\rho) \leq -(8\pi + \delta)\mathcal{K}(\rho) + C.$$

See the proof of Lemma 6.2 of [Caglioti–Lions–Marchioro–Pulvirenti (1995)] for this elementary fact. Then the proof is complete. □

Now we use the *concentration function* of ρ_k defined by

$$Q_k(r) = \sup_{y \in \Omega} \int_{\Omega \cap B(y,r)} \rho_k dx, \quad 0 < r \ll 1.$$

Take $x_k \in \Omega$ as in

$$\int_{\Omega \cap B(x_k, r/2)} \rho_k dx = Q_k(r/2), \tag{4.76}$$

to note

$$1 - Q_k(r) \leq 1 - \int_{\Omega \cap B(x_k, r)} \rho_k dx = \int_{\Omega \setminus B(x_k, r)} \rho_k dx.$$

It holds that

$$\min\{Q_k(r/2), 1 - Q_k(r)\}$$

$$\leq \min \left\{ \int_{\Omega \cap B(x_k, r/2)} \rho_k dx, \int_{\Omega \setminus B(x_k, r)} \rho_k dx \right\}$$

and Lemma 4.5 is applicable to $d = r/2$.

Given $m > 0$, we have thus $\beta > 8\pi$ such that

$$m \leq \min\{Q_k(r/2), 1 - Q_k(r)\} \quad \Rightarrow \quad \mathcal{I}_\beta(\rho_k) \leq C. \tag{4.77}$$

We have, on the other hand,

$$\mathcal{I}_\beta(\rho_k) = \frac{8\pi}{\beta} \left\{ \left(\frac{\beta}{8\pi} - 1 \right) \mathcal{K}(\rho_k) + \mathcal{I}(\rho_k) \right\} \quad \rightarrow +\infty$$

from the assumption, and therefore,

$$\lim_{k \to \infty} \min\{Q_k(r/2), 1 - Q_k(r)\} = 0$$

by (4.77).

Here we apply the following lemma, to obtain

$$\lim_{k \to \infty} \{1 - Q_k(r)\} = 0, \quad 0 < r \ll 1. \tag{4.78}$$

Lemma 4.6. *Each bounded open set* $\Omega \subset \mathbf{R}^2$ *admits* $c_0 > 0$ *such that*

$$Q(r) \equiv \sup_{y \in \Omega} \int_{\Omega \cap B(y,r)} \rho \, dx \geq c_0 r^2, \quad 0 < r \leq 1$$

for any $\rho \in P(\Omega)$.

Proof. Given Ω, we have $\hat{\Omega}$ satisfying

$$\Omega \subset \bigcup_{y \in \Omega} B(y,r) \subset \hat{\Omega}, \quad 0 < \forall r \leq 1.$$

Then *Vitali's covering theorem* assures $\omega \subset \Omega$ such that

$$B(y,r) \cap B(y',r) = \emptyset, \quad y, y' \in \omega, \ y \neq y' \tag{4.79}$$

and

$$\Omega \subset \bigcup_{y \in \omega} B(y, 5r). \tag{4.80}$$

First, (4.79) implies

$$\pi r^2 \cdot \sharp \, \omega \leq |\hat{\Omega}|.$$

It holds, second, that

$$\sharp \, \omega \cdot Q(r) \geq \sum_{y \in \omega} \int_{\Omega \cap B(y,5r)} \rho dx \geq \int_{\Omega} \rho \, dx = 1.$$

Then we obtain

$$Q(r) \geq \frac{1}{\sharp \, \omega} \geq \frac{\pi r^2}{|\hat{\Omega}|} = c_0 r^2$$

and the proof is complete. $\qquad\qquad\qquad\qquad\qquad\qquad\qquad\qquad\quad \square$

We are ready to give the following proof.

Proof of Lemma 4.4. Given $0 < r \ll 1$, we have

$$1 - Q_k(r/2) = \int_{\Omega \setminus B(x_k, r/2)} \rho_k dx \leq r \tag{4.81}$$

for $k \gg 1$ by (4.76) and (4.78). Letting

$$\bar{x}_k = \int_{\Omega} x \rho_k dx \in \Omega,$$

therefore, we obtain

$$
\begin{aligned}
|\overline{x}_k - x_k| &= \left| \int_\Omega (x - x_k) \rho_k dx \right| \\
&\leq \int_{\Omega \cap B(x_k, r)} |x - x_k| \, \rho_k dx + \int_{\Omega \backslash B(x_k, r)} |x - x_k| \, \rho_k dx \\
&\leq r + \operatorname{diam} \Omega \cdot \int_{\Omega \backslash B(x_k, r/2)} \rho_k dx = r + \operatorname{diam} \Omega \cdot (1 - Q_k(r/2)) \\
&\leq (1 + \operatorname{diam} \Omega) \, r.
\end{aligned}
$$

It holds that

$$
\lim_{k \to \infty} |\overline{x}_k - x_k| = 0,
$$

and hence $x_\infty \in \Omega$.

Given $\zeta = \zeta(x) \in C(\Omega)$, similarly, we have

$$
\begin{aligned}
\left| \zeta(x_k) - \int_\Omega \zeta(x) \rho_k dx \right| \\
&\leq \int_{\Omega \cap B(x_k, r)} |\zeta(x_k) - \zeta(x)| \, \rho_k dx + \int_{\Omega \backslash B(x_k, r)} |\zeta(x_k) - \zeta(x)| \, \rho_k dx \\
&\leq \|\zeta - \zeta(x_k)\|_{L^\infty(B(x_k, r))} + 2\|\zeta\|_\infty \int_{\Omega \backslash B(x_k, r/2)} \rho_k dx \\
&\leq \|\zeta - \zeta(x_k)\|_{L^\infty(B(x_k, r))} + 2\|\zeta\|_\infty r
\end{aligned}
$$

by (4.76) and (4.81). Sending $k \to \infty$ and then $r \downarrow 0$, we obtain

$$
\lim_{k \to \infty} \left| \zeta(x_k) - \int_\Omega \zeta(x) \rho_k dx \right| = 0
$$

because $\zeta \in C(\Omega)$ is uniformly continuous, and hence (4.71). □

4.7   Pre-Compactness of the Orbit

Here we show the following theorem.

Theorem 4.1. *The solution $w = w(\cdot, t)$ to (4.18) exists global in time, provided that $0 < \lambda \leq 8\pi$. Its orbit is pre-compact in C^∞ topology, and the ω-limit set*

$$
\omega(w_0) = \left\{ w_\infty \in C^2(\Omega) \mid \exists t_k \uparrow +\infty, \lim_{k \to \infty} \|w(\cdot, t_k) - w_\infty\|_{C^2} = 0 \right\} \quad (4.82)
$$

is non-empty, compact, connected, and contained in the set of stationary solutions:

$$
-\Delta w_\infty = \lambda \left(\frac{e^{w_\infty}}{\int_\Omega e^{w_\infty} dx} - \frac{1}{|\Omega|} \right) \text{ in } \Omega, \quad \int_\Omega e^{w_\infty} dx = \int_\Omega e^{w_0} dx. \quad (4.83)
$$

Proof. Assume $0 < \lambda \leq 8\pi$. It suffices to show that the solution $u = e^w$ to (4.46) satisfies the a priori estimate

$$0 < u(x,t), \ u(x,t)^{-1} \leq C \quad \text{in } \Omega \times [0,T), \tag{4.84}$$

from the parabolic regularity.

First, we confirm

$$\|\nabla w(\cdot,t)\|_2 \leq C, \quad 0 \leq t < T. \tag{4.85}$$

In fact, this (4.85) is obvious for $0 < \lambda < 8\pi$ by (4.55)–(4.56) and

$$J_\lambda(w) = \frac{1}{2}\left(1 - \frac{\lambda}{8\pi}\right)\|\nabla w\|_2^2 + \frac{\lambda}{8\pi}J_{8\pi}(w).$$

In the case of $\lambda = 8\pi$, we use

$$J_{8\pi}(w) = \frac{1}{2}\|\nabla w\|_2^2 - 8\pi \log\left(\int_\Omega e^{w-\overline{w}}dx\right)$$

$$= \frac{1}{2}\|\nabla w\|_2^2 + 8\pi\overline{w} - 8\pi\log(8\pi).$$

Then, equality (4.55) and Lemma 4.2 imply (4.85).

Second, we have

$$|\overline{w}(t)| \leq C. \tag{4.86}$$

It follows from

$$\exp\left(\frac{1}{|\Omega|}\int_\Omega w \, dx\right) \leq \frac{1}{|\Omega|}\int_\Omega e^w dx = \frac{\lambda}{|\Omega|} \equiv \alpha$$

derived from Jensen's inequality and Lemma 4.2 in §4.5.

Inequalities (4.85)–(4.86) then imply

$$\|\log u\|_{H^1(\Omega)} = \|w\|_{H^1(\Omega)} \leq C$$

by *Poincaré–Wirtinger's inequality*, and therefore,

$$\left\|e^{p\log u}\right\|_1, \ \left\|e^{-p\log u}\right\|_1 \leq C_p, \quad p \geq 1$$

by Fontana's inequality in the form of (3.26) in §3.2, which results in

$$\|u(\cdot,t)\|_p, \ \|u^{-1}(\cdot,t)\|_p \leq C_p, \quad p \geq 1. \tag{4.87}$$

To estmate u above, we use (4.46), to deduce

$$\frac{1}{p+1}\frac{d}{dt}\|u\|_{p+1}^{p+1} = -\int_\Omega \nabla u^p \cdot \nabla \log u \, dx + \|u\|_{p+1}^{p+1} - \alpha\|u\|_p^p$$

$$= -\int_\Omega u^{-1}\nabla u^p \cdot \nabla u \, dx + \|u\|_{p+1}^{p+1} - \alpha\|u\|_p^p$$

$$= -\frac{4}{p}\|\nabla u^{\frac{p}{2}}\|_2^2 + \|u\|_{p+1}^{p+1} - \alpha\|u\|_p^p,$$

recalling $\alpha = \lambda/|\Omega|$. We thus obtain

$$\frac{1}{p+1}\frac{d}{dt}\|u\|_{p+1}^{p+1} + \frac{4}{p}\|\nabla u^{\frac{p}{2}}\|_2^2 + \alpha\|u\|_p^p = \|u\|_{p+1}^{p+1} \tag{4.88}$$

for $p \geq 1$. Writing the right-hand side of (4.88) as

$$\|u\|_{p+1}^{p+1} = \|u^{\frac{p}{2}}\|_{2\frac{p+1}{p}}^{2\frac{p+1}{p}},$$

and apply the Gagliardo Nirenberg inequality in two space dimension,

$$\|w\|_q^q \leq C_{q_0}\|w\|_{H^1}^{q-1}\|w\|_1, \quad 1 \leq q \leq q_0 < \infty \tag{4.89}$$

for $q = 2\frac{p+1}{p} \in [2,4]$. From Poincaré–Wirtinger's inequality, (4.86), and $1 < 1 + \frac{2}{p} \leq 3$, it follows that

$$\|u\|_{p+1}^{p+1} \leq C\left(\|\nabla u^{\frac{p}{2}}\|_2 + \|u^{\frac{p}{2}}\|_1\right)^{1+\frac{2}{p}}\|u^{\frac{p}{2}}\|_1$$
$$\leq C\left(\|\nabla u^{\frac{p}{2}}\|_2^{1+\frac{2}{p}}\|u^{\frac{p}{2}}\|_1 + \|u^{\frac{p}{2}}\|_1^{2+\frac{2}{p}}\right). \tag{4.90}$$

Here, we take $p > 2$ and apply Young's inequality to the right-hand side of (4.90) for

$$q = \frac{2p}{p+2} > 1, \quad \frac{1}{q} + \frac{1}{q'} = 1.$$

It follows that

$$\|u\|_{p+1}^{p+1} \leq \frac{1}{q}\left(a\|\nabla u^{\frac{p}{2}}\|_2^{1+\frac{2}{p}}\right)^q + \frac{1}{q'}\left(a^{-1}C\|u^{\frac{p}{2}}\|_1\right)^{q'} + C\|u^{\frac{p}{2}}\|_1^{2+\frac{2}{p}}$$

with $a > 0$, prescribed by

$$\frac{a^q}{q} = \frac{2}{p}.$$

Since

$$q \to 2, \quad q' = \frac{2p}{p-2} \to 2, \quad a^q \sim p^{-1}, \quad a^{q'} \sim p^{-1}$$

as $p \to \infty$ and it holds that $q' > 2 + \frac{2}{p}$ for $p > 2$, we obtain

$$\|u\|_{p+1}^{p+1} \leq \frac{2}{p}\|\nabla u^{\frac{p}{2}}\|_2^2 + C \cdot p\left(\|u^{\frac{p}{2}}\|_1^{\frac{2p}{p-2}} + 1\right). \tag{4.91}$$

It thus holds that

$$\frac{d}{dt}\|u\|_{p+1}^{p+1} + 2\|\nabla u^{\frac{p}{2}}\|_2^2 \leq C \cdot p^2\left(\|u^{\frac{p}{2}}\|_1^{\frac{2p}{p-2}} + 1\right) \tag{4.92}$$

for $p \geq 3$ by (4.88) and (4.91).

Adjusting $a > 0$, differently, we get

$$\|u\|_{p+1}^{p+1} \leq \|\nabla u^{\frac{p}{2}}\|_2^2 + C\left(\|u^{\frac{p}{2}}\|_1^{\frac{2p}{p-2}} + 1\right) \tag{4.93}$$

similarly to (4.91). Adding (4.92)–(4.93) then ensures

$$\frac{d}{dt}\|u\|_{p+1}^{p+1} + \|u\|_{p+1}^{p+1} \leq C \cdot p^2\left(\|u^{\frac{p}{2}}\|_1^{\frac{2p}{p-2}} + 1\right). \tag{4.94}$$

Use

$$\|u^{\frac{p}{2}}\|_1^{\frac{2p}{p-2}} = \left(\int_\Omega u^{\frac{(p+1)(p-2)}{2p}} \cdot u^{\frac{(p+2)}{2p}}\right)^{\frac{2p}{p-2}}$$

on the right-hand side of (4.94), and apply Hölder's inequality for $q = \frac{3}{4}\frac{2p}{p-2} > 1$ and $q' = \frac{3p}{p+4} \in (1,3)$ satisfying $\frac{1}{q} + \frac{1}{q'} = 1$. It follows that

$$\|u^{\frac{p}{2}}\|_1^{\frac{2p}{p-2}} \leq \left(\int_\Omega u^{\frac{3}{4}(p+1)}dx\right)^{\frac{4}{3}}\left(\int_\Omega u^{q'\frac{(p+2)}{2p}}dx\right)^{\frac{1}{q'}\frac{2p}{p-2}}$$

with

$$\left(\int_\Omega u^{q'\frac{(p+2)}{2p}}dx\right)^{\frac{1}{q'}\frac{2p}{p-2}} \leq C$$

from (4.87), to reach

$$\frac{d}{dt}\|u\|_{p+1}^{p+1} + \|u\|_{p+1}^{p+1} \leq Cp^2\left[\left(\int_\Omega u^{\frac{3}{4}(p+1)}dx\right)^{\frac{4}{3}} + 1\right], \quad p \geq 3. \tag{4.95}$$

The power of u is homogeneous in (4.95) similarly to the linear equation, and then the estimate

$$\|u(\cdot, t)\|_\infty \leq C$$

is derived from Moser's iteration scheme as below [Alikakos (1979)].
First, inequality (4.95) implies

$$\sup_{0 \leq t < T}\left[\|u\|_{p+1}^{p+1} + 1\right]$$

$$\leq C\max\left[p^2\sup_{0 \leq t < T}\left[\|u\|_{\frac{3}{4}(p+1)}^{p+1} + 1\right]^{\frac{4}{3}}, |\Omega|\|u_0\|_\infty^{p+1} + 1\right] \tag{4.96}$$

for $p \geq 3$. Second, putting

$$\Phi_k = \sup_{0 \leq t < T}\left[\int_\Omega u^{(\frac{4}{3})^k}dx + 1\right], \quad d = \|u_0\|_\infty + 1 < +\infty,$$

we obtain

$$\Phi_{k+1} \leq C \max \left[\left(\frac{4}{3}\right)^{2(k+1)} \Phi_k^{\frac{4}{3}}, |\Omega| \|u_0\|_{\infty}^{(\frac{4}{3})^{k+1}} + 1 \right]$$

$$\leq C \left(\frac{4}{3}\right)^{2(k+1)} \max \left[\Phi_k^{\frac{4}{3}}, d^{(\frac{4}{3})^{k+1}} \right]$$

by (4.96), which implies

$$\Phi_{k+1} \leq C^{\sum_{j=1}^{k-2}(\frac{4}{3})^{j-1}} \left(\frac{4}{3}\right)^{2\sum_{j=1}^{k-2} j (\frac{4}{3})^{k-2-j}} \max \left[\Phi_3^{(\frac{4}{3})^{k+1}}, d^{(\frac{4}{3})^{k+1}} \right]$$

for $k = 3, 4, 5, ...$, inductively. This inequality induces

$$\sup_{0 \leq t < T} \left[\int_{\Omega} u^{(\frac{4}{3})^{k+1}} dx \right]^{\frac{1}{(\frac{4}{3})^{k+1}}} \leq \Phi_{k+1}^{\frac{1}{(\frac{4}{3})^{k+1}}}$$

$$\leq C \left(\frac{4}{3}\right)^{2\sum_{j=1}^{k-1} j (\frac{4}{3})^{-j-3}} \max \left[\Phi_3, d \right]. \tag{4.97}$$

Making $k \to \infty$ in (4.97), we obtain

$$\sup_{0 \leq t < T} \|u\|_{\infty} \leq C \max \left[\Phi_3, d \right] < +\infty.$$

To estimate u below, we use

$$v_t = v\Delta v - |\nabla v|^2 - v + \alpha v^2 \quad \text{in } \Omega \times (0, T) \tag{4.98}$$

valid to $v = u^{-1}$. Equation (4.98) implies

$$\frac{1}{p+1} \frac{d}{dt} \|v\|_{p+1}^{p+1} = -(p+2) \int_{\Omega} v^p |\nabla v|^2 \, dx - \|v\|_{p+1}^{p+1} + \alpha \|v\|_{p+2}^{p+2}$$

for $p \geq 1$ and hence

$$\frac{1}{p+1} \frac{d}{dt} \|v\|_{p+1}^{p+1} + \frac{4}{p+2} \|\nabla v^{\frac{p+2}{2}}\|_2^2 + \|v\|_{p+1}^{p+1} = \alpha \|v\|_{p+2}^{p+2}. \tag{4.99}$$

We write the right-hand side of (4.99) as

$$\|v\|_{p+2}^{p+2} = \|v^{\frac{p+2}{2}}\|_2^2$$

and apply the Gagliardo Nirenberg inequality (4.89) for $q = 2$, to obtain

$$\|v\|_{p+2}^{p+2} \leq C \left(\|\nabla v^{\frac{p+2}{2}}\|_2 + \|v^{\frac{p+2}{2}}\|_1 \right) \|v^{\frac{p+2}{2}}\|_1$$

$$\leq \frac{1}{2(p+2)} \|\nabla v^{\frac{p+2}{2}}\|_2^2 + \left\{ C^2 \frac{(p+2)}{2} + C \right\} \|v^{\frac{p+2}{2}}\|_1^2.$$

Hence it holds that

$$\frac{1}{p+1}\frac{d}{dt}\|v\|_{p+1}^{p+1} + \|v\|_{p+1}^{p+1} \leq C(p+2)\|v^{\frac{p+2}{2}}\|_1^2 \qquad (4.100)$$

For the right-hand side of (4.100) we have

$$\|v^{\frac{p+2}{2}}\|_1^2 = \left(\int_\Omega v^{\frac{p+1}{2}} v^{\frac{1}{2}} dx\right)^2 \leq \left(\int_\Omega v^{\frac{p+1}{2}\cdot\frac{3}{2}} dx\right)^{\frac{4}{3}} \left(\int_\Omega v^{\frac{3}{2}} dx\right)^{\frac{2}{3}}$$

$$\leq C\left(\int_\Omega v^{\frac{3}{4}(p+1)} dx\right)^{\frac{4}{3}}$$

by (4.87), and therefore, it holds that

$$\frac{d}{dt}\int_\Omega v^{p+1} dx + \int_\Omega v^{p+1} dx \leq C(p+1)^2 \left(\int_\Omega v^{\frac{3}{4}(p+1)} dx\right)^{\frac{4}{3}} \qquad (4.101)$$

Moser's iteration scheme is applicable in this form of (4.101) as in u.

Inequality (4.101) thus implies

$$\sup_{0\leq t<T}\|v\|_{p+1}^{p+1} \leq C\max\left[p^2 \sup_{0\leq t<T}\|v\|_{\frac{3}{4}(p+1)}^{p+1}, \ |\Omega|\cdot\|v_0\|_\infty^{p+1}\right]$$

for $v_0 = u_0^{-1}$. Putting

$$\Phi_k = \sup_{0\leq t<T}\int_\Omega v^{(\frac{4}{3})^k} dx, \quad \gamma = \max\{|\Omega|, 1\}\cdot\|v_0\|_\infty,$$

therefore, we obtain

$$\Phi_{k+1} \leq C\cdot\max\left[\left(\frac{4}{3}\right)^{2(k+1)} \Phi_k^{\frac{4}{3}}, \ |\Omega|\cdot\|v_0\|_\infty^{(\frac{4}{3})^{k+1}}\right]$$

$$\leq C\cdot\left(\frac{4}{3}\right)^{2(k+1)} \max\left[\Phi_k^{\frac{4}{3}}, \ \gamma^{(\frac{4}{3})^{k+1}}\right], \quad k = 0, 1, 2, \cdots.$$

It thus follows that

$$\Phi_{k+1} \leq C^{\sum_{j=1}^{k+1}(\frac{4}{3})^{j-1}} \left(\frac{4}{3}\right)^{2\sum_{j=1}^{k+1} j\left(\frac{4}{3}\right)^{k+1-j}} \max\left[\Phi_0^{(\frac{4}{3})^{k+1}}, \ \gamma^{(\frac{4}{3})^{k+1}}\right]$$

inductively, and hence the inequality

$$\sup_{0\leq t<T_{max}} \left[\int_\Omega v^{(\frac{4}{3})^{k+1}}\right]^{\frac{1}{(\frac{4}{3})^{k+1}}} \leq \Phi_{k+1}^{\frac{1}{(\frac{4}{3})^{k+1}}}$$

$$\leq C\left(\frac{4}{3}\right)^{2\sum_{j=1}^{k+1} j\left(\frac{4}{3}\right)^{-j}} \max\left[\Phi_0, \ \gamma\right] \qquad (4.102)$$

arises. Making $k\to\infty$ in (4.102), we obtain

$$\sup_{0\leq t<T}\|v\|_\infty \leq C\max\left[\Phi_0, \ \gamma\right] < +\infty,$$

and the proof is complete. $\qquad\square$

Remark 4.1. If (4.83) has a unique solution denoted by w_∞, the nonstationary solution w to (4.18) converges to this w_∞ in C^∞ topology by Theorem 4.1 and the parabolic regularity. The first equation of (4.83), on the other hand, is invariant up to additive constant. Hence this uniqueness is equivalent to that of

$$-\Delta v = \lambda \left(\frac{e^v}{\int_\Omega e^v dx} - \frac{1}{|\Omega|} \right) \text{ in } \Omega, \qquad \int_\Omega v \, dx = 0 \qquad (4.103)$$

as is described in §4.1. This property of uniqueness is valid to (4.19), which implies (4.12) except for the rate of convergence. The other example for such Ω is the two dimensional torus

$$\mathbf{T}^2 = \mathbf{R}^2 / a\mathbf{Z} \times b\mathbf{Z}$$

with $a \geq b$ satisfying $\dfrac{\pi}{4} \leq \dfrac{b}{a} \leq 1$ and $0 < \lambda \leq 8\pi$ ([Lin–Lucia (2006)]).

Even without uniqueness of the solution to (4.103), nonstationary solution converges to a stationary solution in C^∞ topology if they are discrete by the connectivity of the ω-limit set $\omega(w_0)$. Convergence to the stationary solution, however, is assured by itself via the method of gradient inequality, regardless of the discreteness of $\omega(w_0)$ as we see below.

4.8   Steady States

Theorem 4.1 in §4.7 guarantees pre-compactness in C^∞ topology of the orbit $\{u(\cdot, t)\}$ of the global-in time solution to (4.40) with (4.45). Then, equalities (4.44)–(4.45) imply the following fact by the theory of dynamical systems [Henry (1981)], where

$$\omega(u_0) \equiv \{u_* = u_*(x) > 0 \mid \exists t_k \to \infty, \ \lim_{k \to \infty} \|u(\cdot, t_k) - u_*\|_{C^2} = 0\}$$

denotes the ω-limit set. In the following, $v = (-\Delta)^{-1}u$ means (4.41) in §4.4:

$$-\Delta v = u - \frac{1}{|\Omega|} \int_\Omega u \, dx, \qquad \int_\Omega v \, dx = 0.$$

Theorem 4.2. *Under the assumption of Theorem 4.1, the above ω-limit set $\omega(u_0)$ is non-empty, connected, compact, and contained in the set of steady states denoted by*

$$F_\lambda = \{u_* = u_*(x) > 0 \mid classical \ solution \ to \ (4.105)\}, \qquad (4.104)$$

where

$$\delta\mathcal{F}(u_*) = \log u_* - (-\Delta)^{-1}u_* = constant \,, \qquad \int_\Omega u_* \, dx = \lambda. \qquad (4.105)$$

Here we note the following property.

Lemma 4.7. *Equation (4.105) is equivalent to*

$$-\Delta \log u_* = u_* - \frac{1}{|\Omega|} \int_\Omega u_* \, dx, \quad \int_\Omega u_* \, dx = \lambda \qquad (4.106)$$

Proof. First we put $v_* = (-\Delta)^{-1} u_*$ in (4.105), to obtain

$$-\Delta v_* = u_* - \frac{1}{|\Omega|} \int_\Omega u_* \, dx, \quad \int_\Omega v_* \, dx = 0, \qquad (4.107)$$

$$\log u_* = v_* + \text{constant}, \quad \int_\Omega u_* dx = \lambda. \qquad (4.108)$$

Then it holds that

$$u_* = \frac{\lambda e^{v_*}}{\int_\Omega e^{v_*} dx} \qquad (4.109)$$

and

$$-\Delta v_* = \lambda \left(\frac{e^{v_*}}{\int_\Omega e^{v_*} \, dx} - \frac{1}{|\Omega|} \right), \quad \int_\Omega v_* \, dx = 0. \qquad (4.110)$$

Hence there arises that (4.106).

If $u_* = u_*(x) > 0$ solves (4.106), conversely, then (4.110) arises for

$$v_* = w_* - \frac{1}{|\Omega|} \int_\Omega w_* \, dx, \quad w_* = \log u_*, \qquad (4.111)$$

and hence (4.105). Thus (4.106) is equivalent to (4.105). □

Remark 4.2. The property

$$F_\lambda = \{\lambda/|\Omega|\} \qquad (4.112)$$

is equivalent to

$$E_\lambda \equiv \{v_* \mid \text{solution to (4.110)}\} = \{0\}. \qquad (4.113)$$

Equality (4.113) holds if $0 < \lambda \le 8\pi$ for $\Omega = S^2$ and $\Omega = \mathbf{R}^2/(a\mathbf{Z} \times b\mathbf{Z})$ with $\frac{\pi}{4} < \frac{a}{b} \le 1$ (Remark 4.1). In these cases we obtain

$$\lim_{t\uparrow+\infty} \|u(\cdot,t) - u_*\|_{C^2} = 0 \qquad (4.114)$$

for $u_* = \lambda/|\Omega|$.

The convergence (4.114) with $u_* \in F_\lambda$ for F_λ defined by (4.104), however, is valid even if E_λ forms a continuum. We can, furthermore, determine the rate of convergence by using a gradient inequality of Lojasiewicz type as in [Haraux–Jendoubi (2001)]. Theorem 4.2 in §4.8 is thus refined as follows.

Theorem 4.3. *If (4.84) holds to (4.40), there is a solution $u_* = u_*(x) > 0$ to (4.106) satisfying (4.114). The rate of this convergence is at least of polynomial order.*

Theorem 4.4. *If the above u_* is non-degenerate, the rate of convergence in (4.114) is exponential.*

To define the non-degeneracy of the steady state u_*, we use the fact that v_* defined by (4.111) is a solution to (4.110), which is the Euler–Lagrange equation of the energy functional

$$J_\lambda(v) = \frac{1}{2}\|\nabla v\|_2^2 - \lambda \log \int_\Omega e^v \, dx, \qquad (4.115)$$

defined for $v \in V_0$, where

$$V_0 = \{v \in H^1(\Omega) \mid \int_\Omega v \, dx = 0\}. \qquad (4.116)$$

Thus we say that $u^* = u^*(x) > 0$ is non-degenerate in Theorem 10.13, if $v^* \in V_0$ defined by (4.111) is non-degenerate as a critical point of J_λ on V_0. Later we show that this non-degeneracy of $u^* = u^*(x) > 0$ means that

$$\psi \in H^2(\Omega), \ -\Delta\psi = u_*\psi \text{ in } \Omega \quad \Rightarrow \quad \psi = 0 \qquad (4.117)$$

(Lemma 4.16 in §4.12).

Remark 4.3. We can regard (4.110) as a nonlinear eigenvalue problem of finding (λ, v_*) simultaneously. If $\Omega = S^2$ then non-trivial solutions bifurcate at $\lambda = 8\pi$ from the branch of trivial solutions in λv_* space,

$$\{(\lambda, v_*) \mid \lambda \in \mathbf{R}, \ v* \in H^1(\Omega), \ \int_\Omega v_* \, dx = 0\}.$$

Hence we cannot apply Theorem 4.4 to (4.19) in §4.2, because the stationary state to (4.105),

$$u_* = \frac{\lambda}{|\Omega|}, \ \lambda = 8\pi,$$

for $\Omega = S^2$ is degenerate by this fact. There is, however, exponential rate of convergence by the result of [Hamilton (1988)] concerning (4.19), indicated by (4.12) in §4.1. We examine this rate in connection with the theory of center manifold in §4.14.

The main tool for the proof of Theorems 4.3 and 4.4 is a gradient inequality which takes the following classical form in the finite dimensional case (§1.11 of [Suzuki (2018)]).

Lemma 4.8 ([Łojasiewicz (1963)]). *Let $E = E(x) : \mathbf{R}^n \to \mathbf{R}$ be real-analytic at $x = 0$, satisfying*

$$E(0) = 0, \quad \delta E(0) = 0.$$

Then there is $0 < \theta \leq \frac{1}{2}$ such that

$$|E(x)|^{1-\theta} \leq C|\delta E(x)|, \quad |x| \ll 1.$$

4.9 Critical Manifolds

This section develops an abstract theory to extend Lemma 4.8 to infinite dimensional spaces.

Let $E : V \to \mathbf{R}$ be a C^2 mapping on the real Hilbert space V, and $\phi \in V$ be its critical point:

$$\delta E(\phi) = 0. \tag{4.118}$$

The linearized operator is defined by

$$L = \delta^2 E(\phi) : V \to V'.$$

It is supposed to be a *Fredholm operator*, where V' denotes the dual space of V. Its kernel

$$\text{Ker } L = \{u \in V \mid Lu = 0\}$$

is thus of finite dimension, and its range

$$\text{Ran } L = \{Lu \mid u \in V\} \subset V'$$

is closed and takes a finite co-dimension. We assume also the Gel'fand triple

$$V \hookrightarrow H \hookrightarrow V'$$

(Chapter 7 of [Suzuki (2022a)]). Hence V is a linear subspace of H with continuous and dense inclusion, and the identification $H \cong H'$ is taken by the representation formula of Riesz. Then the orthogonal projection

$$P : H \to \text{Ker } L \subset H$$

is of finite rank and has an extension

$$P : V' \to \mathrm{Ker}\ L$$

as a bounded linear operator. We call

$$S = \{u \in V \mid (I - P)\delta E(u) = 0\} \tag{4.119}$$

the *critical manifold*, where I denotes the identity operator. It holds that $\phi \in S$ by (4.118).

Theorem 4.5 ([Chill (2006)]). *The set S in (4.119) is a local differentiable manifold near ϕ. If $E : V \to \mathbf{R}$ is C^k, $k \geq 2$, this S is C^{k-1}. If E is analytic, S is analytic.*

For the proof, we regard P as a bounded linear operator $P : V \to V$ and induce the decomposition

$$V = V_0 \oplus V_1, \quad V_0 = \mathrm{Ker}\ L = PV, \quad V_1 = \mathrm{Ker}\ P = \{u \in V \mid Pu = 0\}.$$

Then it holds that

$$V_1 \hookrightarrow H_1 = \{u \in H \mid Pu = 0\} \hookrightarrow V_1' = \{u \in V' \mid Pu = 0\}.$$

We define the C^1 mapping

$$G : V = V_0 \oplus V_1 \to V_1'$$

by

$$G : u = u_0 + u_1 \in V_0 \oplus V_1 \mapsto (I - P)\delta E(u),$$

to obtain

$$\delta G(\phi) = (1 - P)L.$$

Lemma 4.9. *The operator*

$$M = \frac{\delta G}{\delta u_1}(\phi) = (I - P)L|_{V_1} : V_1 \to V_1'$$

is an isomorphism.

Proof. To show the injectivity, let $u \in V_1$ satisfy $(I - P)Lu = 0$. We take $v \in V$ arbitrary, to get

$$(I - P)Lu = 0. \tag{4.120}$$

Hence any $v \in V$ admits the equality

$$\langle PLu, v \rangle_{V',V} = \langle Lu, Pv \rangle_{V',V} = \langle u, LPv \rangle_{V,V'} = 0$$

because $Pv \in \operatorname{Ker} L$. It thus follows that $PLu = 0$, and hence $Lu = 0$ by (4.120). This property impleis $u \in \operatorname{Ker} L \cap V_1$, and hence $u = 0$.

For the surjectivity of $M : V_1 \to V_1'$, we note its symmetry,

$$\langle Mu, v \rangle_{V_1', V_1} = \langle u, Mv \rangle_{V_1, V_1'},$$

which implies that $M(V_1)$ is dense in V_1' becaue of the injectivity of M. We have, on the other hand, the closedness of Ran $L \subset V'$, and hence that of $M(V_1)$ in V_1'. Hence $M(V_1) = V_1'$ follows. $\qquad\square$

Proof of Theorem 4.5. Put

$$\phi = \phi_0 + \phi_1 \in V_0 \oplus V_1.$$

From the implicit function theorem, we have open sets $U_0 \subset V_0$ and $U_1 \subset V_1$ such that $\phi_0 \in U_0$ and $\phi_1 \in U_0$, and the mapping $g \in C^1(U_0, U_1)$ such that

$$g(\phi_0) = \phi_1.$$

Then we obtain

$$\mathcal{S} \cap U = \{u \in U \mid G(u) = 0\} = \{(u_0, g(u_0)) \mid u_0 \in U_0\}$$

for $U = U_0 \otimes U_1$, and therefore, $\mathcal{S}$ is a locally C^1 manifold near ϕ. This g is C^{k-1}, $k \geq 2$, if E is C^k, and is analytic if E is so. We thus obtain the result. $\qquad\square$

4.10 Łojasiewicz Simon Inequality

Under the change of variables (4.51), problem (4.46) is reduced to (4.52). The related variational functional is

$$\mathcal{E}(w) = \int_\Omega \frac{1}{2}|\nabla w|^2 - e^w + \frac{\lambda}{|\Omega|}w \, dx, \quad w \in H^1(\Omega) = V. \tag{4.121}$$

In relation to the Gel'fand triple

$$V = H^1(\Omega) \hookrightarrow X = L^2(\Omega) \cong X' \hookrightarrow V',$$

the first variation of $\mathcal{E}(w)$ is given by

$$\delta\mathcal{E}(w) = -\Delta w - e^w + \frac{\lambda}{|\Omega|} \tag{4.122}$$

and thus (4.52) is reduced to

$$\frac{\partial e^w}{\partial t} = -\delta\mathcal{E}(w) \quad \text{in } \Omega \times (0, +\infty). \tag{4.123}$$

Hence it holds that

$$\frac{d}{dt}\mathcal{E}(w) = -\int_\Omega e^w w_t^2 \, dx \leq 0. \tag{4.124}$$

Each steady state $u_* \in F_\lambda$ to (4.46) is a solution to (4.106), and hence $w_* = \log u_*$ satisfies

$$-\Delta w_* = e^{w_*} - \frac{\lambda}{|\Omega|} \quad \text{in } \Omega,$$

or equivalently,

$$\delta\mathcal{E}(w_*) = 0.$$

Here we show an inequality which takes a fundamental role in the proof of Theorem 10.7, following the theory of gradient inequality [Simon (1983)].

Theorem 4.6. *Given $w_* \in V$ satisfying*

$$\delta\mathcal{E}(w_*) = 0,$$

there exist

$$0 < \theta \leq \frac{1}{2}, \quad \varepsilon_0 > 0$$

such that

$$w \in V, \ \|w - w_*\|_V < \varepsilon_0$$
$$\Rightarrow |\mathcal{E}(w) - \mathcal{E}(w_*)|^{1-\theta} \leq C\|\delta\mathcal{E}(w)\|_{V'}. \tag{4.125}$$

To prove this result we decompose $\mathcal{E}(w)$ as in

$$\mathcal{E}(w) = \mathcal{E}_1(w) - \mathcal{E}_2(w),$$

for

$$\mathcal{E}_1(w) = \int_\Omega \frac{1}{2}|\nabla w|^2 + \frac{\lambda}{|\Omega|}w \, dx$$

and

$$\mathcal{E}_2(w) = \int_\Omega e^w \, dx.$$

The first functional $\mathcal{E}_1 : V \to \mathbf{R}$ is analytic, and it holds that

$$\delta\mathcal{E}_1(w_*)[w] = \int_\Omega \nabla w \cdot \nabla w_* + \frac{\lambda}{|\Omega|} w \, dx$$

$$\delta^2\mathcal{E}_1(w_*)[w, w] = \int_\Omega |\nabla w|^2 \, dx = (\nabla w, \nabla w)$$

$$\delta^k\mathcal{E}_1(w_*)[\overbrace{w, w, \cdots, w}^{k}] = 0, \quad k \geq 3, \quad w, w_* \in V,$$

where $(\ ,\ )$ denotes the L^2-inner product. The second functional $\mathcal{E}_2 : V \to \mathbf{R}$ is also analytic by Fontana's inequality (3.25), which assures the convergence

$$\sum_{k=0}^{\infty} \frac{1}{k!} \int_\Omega e^{w_*} |w|^k \, dx = \int_\Omega e^{w_* + |w|} \, dx < +\infty, \quad w, w_* \in V.$$

Then it holds that

$$\mathcal{E}_2(w + w_*) - \mathcal{E}_2(w_*) = \sum_{k=1}^{\infty} \frac{1}{k!} \int_\Omega e^{w_*} w^k \, dx$$

and hence

$$\delta^k\mathcal{E}_2(w_*)[\overbrace{w, w, \cdots, w}^{k}] = \int_\Omega e^{w_*} w^k \, dx, \quad k \geq 1.$$

Given

$$w_* \in V, \ \delta\mathcal{E}(w_*) = 0,$$

the linearized operator

$$\mathcal{L} \equiv \delta^2\mathcal{E}(w_*) = -\Delta - e^{w_*} : V \to V'$$

is realized as a self-adjoint operator in $X = L^2(\Omega)$ with domain

$$D(\mathcal{L}) = H^2(\Omega).$$

To develop the theory of critical manifold [Chill (2006)], we introduce

$$X_1 \equiv \operatorname{Ker} \mathcal{L} = \{v \in D(\mathcal{L}) \mid \mathcal{L}v = 0\} \subset V = H^1(\Omega).$$

Put $n = \dim X_1 < +\infty$ and let $\langle \phi_1, \ldots, \phi_n \rangle$ be an orthonormal basis of X_1. We define the orthogonal projection $\mathcal{P} : X \to X_1$, which can be extended to $\mathcal{P} : V' \to X_1$, by

$$\mathcal{P}v = \sum_{i=1}^{n} (v, \phi_i)\phi_i = \sum_{i=1}^{n} \langle \phi_i, v \rangle_{V, V^*} \phi_i.$$

Here we recall Theorem 4.5 in the previous section, derived from the implicit function theorem applied to

$$(I - \mathcal{P})\delta\mathcal{E}(v) = 0.$$

The local manifold S defined by (4.126) below is analytic because $\mathcal{E} : V \to \mathbf{R}$ is so.

Theorem 4.7. *Each $w_* \in V$ with $\delta\mathcal{E}(w_*) = 0$ admits a neighbourhood $U \subset V$ such that*

$$S = \{w \in U \mid (I - \mathcal{P})\delta\mathcal{E}(w) = 0\}, \tag{4.126}$$

is a local analytic manifold around w_* with dimension equal to n.

More precisely, we have the analytic mapping

$$g : U_1 = U \cap X_1 \to U_2 = (I - \mathcal{P})U$$

such that $g(w_1^*) = w_2^*$ for

$$w_* = w_1^* + w_2^* \in U_1 \oplus U_2.$$

The critical manifold S takes the form

$$S = \{w_1 + g(w_1) \mid w_1 \in U_1\},$$

and therefore, the following decomposition is valid:

$$w = w_1 + w_2 \in S = U_1 \oplus U_2, \quad w_1 = \mathcal{P}w, \quad w_2 = g(w_1). \tag{4.127}$$

The analytic mapping

$$Q : U \to S$$

is now defined by

$$Qw = w_1 + g(w_1) \in S, \quad w = w_1 + w_2 \in U_1 \oplus U_2, \tag{4.128}$$

and it holds that

$$w - Qw = w_2 - g(w_1) \in U_2 \tag{4.129}$$

and

$$Qw = w, \ w \in S, \tag{4.130}$$

by (4.127)–(4.128).

In the sequel we confirm several lemmas concerning the mapping Q.

Lemma 4.10. *It holds that*

$$|\mathcal{E}(w) - \mathcal{E}(Qw)| \leq C\|w - Qw\|_V^2, \ w \in U.$$

Proof. First, we have

$$\mathcal{E}(w) - \mathcal{E}(Qw) = \langle w - Qw, \delta\mathcal{E}(Qw)\rangle$$
$$+ \frac{1}{2}\delta^2\mathcal{E}(Qw)[w - Qw, w - Qw] + o\left(\|w - Qw\|_V^2\right). \qquad (4.131)$$

It also arises

$$(I - \mathcal{P})(w - Qw) = w - Qw,$$

by (4.129), and therefore,

$$\langle w - Qw, \delta\mathcal{E}(Qw)\rangle = \langle (I - \mathcal{P})(w - Qw), \delta\mathcal{E}(Qw)\rangle$$
$$= \langle w - Qw, (I - \mathcal{P})\delta\mathcal{E}(Qw)\rangle = 0$$

by $Qw \in \mathcal{S}$. Then (4.131) entails the desired estimate

$$|\mathcal{E}(w) - \mathcal{E}(Qw)| \leq C\|w - Qw\|_V^2.$$

$\square$

Lemma 4.11. *Any $\varepsilon > 0$ admits $\delta > 0$ such that*

$$w \in V, \ \|w - w_*\|_V < \delta \quad \Rightarrow \quad \|w - Qw\|_V < \varepsilon.$$

Proof. We may assume $w \in U$ by making $0 < \delta \ll 1$. Since $w_* \in \mathcal{S}$ it holds that

$$Qw_* = w_*$$

by (4.130), which implies

$$\|w - Qw\|_V \leq \|w - w_*\|_V + \|Qw_* - Qw\|_V.$$

The result is now obvious because $Q : U \to \mathcal{S}$ is continuous. $\square$

Lemma 4.12. *There is $\varepsilon_0 > 0$ such that*

$$w \in V, \ \|w - w_*\|_V < \varepsilon_0 \quad \Rightarrow \quad \|w - Qw\|_V \leq C\|\delta\mathcal{E}(w)\|_{V'}.$$

Proof. We have

$$\delta\mathcal{E}(w) - \delta\mathcal{E}(Qw) = \delta^2\mathcal{E}(Qw)(w - Qw) + o\left(\|w - Qw\|_V\right). \qquad (4.132)$$

Since $Qw \in \mathcal{S}$ implies

$$(I - \mathcal{P})\delta\mathcal{E}(Qw) = 0,$$

it holds also that

$$(I - \mathcal{P})\delta\mathcal{E}(w) = (I - \mathcal{P})\delta^2\mathcal{E}(Qw)(w - Qw) + o\left(\|w - Qw\|_V\right). \qquad (4.133)$$

Let

$$V_2 = (I - \mathcal{P})(V)$$

and recall $\mathcal{L} = \delta^2 \mathcal{E}(w_*)$. Then,

$$(I - \mathcal{P})\mathcal{L} : V_2 \to V_2'$$

is an isomorphism. By Lemma 4.11, therefore, there is $\varepsilon_1 > 0$ such that

$$(I - \mathcal{P})\delta^2 \mathcal{E}(Qw) : V_2 \to V_2',$$

is an isomorphism, provided that

$$\|w - w_*\|_V < \varepsilon_1, \quad w \in V.$$

More precisely, we have $C_1 > 0$ such that

$$\|z\|_V \le C_1 \left|\left| (I - \mathcal{P})\delta^2 \mathcal{E}(Qw)(z) \right|\right|_{V'}, \quad z \in V_2, \tag{4.134}$$

for any $w \in V$ in $\|w - w_*\|_V < \varepsilon_1$. Putting

$$z = w - Qw \in V_2 = (I - \mathcal{P})V,$$

in (4.134), we deduce

$$\begin{aligned}\|w - Qw\|_V &\le C_1 \left|\left| (I - \mathcal{P})\delta^2 \mathcal{E}(Qw)(w - Qw) \right|\right|_{V'} \\ &= C_1 \left|\left| (I - \mathcal{P})\delta \mathcal{E}(w) \right|\right|_{V'} + o\left(\|w - Qw\|_V \right)\end{aligned}$$

by (4.133). Hence there is $\varepsilon_0 > 0$ such that

$$\begin{aligned} w \in V, \ &\|w - w^*\|_V < \varepsilon_0 \\ &\Rightarrow \|w - Qw\|_V \le C_2 \left|\left| (I - \mathcal{P})\delta \mathcal{E}(w) \right|\right|_{V'} \le C_3 \left|\left| \delta \mathcal{E}(w) \right|\right|_{V'} \end{aligned}$$

by Lemma 4.11. $\qquad\square$

Lemma 4.13. *There is $\varepsilon_0 > 0$ such that*

$$\begin{aligned} w \in V, \ &\|w - w_*\|_V < \varepsilon_0 \\ &\Rightarrow \quad \|\delta \mathcal{E}(Qw)\|_{V'} \le C\|\delta \mathcal{E}(w)\|_{V'}. \end{aligned}$$

Proof. Equality (4.132), combined with $Qw \in \mathcal{S}$, implies

$$\begin{aligned} \delta \mathcal{E}(Qw) &= \mathcal{P}\delta \mathcal{E}(Qw) \\ &= \mathcal{P}\delta \mathcal{E}(w) - \mathcal{P}\delta^2 \mathcal{E}(Qw)(w - Qw) + o\left(\|w - Qw\|_V \right). \end{aligned}$$

Hence it follows that

$$\|\delta \mathcal{E}(Qw)\|_{V'} \le \|\delta \mathcal{E}(w)\|_{V'} + C\|w - Qw\|_V + o\left(\|w - Qw\|_V \right). \tag{4.135}$$

Then Lemma 4.11 assures the existence of $\varepsilon_1 > 0$ such that

$$w \in V, \ \|w - w_*\|_V < \varepsilon_1$$
$$\Rightarrow \quad \|\delta\mathcal{E}(Qw)\|_{V'} \le \|\delta\mathcal{E}(w)\|_{V'} + C\|w - Qw\|_V.$$

We finally obtain $\varepsilon_0 > 0$ such that

$$w \in V, \ \|w - w_*\|_V < \varepsilon_0 \quad \Rightarrow \quad \|\delta\mathcal{E}(Qw)\|_{V'} \le C\|\delta\mathcal{E}(w)\|_{V'}$$

by Lemma 4.12. $\qquad\qquad\square$

Lemma 4.14. *There exist $0 < \theta \le \frac{1}{2}$ and $\varepsilon_0 > 0$ such that*

$$w \in \mathcal{S}, \ \|w - w^*\|_V < \varepsilon_0$$
$$\Rightarrow |\mathcal{E}(w) - \mathcal{E}(w_*)|^{1-\theta} \le C\|\delta\mathcal{E}(w)\|_{V'}. \tag{4.136}$$

Proof. Since $\mathcal{S}$ is a finite dimensional analytic manifold and $\mathcal{E} : S \to \mathbf{R}$ is analytic, the result follows from Lemma 4.8. $\qquad\qquad\square$

We are ready to complete the following proof.

Proof of Theorem 4.6. Given $w \in V$ in $\|w - w_*\|_V \ll 1$, we have

$$|\mathcal{E}(w) - \mathcal{E}(w_*)| \le |\mathcal{E}(w) - \mathcal{E}(Qw)| + |\mathcal{E}(Qw) - \mathcal{E}(w_*)|$$
$$\le C(\|w - Qw\|_V^2 + \|\delta\mathcal{E}(Qw)\|_{V'}^{\frac{1}{1-\theta}})$$

by Lemma 4.10, Lemma 4.14, and $Qw \in \mathcal{S}$, where $0 < \theta \le \frac{1}{2}$. Then Lemmas 4.12 and 4.13 imply

$$\mathcal{E}(w) - \mathcal{E}(w_*)| \le C(\|\delta\mathcal{E}(w)\|_{V'}^2 + \|\delta\mathcal{E}(w)\|_{V'}^{\frac{1}{1-\theta}})$$

and hence the desired property (4.136). $\qquad\qquad\square$

4.11   Convergence to the Steady State

If $0 < \lambda \le 8\pi$ holds in (4.52), there arises that $T = +\infty$ and the orbit $\mathcal{O} = \{w(\cdot, t)\}$ is pre-compact in C^∞ topology by Theorem 4.1 in §4.7. The ω-limit set $\omega(w_0)$ defined by (4.82) is compact, connected, non-empty and contained in the set of steady states (Theorem 4.2 in §4.8). Since the steady state is defined by (4.83), it holds that

$$\omega(w_0) \subset D \equiv \{w_* \in V \mid \delta\mathcal{E}(w_*) = 0\}. \tag{4.137}$$

Here we show

$$\sharp \, \omega(w_0) = 1, \tag{4.138}$$

regardless of the structure of the set of D, and hence there is $w_* \in D$ such that

$$\lim_{t\uparrow+\infty} \|w(\cdot,t) - w_*\|_\infty = 0.$$

This rate is, furthermore, at least of polynomial order, provided that $\mathcal{O}$ is pre-compact in C^∞ topology:

$$T = +\infty, \ \|w(\cdot,t)\|_\infty \le C.$$

Thus we prove an equivalent form of Theorem 4.3.

To apply Theorem 4.6 for the proof, we use the parabolic regularity in the following form.

Lemma 4.15. *Each $w_* \in V$ in $\delta\mathcal{E}(w_*) = 0$ admits $C > 0$ such that*

$$\sup_{t_0 \le t < t_0 + T} \|w(\cdot,t) - w_*\|_V \le C(\|w(\cdot,t_0) - w_*\|_V$$
$$+ \sup_{t_0 \le t < t_0 + T} \|w(\cdot,t) - w_*\|_2), \quad t_0 \ge 0, \ T > 0. \tag{4.139}$$

Proof. By

$$w_t = e^{-w}\Delta w + 1 - \frac{\lambda}{|\Omega|}e^{-w}, \quad 0 = e^{-w_*}\Delta w_* + 1 - \frac{\lambda}{|\Omega|}e^{-w_*},$$

the function

$$z = w - w_*$$

solves

$$z_t = e^{-w}\Delta z + \left(e^{-w} - e^{-w_*}\right)\Delta w_* - \frac{\lambda}{|\Omega|}\left(e^{-w} - e^{-w_*}\right)$$
$$= \left(e^{-w}\Delta - 1\right)z + bz$$

with $b = b(x,t)$ uniformly bounded.

Since w is uniformly bounded including its derivatives of any order, $e^{-w}\Delta$ generates an evolution operator, denoted by $\{U(t,s)\}$, satisfying

$$\|U(t,s)z_0\|_V \le C\|z_0\|_V$$
$$\|U(t,s)z_0\|_V \le C(t-s)^{-1/2}\|z_0\|_2, \quad 0 \le s < t < \infty \tag{4.140}$$

([Yagi (2010)]). Hence

$$z(t) = U(t,s)z_0$$

is the solution to

$$z_t = e^{-w}\Delta z \quad \text{in } \Omega \times (s, +\infty), \quad z|_{t=s} = z_0.$$

If $\{\widetilde{U}(t,s)\}$ denotes the evolution operator associated with

$$e^{-w}\Delta - 1,$$

therefore, it holds that

$$\||\widetilde{U}(t,s)\|_{V \to V} \leq Ce^{-(t-s)}$$
$$\|\widetilde{U}(t,s)\|_{X \to V} \leq C(t-s)^{-1/2}e^{-(t-s)}, \ 0 \leq s < t < \infty,$$

and furthermore,

$$z(t) = \widetilde{U}(t,t_0)z(t_0) + \int_{t_0}^{t} \widetilde{U}(t,r)(bz)(r) \, dr, \ t \geq t_0.$$

Thus we obtain

$$\|z(t)\|_V \leq C\left(\|z(t_0)\|_V + \int_{t_0}^{t} (t-r)^{-1/2}e^{-(t-r)} \, dr \sup_{t_0 \leq r < t} \|z(r)\|_2 \right)$$

which finally entails

$$\sup_{t_0 \leq t < t_0 + T} \|z(t)\|_V \leq C(\|z(t_0)\|_V + \int_{0}^{\infty} s^{-1/2}e^{-s} \, ds \sup_{t_0 \leq t < t_0 + T} \|z(t)\|_2)$$
$$\leq C(\|z(t_0)\|_V + \sup_{t_0 \leq t < t_0 + T} \|z(t)\|_2).$$

$\square$

Remark 4.4. The second inequality of (4.140) implies

$$\|w(\cdot, t+1) - w_*\|_V \leq C \sup_{t \leq s < t+1} \|w(\cdot, s) - w_*\|_2$$

for any $t \geq 0$ by

$$z(t+1) = \widetilde{U}(t+1,t)z(t) + \int_{t}^{t+1} \widetilde{U}(t+1,r)(bz)(r) \, dr.$$

We are ready to complete the following proof.

Proof of Theorem 4.3. We prescribe the constant C in (4.139) as $C = C_1 \geq 1$ and thus:

$$\sup_{t_0 \leq t < t_0 + T} \|w(\cdot, t) - w_*\|_V \leq C_1(\|w(\cdot, t_0) - w_*\|_V$$
$$+ \sup_{t_0 \leq t < t_0 + T} \|w(\cdot, t) - w_*\|_2). \tag{4.141}$$

Recall the ω-limit set (4.82) in §4.7 of (4.52) in §4.5. By (4.137), we have $w_* \in V$ with $\delta\mathcal{E}(w_*) = 0$ and $t_k \to +\infty$ such that

$$w(\cdot, t_k) \to w_* \quad \text{in } C^\infty \text{ topology} \tag{4.142}$$

and in particular,

$$\|w(\cdot, t_k) - w_*\|_V \leq \frac{\varepsilon_0}{4C_1}, \quad k \gg 1, \tag{4.143}$$

where $\varepsilon_0 > 0$ and $C_1 \geq 1$ are constants prescribed in Theorem 4.6 and (4.141), respectively.

We have

$$\frac{d}{dt}\mathcal{E}(w) = -\langle w_t, \delta\mathcal{E}(w)\rangle_{V,V'} = -(w_t, e^w w_t) \leq 0,$$

by (4.52), and hence the existence of

$$\lim_{t \to +\infty} \mathcal{E}(w(\cdot, t)) = \mathcal{E}_\infty = \mathcal{E}(w_*), \tag{4.144}$$

where the second equality follows from $w_* \in \omega(w_0)$. In particular,

$$\mathcal{H}(t) = (\mathcal{E}(w(\cdot, t)) - \mathcal{E}(w_*))^\theta \geq 0$$

is well-defined, and it holds that

$$\lim_{t \to +\infty} \mathcal{H}(t) = 0. \tag{4.145}$$

Since

$$C_2^{-1} \leq e^w \leq C_2 \quad \text{in } \Omega \times (0, \infty)$$

is valid with $C_2 \geq 1$, we obtain

$$\begin{aligned}
-\frac{d\mathcal{H}}{dt} &= -\theta\left(\mathcal{E}(w) - \mathcal{E}(w_*)\right)^{\theta-1}\langle w_t, \delta\mathcal{E}(w)\rangle_{V,V'} \\
&= \theta\left(\mathcal{E}(w) - \mathcal{E}(w_*)\right)^{\theta-1}(e^w, w_t^2) \\
&\geq \theta C_2^{-1}\left(\mathcal{E}(w) - \mathcal{E}(w_*)\right)^{\theta-1} V\|w_t\|_2^2 \\
&\geq \theta C_2^{-3/2}\left(\mathcal{E}(w) - \mathcal{E}(w_*)\right)^{\theta-1}\|w_t\|_2\left(\int_\Omega e^w w_t^2\, dx\right)^{1/2} \\
&= \theta C_2^{-3/2}\left(\mathcal{E}(w) - \mathcal{E}(w_*)\right)^{\theta-1}\|w_t\|_2\|\delta\mathcal{E}(w)\|_2
\end{aligned}$$

by (4.124) in §4.10. Therefore, there is $C_3 > 0$ such that

$$-\frac{d\mathcal{H}}{dt} \geq \frac{1}{C_3}\left(\mathcal{E}(w) - \mathcal{E}(w_*)\right)^{\theta-1}\|w_t\|_2\|\delta\mathcal{E}(w)\|_{V'}. \tag{4.146}$$

Given k, we assume the existence of $t_0 > t_k$ such that

$$\|w(\cdot, t) - w_*\|_V < \varepsilon_0, \quad t_k \leq t \leq t_0. \tag{4.147}$$

Then inequality (4.146) implies

$$\|w_t\|_2 \le -C_4 \frac{d\mathcal{H}}{dt}, \quad t_k \le t \le t_0, \tag{4.148}$$

where $C_4 > 0$ is a constant. It follows that

$$\|w(\cdot,t) - w(\cdot,t_k)\|_2 \le C_4 \mathcal{H}(t_k), \quad t_k \le t \le t_0,$$

and thus we obtain

$$\|w(\cdot,t) - w(\cdot,t_k)\|_V \le \frac{\varepsilon_0}{4} + C_1 C_4 \mathcal{H}(t_k) \quad t_k \le t \le t_0 \tag{4.149}$$

by (4.141) and (4.143) with $C_1 \ge 1$.

Equality (4.145) assures $k \gg 1$ satisfying

$$\mathcal{H}(t_k) < \frac{\varepsilon_0}{4C_1 C_4}. \tag{4.150}$$

Fix such k. By the above argument, if there is $t_0 > t_k$ provided with (4.147), it holds that (4.149) and hence

$$\|w(\cdot,t) - w(\cdot,t_k)\|_V < \varepsilon_0/2, \quad t_k \le t \le t_0 \tag{4.151}$$

by (4.150). Since we have (4.143) with $C_1 \ge 1$, inequality (4.151) implies

$$\|w(\cdot,t) - w_*\|_V < 3\varepsilon_0/4, \quad t_k \le t \le t_0. \tag{4.152}$$

We have thus observed that (4.147) implies (4.152). Regarding (4.143) with $C_1 \ge 1$ again, we conclude

$$\|w(\cdot,t) - w_*\|_V < \varepsilon_0, \quad t \ge t_k. \tag{4.153}$$

Consequently, by (4.153) inequality (4.148) is improved as

$$\|w_t\|_2 \le -C_4 \frac{d\mathcal{H}}{dt}, \quad t \ge t_k, \tag{4.154}$$

which implies

$$\int_0^\infty \|w_t\|_2 dt < \infty.$$

Then we obtain

$$\lim_{t \to \infty} \|w(\cdot,t) - w_*\|_2 = 0$$

by (4.142), and hence $\omega(w_0) = \{w_*\}$ from the uniqueness of the limit. It thus follows that

$$w(\cdot,t) \to w_* \quad \text{in } C^\infty \text{ topology} \tag{4.155}$$

as $t \to +\infty$.

Turning to the rate of convergence, we use

$$|\mathcal{E}(w(\cdot,t)) - \mathcal{E}(w_*)|^{1-\theta} \le C_5 \|\delta\mathcal{E}(w(\cdot,t))\|_{V^*}, \quad t \ge t_k$$

derived from Theorem 4.6 and (4.153). Using (4.123) in §4.10, we derive

$$
\begin{aligned}
-\frac{d\mathcal{H}}{dt} &= \theta(\mathcal{E}(w) - \mathcal{E}(w_*))^{\theta-1}\langle w_t, -\delta\mathcal{E}(w)\rangle \\
&= \theta(\mathcal{E}(w) - \mathcal{E}(w_*))^{\theta-1}\langle e^{-w}\delta\mathcal{E}(w), \delta\mathcal{E}(w)\rangle \\
&\ge \theta(\mathcal{E}(w) - \mathcal{E}(w_*))^{\theta-1}\|\delta\mathcal{E}(w)\|_2^2 \\
&\ge \frac{1}{C_6}(\mathcal{E}(w) - \mathcal{E}(w_*))^{\theta-1}\|\delta\mathcal{E}(w)\|_{V'}^2 \\
&\ge \frac{1}{C_6 C_5^2}(\mathcal{E}(w) - \mathcal{E}(w_*))^{1-\theta} = \gamma\mathcal{H}^{\frac{1}{\theta}-1}, \quad t \ge t_k,
\end{aligned}
$$

where $\gamma = \frac{1}{C_6 C_5^2}$. We thus obtain

$$\mathcal{H}(t) \le C\Phi(t), \quad t \ge t_k$$

for

$$\Phi(t) = \begin{cases} t^{-\frac{\theta}{1-2\theta}}, & 0 < \theta < \frac{1}{2} \\ e^{-\gamma t}, & \theta = \frac{1}{2}. \end{cases} \tag{4.156}$$

Inequality (4.154) now implies

$$\|w(\cdot,t) - w(\cdot,s)\|_2 \le C\Phi(s), \quad t \ge s \ge t_k,$$

and sending $t \to \infty$, we get

$$\|w_* - w(\cdot,s)\|_2 \le C\Phi(s), \quad s \ge t_k,$$

or

$$\|w(\cdot,t) - w_*\|_2 \le C\Phi(t), \quad t \ge t_k, \tag{4.157}$$

Then, Remark 4.4 entails

$$\|w(\cdot,t) - w_*\|_V \le C\Phi(t), \quad t \to +\infty. \tag{4.158}$$

We have a parabolic equation with smooth, uniformly bounded coefficients for

$$z_i = \partial_i z = \frac{\partial}{\partial x_i}(w - w_*), \quad i = 1, 2$$

by (4.52) in §4.5. Then we obtain

$$\|z_i(\cdot,t)\|_V \le C\Phi(t), \quad t \to +\infty.$$

similarly to (4.158), by the second estimate of (4.140). Given multi-index α, we can derive analogous inequaliy for $z_\alpha = D^\alpha(w - w_*)$ by an induction. Then we obtain (4.114) in §4.8. $\qquad\square$

4.12 Non-Degeneracy

In §4.12–§4.13 we show Theorem 4.4, recalling that $u_* = u_*(x) > 0$ is called a steady-state to (4.46) in §4.4 when it solves (4.106) in §4.8. Then $v_* = v_*(x)$ defined by (4.111) satisfies (4.110), which is the Euler–Lagrange equation for the functional $J_\lambda = J_\lambda(v)$, $v \in V_0$, defined by (4.115)–(4.116):

$$J_\lambda(v) = \frac{1}{2}\|\nabla v\|_2^2 - \lambda \log \int_\Omega e^v dx, \quad V_0 = \{v \in H^1(\Omega) \mid \int_\Omega v\, dx = 0\}.$$

We say that u_* is non-degenerate if this $v_* \in V_0$ is a non-degenerate critical point of J_λ on V_0. Here, u_* is reproduced by v_* through (4.109).

More precisely, first, we notice

$$\delta J_\lambda(v)[\phi] = \frac{d}{ds} J_\lambda(v + s\phi)\Big|_{s=0}$$
$$= (\nabla v, \nabla \phi) - \frac{\lambda \int_\Omega e^v \phi\, dx}{\int_\Omega e^v\, dx}, \quad \phi \in V_0,$$

to identify

$$\delta J_\lambda(v) = -\Delta v - \lambda \left(\frac{e^v}{\int_\Omega e^v\, dx} - \frac{1}{|\Omega|} \right) \in V_0', \quad v \in V_0.$$

Hence the above v_*, realized as a solution to (4.110), belongs to V_0 and is a critical point of J_λ on V_0.

Second, the quadratic form

$$\mathcal{Q} : V_0 \times V_0 \to \mathbf{R}$$

defined by

$$\mathcal{Q}(\phi, \phi) = \frac{d^2}{ds^2} J_\lambda(v_* + s\phi)\Big|_{s=0}$$
$$= (\nabla \phi, \nabla \phi) - \lambda \frac{\int_\Omega e^{v_*} \phi^2\, dx}{\int_\Omega e^{v_*} dx} + \lambda \left(\frac{\int_\Omega e^{v_*} \phi\, dx}{\int_\Omega e^{v_*}\, dx} \right)^2$$

is associated with the linearized operator

$$\delta^2 J_\lambda(v_*) : V_0 \to V_0'$$

through

$$\mathcal{Q}(\phi, \phi) = \langle \phi, \delta^2 J_\lambda(v_*)\phi \rangle_{V,V'}.$$

This $\delta^2 J_\lambda(v_*)$ is realized as a self-adjoint operator in

$$X_0 = L^2(\Omega) \cap V_0,$$

denoted by $\mathcal{B}$, with the domain

$$D(\mathcal{B}) = H^2(\Omega) \cap V_0,$$

satisfying

$$(\mathcal{B}\phi, \psi) = \mathcal{Q}(\phi, \psi), \quad \phi \in D(\mathcal{B}) \subset V_0, \ \psi \in V_0.$$

Hence it holds that

$$\mathcal{B}\phi = -\Delta\phi - \frac{\lambda e^{v_*}}{\int_\Omega e^{v_*}\, dx}\phi + \frac{\lambda \int_\Omega e^{v_*}\phi\, dx}{\left(\int_\Omega e^{v_*}\, dx\right)^2} e^{v_*}$$

$$= -\Delta\phi - u_*\phi + \frac{1}{\lambda}(\phi, u_*)u_* \tag{4.159}$$

for

$$\phi \in D(\mathcal{B}) = H^2(\Omega) \cap V_0$$

by (4.109) in §4.8.

Now we show the following lemma.

Lemma 4.16. *The stationary solution $u_* = u_*(x) > 0$ to (4.46) is non-degenerate if and only if the property (4.117) holds.*

Proof. By the definition, the non-degeneracy of u_* means that of $\mathcal{B}$ in $X_0 = L^2(\Omega) \cap V_0$, which is equivalent to

$$\phi \in D(\mathcal{B}), \ \mathcal{B}\phi = 0 \quad \Rightarrow \quad \phi = 0. \tag{4.160}$$

Assume, first, $\phi \in D(\mathcal{B}) \setminus \{0\}$ with $\mathcal{B}\phi = 0$, and let

$$\psi = \phi - \frac{1}{\lambda}\int_\Omega u_*\phi\, dx \in H^2(\Omega).$$

Then we have

$$-\Delta\psi = u_*\psi.$$

It also holds that $\psi \neq 0$ by $\phi \in V_0 \setminus \{0\}$. Hence if u_* is degenerate there is $\psi \in V \setminus \{0\}$ satisfying (4.117).

If problem (4.117) admits $\psi \in H^2(\Omega) \setminus \{0\}$, conversely, we take

$$\phi = \psi - \frac{1}{|\Omega|}\int_\Omega \psi\, dx \in H^2(\Omega) \cap V_0 = D(\mathcal{B}).$$

It holds that

$$\int_\Omega \psi u_*\, dx = 0,$$

and therefore,

$$(\phi, u_*) = -\frac{\lambda}{|\Omega|} \int_\Omega \psi \, dx.$$

Then we obtain

$$\mathcal{B}\phi = -\Delta\phi - u_*\phi + \frac{1}{\lambda}(\phi, u_*)u_*$$

$$= -\Delta\psi - u_*\psi + \frac{u_*}{|\Omega|} \int_\Omega \psi \, dx + \frac{1}{\lambda}(\phi, u_*)u_*$$

$$= -\Delta\psi - u_*\psi = 0$$

by the assumption of (4.117). If $\phi = 0$ it holds that $\psi = $ constant and by virtue of the conclusion of (4.117), there arises $\psi = 0$, a contradiction. Thus $\mathcal{B}$ has the eigenvalue 0, and therefore, this operator is degenerate. $\square$

Lemma 4.17. *Let $u_* = u_*(x) > 0$ be a steady state to (4.46) in §4.4 and define $w_* \in V = H^1(\Omega)$ by (4.111) in §4.8:*

$$w_* = \log u_*. \tag{4.161}$$

Set

$$\mathcal{M} = -\Delta - u_* : V \to V'. \tag{4.162}$$

Then the following properties on u_* and w_* are equivalent:

(1) There exists $C > 0$ such that

$$\phi \in V, \ \int_\Omega u_*\phi \, dx = 0 \quad \Rightarrow \quad \|\phi\|_V \le C\|\mathcal{M}\phi\|_{V'}. \tag{4.163}$$

(2) There exists $\varepsilon_0 > 0$ and $C > 0$ such that

$$w \in V, \ \int_\Omega e^w \, dx = \lambda, \ \|w - w_*\|_V < \varepsilon_0$$

$$\Rightarrow \quad \|w - w_*\|_V \le C\|\mathcal{M}(w - w_*)\|_{V'}. \tag{4.164}$$

Proof. $(i) \implies (ii)$: Assume (i), take

$$w \in V, \ \int_\Omega e^w \, dx = \lambda, \ \|w - w_*\|_V < \varepsilon_0, \tag{4.165}$$

and let

$$\phi_* = \frac{u_*}{\|u_*\|_2}, \quad z = w - w_*, \quad \mathcal{P}z = z - (\phi_*, z)\phi_*.$$

It holds that $(\mathcal{P}z, u_*) = 0$, and hence

$$\|\mathcal{P}z\|_V \le C \|\mathcal{M}(\mathcal{P}z)\|_{V'} \tag{4.166}$$

by (4.163). Then, we obtain

$$
\begin{aligned}
\|z\|_V &\le \|\mathcal{P}z\|_V + |(\phi_*, z)| \, \|\phi_*\|_V \\
&\le C \, \|\mathcal{M}(\mathcal{P}z)\|_{V'} + |(\phi_*, z)| \|\phi_*\|_V \\
&= C \, \big\|\mathcal{M}z - (u_*, z)\|u_*\|_2^{-1}\mathcal{M}\phi_*\big\|_{V'} + |(u_*, z)|\|\phi_*\|_V \cdot \|u_*\|_2^{-1} \\
&\le C \, \|\mathcal{M}(z)\|_{V'} + |(u_*, z)|(C\|\mathcal{M}\phi_*\|_{V'} + \|\phi_*\|_V)\|u_*\|_2^{-1}. \quad (4.167)
\end{aligned}
$$

Here we have

$$
\int_\Omega e^w \, dx = \int_\Omega e^{w_*} \, dx = \lambda
$$

by (4.161), $\|u_*\|_1 = \lambda$, and (4.165). Hence it holds that

$$
0 = \int_0^1 \int_\Omega e^{sw+(1-s)w_*}(w - w_*) \, dx \, ds \qquad (4.168)
$$

by

$$
e^w - e^{w_*} = \int_0^1 \frac{d}{ds} e^{sw+(1-s)w_*} \, ds = \int_0^1 e^{sw+(1-s)w_*}(w - w_*) \, ds.
$$

Then (4.168) implies

$$
\begin{aligned}
(u_*, z) &= \int_\Omega e^{w_*}(w - w_*) \, dx \\
&= \int_0^1 \int_\Omega (e^{w_*} - e^{sw+(1-s)w_*})(w - w_*) \, dx \, ds \\
&= \int_0^1 \int_\Omega (e^{w_*} - e^{sw+(1-s)w_*})z \, dx \, ds. \qquad (4.169)
\end{aligned}
$$

In (4.169) we have

$$
\begin{aligned}
e^{w_*} - e^{sw+(1-s)w_*} &= \int_0^1 \frac{d}{dr} e^{rw_*+(1-r)(sw+(1-s)w_*)} \, dr \\
&= \int_0^1 e^{rw_*+(1-r)(sw+(1-s)w_*)}(-s)z \, dr,
\end{aligned}
$$

and therefore,

$$
|(u_*, z)| \le \int_0^1 \int_0^1 \int_\Omega e^{rw_*+(1-r)(sw+(1-s)w_*)} sz^2 \, dx \, dr \, ds.
$$

By Fontana's inequality, any $K > 0$ admits $C_1(K)$ such that

$$
\|w\|_V \le K \quad \Rightarrow \quad \|e^{rw_*+(1-r)(sw+(1-s)w_*)}\|_2 \le C_1(K), \ 0 \le r, s \le 1,
$$

and hence we find $C_2(K) > 0$ such that

$$
|(u_*, z)| \le C_1(K)\|z\|_4^2 \le C_2(K)\|z\|_V^2. \qquad (4.170)
$$

Combining (4.167) and (4.170), we reach

$$\|z\|_V \le C_3(K)(\|\mathcal{M}z\|_V + \|z\|_V^2)$$

for $\|z\|_V \le K$. Then (4.164) folllows for $\varepsilon_0 = \frac{1}{2C_0(K)}$, because then we have

$$\|z\|_V = \|w - w_*\|_V < \varepsilon_0 \quad \Rightarrow \quad C_3(K)\|z\|_V^2 \le \frac{1}{2}\|z\|_V$$

and hence (4.164) with $C = 2C_3(K)$.

$(ii) \implies (i)$: Given

$$\phi \in V, \quad \int_\Omega u_* \phi \, dx = 0, \tag{4.171}$$

we show the conclusion of (4.163):

$$\|\phi\|_V \le C\|\mathcal{M}\phi\|_{V'}. \tag{4.172}$$

For this purpose, it suffices to assume

$$\phi \ne 0. \tag{4.173}$$

Define

$$\Phi(s,z) = \begin{cases} \frac{1}{s}\int_\Omega e^{s\phi+s^2 z + w_*} - e^{w_*} \, dx, & s \ne 0 \\ 0, & s = 0, \end{cases} \quad (s,z) \in \mathbf{R} \times V.$$

We note also that

$$e^{s\phi+s^2 z + w_*} - e^{w_*} = e^{w_*}(e^{s\phi+s^2 z} - 1)$$

$$= e^{w_*}\{(s\phi + s^2 z) + \frac{1}{2}(s\phi + s^2 z)^2 + o(s^2)\}$$

$$= \left\{ s\phi + s^2 \left(z + \frac{1}{2}\phi^2 \right) \right\} e^{w_*} + o(s^2), \quad s \to 0,$$

to deduce

$$\Phi(s,z) = \int_\Omega \{\phi + s(z + \frac{1}{2}\phi^2)\} e^{w_*} \, dx + o(s), \quad s \to 0.$$

First, this $\Phi = \Phi(s,z)$ is continuous in $(s,z) \in \mathbf{R} \times V$ because

$$\lim_{s \to 0} \Phi(s,z) = 0$$

follows from (4.161) and (4.171):

$$\int_\Omega e^{w_*} \phi \, dx = 0. \tag{4.174}$$

Second, the following limit arises

$$\lim_{s \to 0} \Phi_s(s,z) = \int_\Omega (z + \frac{1}{2}\phi^2) e^{w_*} \, dx$$

and hence Φ is C^1 in $\mathbf{R} \times V$. It holds, in particular, that

$$\Phi_s(0,0) = \frac{1}{2} \int_\Omega e^{w_*} \phi^2 \, dx \neq 0$$

by (4.173), and therefore, the implicit function theorem guarantees the existence of a C^1 function $z = z(s)$ of s such that

$$z(0) = 0, \quad \Phi(s, z(s)) = 0, \ |s| \ll 1.$$

Accordingly,

$$w(s) = s\phi + s^2 z(s) + w_*$$

satisfies

$$w(0) = w_*, \quad \dot{w}(0) = \phi, \quad \int_\Omega e^{w(s)} \, dx = \int_\Omega e^{w_*} \, dx = \lambda, \ |s| \ll 1 \quad (4.175)$$

and hence

$$\|w(s) - w_*\|_V \leq C\|\mathcal{M}(w(s) - w_*)\|_{V'}, \quad |s| \ll 1 \qquad (4.176)$$

by (4.164). Then, (4.172) follows from (4.175)–(4.176). $\qquad\qquad\square$

4.13 Exponential Rate of Convergence

Given a non-degenerate steady-state $u_* = u_*(x) > 0$ of (4.46) in §4.4, define $w_* \in V$ by (4.161) in §4.12. Then it holds that

$$\delta\mathcal{E}(w_*) = 0, \quad \int_\Omega e^{w_*} \, dx = \lambda. \qquad (4.177)$$

By Lemma 4.16, the operator $\mathcal{M} : V \to V'$ defined by (4.162) is provided with the property (4.163). Then we obtain $\varepsilon_0 > 0$ satisfying (4.164) by Lemma 4.17.

Having these properties, we see that Theorem 4.4 in §4.8 is reduced to the following lemma by the proof of Theorem 4.3 in §4.11.

Lemma 4.18. *Let $w_* \in V$ satisfy (4.177), and assume the propery (4.164). Then, there arises that $\theta = \frac{1}{2}$ in the conclusion of (4.125) in §4.10 for w satisfying*

$$w \in V, \ \|w - w_*\|_V < \varepsilon_1, \quad \int_\Omega e^w \, dx = \lambda \qquad (4.178)$$

with $\varepsilon_1 > 0$ sufficiently small.

For the proof of this lemma, we verify several facts derived from Fontana's inequality (3.25) in §3.2.

Lemma 4.19. *Any $K > 0$ admits $C(K) > 0$ such that*

$$w_1, w_2 \in V, \quad \|w_1\|_V, \; \|w_2\|_V \leq K$$

$$\Rightarrow \quad \|\delta\mathcal{E}(w_1) - \delta\mathcal{E}(w_2)\|_{V'} \leq C(K)\|w_1 - w_2\|_V. \qquad (4.179)$$

Proof. Given $w \in V = H^1(\Omega)$, let

$$\overline{w} = \frac{1}{|\Omega|} \int_\Omega w \, dx, \quad [w] = w - \overline{w} \in V_0.$$

Take $z \in V$. Then we have

$$\langle z, \delta\mathcal{E}(w_1) - \delta\mathcal{E}(w_2) \rangle_{V,V'}$$

$$= \int_\Omega \nabla z \cdot \nabla(w_1 - w_2) - z(e^{w_1} - e^{w_2}) \, dx \qquad (4.180)$$

by (4.122), where

$$e^{w_1} - e^{w_2} = \int_0^1 \frac{d}{ds} e^{sw_1 + (1-s)w_2} \, ds = \int_0^1 e^{sw_1 + (1-s)w_2} \, ds \cdot (w_1 - w_2)$$

$$= \int_0^1 e^{s\overline{w_1} + (1-s)\overline{w_2}} \cdot e^{[sw_1 + (1-s)w_2]} \, ds \cdot (w_1 - w_2).$$

Hence it follows that

$$|e^{w_1} - e^{w_2}| \leq e^{|\overline{w_1}| + |\overline{w_2}|} \cdot \int_0^t e^{[sw_1 + (1-2)w_2]} ds \cdot |w_1 - w_2|. \qquad (4.181)$$

Letting $w \in V \setminus \mathbf{R}$, on the other hand, we use

$$[w] \leq \frac{4\pi[w]^2}{\|\nabla[w]\|_2^2} + \frac{1}{\pi}\|\nabla[w]\|_2^2$$

to deduce

$$\int_\Omega e^{[w]} \, dx \leq C \cdot \exp\left(\frac{1}{\pi}\|\nabla[w]\|_2^2\right), \quad w \in V, \qquad (4.182)$$

by (3.25).

Inequalities (4.181)–(4.182) imply

$$\left| \int_\Omega z\left(e^{w_1} - e^{w_2}\right) dx \right|$$

$$\leq \|z\|_4 \exp\left(|\overline{w_1}| + |\overline{w_2}|\right) \left\| \int_0^1 e^{[sw_1 + (1-s)w_2]} \, ds \right\|_4 \|w_1 - w_2\|_2$$

$$\leq C'(K)\|z\|_4\|w_1 - w_2\|_2 \leq C(K)\|z\|_V\|w_1 - w_2\|_V,$$

provided that

$$\|w_1\|_V, \|w_2\|_V \leq K.$$

Hence (4.179) is valid by (4.180). $\qquad \square$

Lemma 4.20. *Given $w_* \in V$ with $\delta\mathcal{E}(w_*) = 0$, any $K > 0$ admits $C = C(K) > 0$ such that*

$$w \in V, \ \|w\|_V \leq K \quad \Rightarrow \quad |\mathcal{E}(w) - \mathcal{E}(w_*)| \leq C\|w - w_*\|_V^2.$$

Proof. Since

$$\mathcal{E}(w) - \mathcal{E}(w_*) = \int_0^1 \frac{d}{ds}\mathcal{E}(sw + (1-s)w_*) \, ds$$

$$= \int_0^1 \langle w - w_*, \delta\mathcal{E}(sw + (1-s)w_*)\rangle_{V,V'} \, ds$$

$$= \int_0^1 \langle w - w_*, \delta\mathcal{E}(sw + (1-s)w_*) - \delta\mathcal{E}(w_*)\rangle_{V,V'} \, ds$$

we obtain

$$|\mathcal{E}(w) - \mathcal{E}(w_*)| \leq \|w - w_*\|_V \int_0^1 \|\delta\mathcal{E}(sw + (1-s)w_*) - \delta\mathcal{E}(w_*)\|_{V'} \, ds$$

$$\leq C(K)\|w - w_*\|_V \|w - w_*\|_V \int_0^1 s \, ds = \frac{C(K)}{2}\|w - w_*\|_V^2$$

by the previous lemma. $\qquad\square$

We are ready to complete the following proof.

Proof of Theorem 4.4. It suffices to show Lemma 4.18. We take w as in (4.178). Recall $\delta\mathcal{E}(w_*) = 0$, and deduce from (4.122) that

$$-\delta\mathcal{E}(w) = -\delta\mathcal{E}(w) + \delta\mathcal{E}(w_*)$$

$$= \Delta(w - w_*) + (e^w - e^{w_*})$$

$$= \Delta(w - w_*) + \int_0^1 \frac{d}{ds}e^{sw+(1-s)w_*} \, ds$$

$$= \Delta(w - w_*) + \int_0^1 e^{sw+(1-s)w_*}(w - w_*) \, ds$$

$$= \Delta(w - w_*) + e^{w_*}(w - w_*) + \int_0^1 (e^{sw+(1-s)w_*} - e^{w_*})(w - w_*) \, d\zeta$$

$$= -\mathcal{M}(w - w_*) + z, \qquad\qquad (4.183)$$

where

$$z = \int_0^1 (e^{sw+(1-s)w_*} - e^{w_*})(w - w_*) \, ds.$$

Here we use

$$e^{sw+(1-s)w_*} - e^{w_*} = \int_0^1 \frac{d}{d\zeta} e^{\zeta(sw+(1-s)w_*)+(1-\zeta)w_*} \, d\zeta$$

$$= \int_0^1 e^{\zeta(sw+(1-s)w_*)+(1-\zeta)w_*} \, s(w - w_*) \, d\zeta,$$

to derive

$$|z| \le |w - w_*|^2 e^{|w|+|w_*|}.$$

Hence it holds that

$$\|z\|_2 \le \| \exp(|w|)\|_8 \cdot \| \exp(|w_*|)\|_8 \cdot \|w - w_*\|_8^2,$$

and therefore, the assumption (4.178) ensures

$$\|z\|_{V'} \le C'\|z\|_2 \le C''\|w - w_*\|_8^2 \le C\|w - w_*\|_V^2 \tag{4.184}$$

by Fontana's inequality.

Since $\mathcal{M} : V \to V'$ is provided with (4.164), it follows that

$$\|w - w_*\|_V \le C\|\mathcal{M}(w - w_*)\|_{V'}$$

from (4.178) if $\varepsilon_1 \le \varepsilon_0$. By (4.183)–(4.184), therefore, we obtain

$$\|w - w_*\|_V \le C(\|\delta\mathcal{E}(w)\|_{V'} + \|w - w_*\|_V^2),$$

which implies

$$\|w - w_*\|_V \le C\|\delta\mathcal{E}(w)\|_{V'} \tag{4.185}$$

for $0 < \varepsilon_1 \ll 1$ in (4.178).

Lemma 4.20 now guarantees

$$|\mathcal{E}(w) - \mathcal{E}(w_*)| \le C'\|w - w_*\|_V^2 \le C\|\delta\mathcal{E}(w)\|_{V'}^2,$$

which entails the conclusion of (4.125) for $\theta = 1/2$ under the presence of (4.178). $\qquad \square$

4.14   Vanishing of the Center Manifold

This section is devoted to a remark on the rate of convergence of 2D-NRF, that is, (4.40) in §4.4 for (Ω, λ) given by (4.19) in §4.2. We have readily noticed the uniqueness of steady state in this case, which is given by

$$u_* = \frac{\lambda}{|\Omega|} = 2,$$

assuming

$$S^2 = \{x \in \mathbf{R}^3 \mid |x| = 1\}$$

without loss of generality. Then the geometric result [Hamilton (1988)] ensures

$$u(\cdot, t) \to u_* \quad \text{exponentially in } C^\infty \text{ topology} \tag{4.186}$$

as $t \to +\infty$. This rate of convergence does not follow from Theorem 4.4 in §4.8 because u_* is degenerate.

To examine this property from the view point of analysis, we put $z = u - u_*$ to derive

$$z_t = \Delta\varphi(z) + z \quad \text{in } \Omega \times (0, +\infty), \tag{4.187}$$

where

$$\varphi(z) = \log u - \log u_* = \log(z + 2) - \log 2 = \frac{z}{2} + g(z)$$

for

$$g(z) = -\frac{z^2}{8} + R(z), \quad |R(z)| \le C|z|^3. \tag{4.188}$$

Then equation (4.187) is reduced to

$$z_t + Lz = \Delta g(z), \quad z \in E$$

for

$$E = \{v \in H^1(\Omega) \mid \int_\Omega v \, dx = 0\}$$

and

$$L = -\frac{1}{2}(\Delta + 2),$$

which is realized as a self-adjoint operator in

$$H = \{v \in L^2(\Omega) \mid \int_\Omega v \, dx = 0\}$$

with the domain $D(L) = H^2(\Omega) \cap E$.

The eigenvalues of the Laplace Beltrami operator $-\Delta$ on S^2 is given by

$$\sigma(-\Delta) = \{\mu_\ell \mid \ell = 0, 1, 2, \cdots, \}, \ \mu_\ell = \ell(\ell+1),$$

and each μ_ℓ takes the multiplicity $2\ell + 1$ (Chapter 7 of [Suzuki (2022a)]). The space H is generated by the eigenfunctions corresponding to the non-zero eigenvalue of $-\Delta$, and hence the eigenvalues of L is given by

$$\sigma(L) = \{0\} \cup \{1 - \ell(\ell+1)/2 \mid \ell = 2, 3, \cdots\}.$$

The operator L is thus actually degenerate with $\dim E_0 = 2 \cdot 1 + 1 = 3$, where

$$E_0 = \operatorname{Ker} L.$$

It is actually generated by x_1, x_2, and x_3 for $x = (x_1, x_2, x_3) \in S^2$:

$$E_0 = \langle x_1, x_2, x_3 \rangle. \tag{4.189}$$

Writing

$$G(z) = g(z) - \overline{g(z)}, \quad \overline{g(z)} = \frac{1}{|\Omega|} \int_\Omega g(z) \, dx,$$

we obtain

$$z_t + Lz = \Delta G(z), \quad z \in E, \tag{4.190}$$

where $G : E \to E$ is a C^∞ mapping. Here, $\Delta : E \to E'$ is a bounded linear mapping, and there holds that

$$\|z\|_E \leq K \quad \Rightarrow \quad \|G(z)\|_E \leq C(K)\|z\|_E^2.$$

Concerning the semilinear equation,

$$z_t + Lz = G(z), \quad z \in E \tag{4.191}$$

with $G(0) = G'(0) = 0$, the center manifold theory is available. By Theorem 3.22 in Chapter 2 of [Hargus–Iooss (2011)], there is a C^∞ mapping

$$h = h(y) : E_0 \cap U \to E_1, \quad E = E_0 \oplus E_1,$$

with

$$h(0) = 0, \ h'(0) = 0, \tag{4.192}$$

such that

$$\mathcal{M} = \{y + h(y) \mid y \in E_0 \cap U\}$$

is invariant to (4.191), where $U \subset E$ is an open set containing 0. If

$$z(\cdot, t) \to 0$$

holds in (4.191), there is $w = w(\cdot, t) \in \mathcal{M}$ such that

$$z(\cdot, t) - w(\cdot, t) \to 0 \quad \text{exponentially in } C^\infty \text{ topology} \qquad (4.193)$$

in E.

Equation (4.190), however, does not admit such a local center manifold $\mathcal{M}$. In fact, if it exists, there arises that

$$w(\cdot, t) \to 0 \quad \text{exponentially in } C^\infty \text{ topology}$$

by (4.186)–(4.193). If $P : E \to E_0$ denotes the L^2 orthogonal projection, we have $PL = 0$, and hence

$$-P\Delta v = 2Pv, \quad v \in E.$$

Writing $y = Pw$, therefore, we obtain

$$\dot{y} = -2Pg(y + h(y)), \quad y(t) \in E_0$$

by $P\bar{z} = 0$, with

$$y(t) \to 0 \quad \text{exponentially.} \qquad (4.194)$$

By (4.188) and (4.192) it holds that

$$\dot{y} = \frac{1}{4}Py^2 - 2PQ(y), \quad |Q(y)| \le C|y|^3. \qquad (4.195)$$

Here we use (4.189), to obtain

$$y \in E_0 \;\Rightarrow\; Py^2 = 0$$

by

$$\int_{S^2} x_i x_j x_k \, dS = 0, \quad 1 \le i, j, k \le 3.$$

Hence (4.195) is reduced to

$$\dot{y} = -2\sum_{i=1}^{3}(Q(y), \psi_i)\psi_i \qquad (4.196)$$

for

$$\psi_i = (\frac{3}{4\pi})^{1/2}x_i, \quad 1 \le i \le 3,$$

or

$$\dot{y}_i = -2(Q(y), \psi_i), \quad y_i = (y, \psi_i), \quad 1 \le i \le 3.$$

We have

$$\frac{1}{4}\frac{d}{dt}\sum_{i=1}^{3}\int_{S^2}y_i^4 dS = -2\int_{S^2}\sum_{i=1}^{3}(Q(y), \psi_i)y_i^3 \, dS,$$

and hence

$$\left| \frac{d}{dt} \sum_{i=1}^{3} \int_{S^2} y_i^4 dS \right| \le C' \left(\int_{S^2} |y|^3 dS \right)^2 \le C \left(\sum_{i=1}^{3} \int_{S^2} y_i^4 dS \right)^{3/2},$$

which implies

$$\left| \frac{dx}{dt} \right| \le C x^2 \tag{4.197}$$

for

$$x = \left(\sum_{i=1}^{3} \int_{S^2} y_i^4 dS \right)^{1/2} \ge 0.$$

Equation (4.196) takes the form

$$\dot{y} = F(y)$$

with locally Lipschitz continuous $F : \mathbf{R}^3 \to \mathbf{R}^3$. Since it holds that $F(0) = 0$, we have either $y(t) \equiv 0$ or $y(t) \ne 0$ for any t, and in the latter case it holds that

$$\left| \frac{d}{dt} \frac{1}{x} \right| \le C$$

by (4.197). Then we obtain

$$x(t) \ge \frac{c_0}{1+t}$$

with $c_0 > 0$, a contradiction to (4.194).

PART 2
Differential Forms and Singularities

Chapter 5

Systems of Multiple Components

In accordance with the classical mechanics there arises the integrability of Hamilton system realized by the transformation of symplectic variables, which results in the Hamilton Jacobi theory. Geometric structure is hidden there, and is formulated by differential forms. This integrability induces the relaxation of singularities caused by the skew symmetric interaction of multiple species. This principle of relaxation is different from the one described in Part I, particularly the Smoluchowski Poisson (SP) equation realized by the symmetry of the Green function derived from action reaction law at the occasion of the creation of potential field by particle density. A typical example of the relaxation of skew-symmetric multi-species, on the other hand, is observed in the reaction diffusion system with dissipation, where quadratic nonlinearity admits the solution global in time in any space dimension, in spite of a similar thermo-dynamical structure to that of the SP equation. In this chapter, first, we confirm the notion of Poisson manifolds in accordance with the analytic mechanics to reach integrability of the Hamilton system (§5.1–§5.4). Then we describe several examples of integrable systems in mathematical biology (§5.5) and turn to recent developments in the theory of reaction diffusion system (§5.6). Integrability of the skew-symmetric Lotka Volterra ODE system is then confirmed (§5.7–§5.9), to formulate a class of integrable systems formulated by differential forms (§5.10-§5.13, which results in the persistency of multiple species derived from the foliation made by nested periodic orbits.

5.1 Languages of Geometry

Many mathematical models are concerned with the interaction of multiple species. They are sometimes reduced to the dynamics on lower-dimensional manifolds, because of several conservation laws derived from skew-symmetric interactions. By this structure, formation of singularities of the solution is avoided, resulting in the spatial homogenization in infinite time. Languages of geometry, on the other hand, are useful to describe this reduction to lower dimensional manifolds of multiple components.

A typical example of the formation of the first integral is Liouville's theorem on the integrability of Hamilton systems in analytic mechanics, where symplectic geometry takes a role. This chapter begins with a review of several concepts of geometry, to understand competitive and cooperative interactions of multiple species. Hence we generalize the theories of differential forms, tangent and cotangent spaces, and connections discussed in §1.2, §1.5, §1.6, and §1.8, respectively, for surfaces in $\mathbf{R}^3$.

A *manifold* M is thus a Hausdorff space covered by a family of open sets as in

$$M = \bigcup_{\alpha \in \mathcal{A}} U_\alpha,$$

provided with the homeomorphism

$$\psi_\alpha : U_\alpha \to \psi_\alpha(U_\alpha) \subset \mathbf{R}^n = \{x^1, \cdots, x^n\}, \ \alpha \in \mathcal{A},$$

where

$$\psi_\beta \circ \psi_\alpha^{-1} : \psi_\alpha(U_\alpha \cap U_\beta) \to \psi_\beta(U_\alpha \cap U_\beta)$$

is C^∞ if $U_\alpha \cap U_\beta \neq \emptyset$. The *tangent space* T_pM at $p \in M$ is defined as a vector space formed by infinitesimal directions of curves on M at p. The basis of this vector space is denoted by

$$\left\{ \frac{\partial}{\partial x^1}, \cdots, \frac{\partial}{\partial x^n} \right\} \tag{5.1}$$

according to the *local chart* $(U_\alpha; x^1, \cdots, x^n)$ such that $p \in U_\alpha$. Then the *tangent bundle* of M is induced as

$$TM = \bigcup_{p \in M} T_pM.$$

The *cotangent space* T_p^*M is defined by the dual space of the tangent space, $(T_pM)^*$. The dual basis of (5.1) is denoted by

$$\{dx^1, \cdots, dx^n\}.$$

Then the *cotangent bundle* is formed as

$$T^*M = \bigcup_{p \in M} T_p^*M.$$

The smooth mappings

$$p \in M \mapsto X_p \in T_pM, \quad p \in M \mapsto \omega_p \in T_p^*M$$

are called *vector field* and *1-form*, respectively. The set of vector fields and 1-forms are, furthermore, denoted by $\mathcal{X}(M)$ and $\mathcal{D}^1(M)$, respectively.

If $\mathcal{M} \hookrightarrow \mathbf{R}^3$ is a surface, the connection ∇ formulated in §1.8 becomes a *co-variant derivative*. The co-variant derivative is actually formulated as a mapping

$$(X, Y) \in \mathcal{X}(M) \times \mathcal{X}(M) \ \to \nabla_X Y \in \mathcal{X}(M)$$

satisfying

$$\nabla_X(Y + Z) = \nabla_X Y + \nabla_X Z$$
$$\nabla_X(fY) = (Xf)Y + f\nabla_X Y$$
$$\nabla_{X+Y} Z = \nabla_X Z + \nabla_Y Z$$
$$\nabla_{fX} Y = f\nabla_X Y$$

for any $X, Y, Z \in \mathcal{X}(M)$ and $f \in C^\infty(M)$. This axiom determines the connection coefficient, or the Christoffel symbol Γ_{ij}^k by

$$\nabla_{\frac{\partial}{\partial x^i}} \frac{\partial}{\partial x^j} = \Gamma_{ij}^k \frac{\partial}{\partial x^k}$$

through the local chart under the Einstein convention. Thus ∇ is represented by Γ_{ij}^k subject to the transformation of local charts, and by this reason we call the co-variant derivative the connection as in §1.8.

The *curvature tensor* is a mapping

$$R : \mathcal{X}(M) \times \mathcal{X}(M) \times \mathcal{X}(M) \to \mathcal{X}(M)$$

defined by

$$R(X, Y, Z) = \nabla_X(\nabla_Y Z) - \nabla_Y(\nabla_X Z) - \nabla_{[X,Y]}Z,$$

where

$$[X, Y] = XY - YX$$

denotes the *commutator*. The *torsion tensor*, on the other hand, is a mapping

$$T : \mathcal{X}(M) \times \mathcal{X}(M) \to \mathcal{X}(M)$$

defined by

$$T(X, Y) = \nabla_X Y - \nabla_Y X - [X, Y].$$

A manifold M provided with the connection ∇ is said to be *flat* if curvature and torsion tensors are free, that is, $R \equiv 0$ and $T \equiv 0$. In this case there is a local chart satisfying $\Gamma_{ij}^k = 0$ for any k, i, j.

If M is a manifold provided with the connection ∇, and

$$C : \ p(t) \in M, \quad a \leq t \leq b$$

is a curve on M, we say that $Z \in \mathcal{X}(M)$ is *parallel* to ∇ along C if

$$\nabla_{\dot{p}} Z = 0$$

for $a \leq t \leq b$. Here, $\dot{p} = \dot{p}(t)$ is regarded as a vector field on C. If

$$\nabla_{\dot{p}} \dot{p} = 0, \quad a \leq t \leq b, \tag{5.2}$$

this C is called *geodesic*. If this M is a Riemann manifold provided with the metric $g(\cdot, \cdot)$, and ∇ is the Levi Chivita connection, equality (5.2) means

$$\frac{d^2 x^i}{dt^2} + \Gamma_{jk}^i \frac{dx^j}{dt} \frac{dx^k}{dt} = 0, \quad 1 \leq i \leq n \tag{5.3}$$

in the local chart. Here, (5.3) is the Euler–Lagrange equation of the minimizing the length of $\mathcal{C}$,

$$L(\mathcal{C}) = \int_a^b \|\dot{p}(t)\| \, dt,$$

where

$$\|X\| = g(X, X)^{1/2}, \quad X \in TM.$$

We recall that the Levi Chivita connection is formulated in §1.8 for surfaces in $\mathbf{R}^3$.

Given a vector space V, let V^* be its dual space. A multi-linear form

$$F : \overbrace{V^* \times \cdots \times V^*}^{r} \times \underbrace{V \times \cdots V}_{s} \to \mathbf{R}$$

is called an (r, s) *tensor* on V. If F and G are (r, s) and (r', s') tensors, respectively, their tensor product $F \otimes G$ is an $(r + r', s + s')$ tensor defined by

$$(F \otimes G)(x, x' : y, y') = F(x, x') \cdot G(y, y'),$$

where $(x, x', y, y') \in V^r \times V^{r'} \times (V^*)^r \times (V^*)^{r'}$.

A *Riemann metric* g on the manifold M is a smooth mapping

$$p \in M \mapsto g_p,$$

where g_p is a $(0, 2)$ tensor on $V = T_pM$, symmetric and positive definite:

$$g_p(X, Y) = g_p(Y, X), \quad g_p(X, X) \geq \delta |X|^2, \quad X, Y \in T_pM, \ p \in M,$$

where $\delta > 0$ is a constant. This metric is thus represented by

$$g = g_{ij} dx^i \otimes dx^j$$

under the local chart, where (g_{ij}), defined for

$$g_{ij} = g\left(\frac{\partial}{\partial x^i}, \frac{\partial}{\partial x^j}\right),$$

is a positive definite symmetric matrix.

A manifold provided with a Riemann metric is called a *Riemann manifold*, and a Riemann manifold provided with connection is called the *Riemann Cartan manifold*. Any Riemann manifold can be a Riemann Cartan manifold by Levi Chivita connection, while the other connection can be provided to form a Riemann Cartan manifold.

The Riemann metric $g(\ ,\ )$ thus always induces a connection ∇, called the *Levi Chivita connection*, satisfying

$$Xg(Y, Z) = g(\nabla_X Y, Z) + g(Y, \nabla_X Z), \quad X, Y, Z \in \mathcal{X}(M). \tag{5.4}$$

Such a connection is called a *metric*. A connection ∇ on a Riemann manifold (M, g), furthermore, is Levi Chivita if and only if it is a metric and torsion free.

Let (M, g) be a Riemann manifold and ∇, ∇^* be connections. We say that ∇^* is a *dual connection* of ∇ and that (∇, ∇^*) forms a *dual affine structure* to (M, g) if there arises that

$$Xg(Y, Z) = g(\nabla_X Y, Z) + g(Y, \nabla_X^* Z), \quad X, Y, Z \in \mathcal{X}(M).$$

By (5.4), therefore, the Levi Chivita connection $\nabla = \nabla^*$ provides a dual affine structure to (M, g).

A Riemann manifold M is *dual flat* if there is a dual affine structure (∇, ∇^*) of (M, g) such that

$$R = R^* = 0, \quad T = T^* = 0,$$

where R and R^* are curvatures with respect to ∇ and ∇^*, respectively, and T and T^* are torsions with respect to ∇ and ∇^*, respectively. This dual flat property is a fundamental structure of *information geomerty*.

5.2   Analytic Mechanics

If mass particles $x_i = x_i(t) \in \mathbf{R}^3$, $i = 1, 2, \cdots, f$, are subject to the *potential energy* $U = U(x_1, x_2, \cdots, x_f)$, then *Newton's equation of motion* takes the form

$$\frac{dp_i}{dt} = -\frac{\partial U}{\partial x_i}, \qquad i = 1, 2, \cdots, f. \tag{5.5}$$

Kinetic energy then arises,

$$K(\dot{x}_1, \dot{x}_2, \cdots, \dot{x}_f) = \frac{1}{2} \sum_{i=1}^{f} m_i \dot{x}_i^2,$$

and it holds that

$$p_i = \frac{\partial K}{\partial \dot{x}_i}$$

where $p_i = m_i \dot{x}_i$ is the *momentum*.

If $(q_1, \cdots, q_f)$ denotes the *generalized coordinate* then it holds that

$$x_1 = x_1(q_1, \cdots, q_f; t), \ \ \ldots \ x_f = x_f(q_1, \cdots, q_f; t)$$

and hence

$$\dot{x}_i = \sum_j \frac{\partial x_i}{\partial q_j} \dot{q}_j + \frac{\partial x_i}{\partial t}. \tag{5.6}$$

Regarding $(q, \dot{q}, t)$ as independent variables, we differentiate (5.6) with respect to $\dot{q}_j$ to obtain

$$\frac{\partial \dot{x}_i}{\partial \dot{q}_j} = \frac{\partial x_i}{\partial q_j}.$$

Hence we have

$$\frac{\partial K}{\partial \dot{q}_j} = \sum_k \frac{\partial K}{\partial \dot{x}_k} \frac{\partial \dot{x}_k}{\partial \dot{q}_j} = \sum_k p_k \frac{\partial x_k}{\partial q_j},$$

and then it follows that

$$\frac{d}{dt} \frac{\partial K}{\partial \dot{q}_j} = \sum_k \left(\frac{dp_k}{dt} \frac{\partial x_k}{\partial q_j} + p_k \frac{d}{dt} \frac{\partial x_k}{\partial q_j} \right)$$

$$= \sum_k \left(-\frac{\partial U}{\partial x_k} \frac{\partial x_k}{\partial q_j} + \frac{\partial K}{\partial \dot{x}_k} \frac{\partial \dot{x}_k}{\partial q_j} \right) = -\frac{\partial U}{\partial q_j} + \frac{\partial K}{\partial q_j} \tag{5.7}$$

from

$$\frac{d}{dt} \frac{\partial x_k}{\partial q_j} = \frac{\partial \dot{x}_k}{\partial q_j}.$$

If we regard K as a function of $(q, \dot{q}, t)$ by (5.6) and take the *Lagrangian*,

$$L(q, \dot{q}, t) = K(q, \dot{q}, t) - U(q, t), \tag{5.8}$$

it follows that

$$\frac{d}{dt}\frac{\partial L}{\partial \dot{q}_j} = \frac{\partial L}{\partial q_j}, \quad j = 1, \cdots, f \tag{5.9}$$

from (5.7). Equation (5.9), called *Lagrange's equation of motion*, is nothing but the *Euler equation* of the variational problem $\delta S = 0$ under the constraint

$$\delta q(t_1) = \delta q(t_2) = 0$$

where

$$S = \int_{t_1}^{t_2} L\left(q(t), \dot{q}(t), t\right) dt$$

denotes the action integral defined for $q = q(t) \in \mathbf{R}^{3f}$ with $t \in [t_1, t_2]$. This fact is called *Hamilton's principle of least action*.

The *Legendre transformation* of

$$\dot{q} \mapsto L = L(q, \dot{q}, t)$$

is given by

$$L^*(p) = \sup_{\dot{q}}\{p \cdot \dot{q} - L(\dot{q})\} = p \cdot \dot{q} - L, \quad p = \frac{\partial L}{\partial \dot{q}}.$$

Writing this L^* as

$$H = H(p, q, t) = \sum_j p_j \dot{q}_j - L, \quad p_j = \frac{\partial L}{\partial \dot{q}_j}, \tag{5.10}$$

we call it the *Hamiltonian* associated with the generalized momentum p and the generalized coordinate q.

From (5.9) we have

$$\begin{aligned}
dL &= \frac{\partial L}{\partial q}dq + \frac{\partial L}{\partial \dot{q}}d\dot{q} + \frac{\partial L}{\partial t}dt \\
&= \frac{d}{dt}\left(\frac{\partial L}{\partial \dot{q}}\right)dq + \frac{\partial L}{\partial \dot{q}}d\dot{q} + \frac{\partial L}{\partial t}dt \\
&= \dot{p}dq + pd\dot{q} + \frac{\partial L}{\partial t}dt,
\end{aligned}$$

which implies

$$dH = \dot{q}dp + pd\dot{q} - dL = \dot{q}dp - \dot{p}dq - \frac{\partial L}{\partial t}dt. \tag{5.11}$$

Equality (5.11) induces *Hamilton's canonical equation*, or *Hamilton system* in §2.2,

$$\dot{q} = \frac{\partial H}{\partial p}, \quad \dot{p} = -\frac{\partial H}{\partial q}, \tag{5.12}$$

by

$$dH = \frac{\partial H}{\partial p} dp + \frac{\partial H}{\partial q} dq + \frac{\partial H}{\partial t} dt.$$

Here we obtain

$$\dot{H} \equiv \frac{d}{dt} H(p, q) = \frac{\partial H}{\partial q} \cdot \dot{q} + \frac{\partial H}{\partial p} \cdot \dot{p}$$

$$= \frac{\partial H}{\partial q} \cdot \frac{\partial H}{\partial p} - \frac{\partial H}{\partial p} \cdot \frac{\partial H}{\partial q} = 0 \tag{5.13}$$

for the solution

$$(p, q) = (p(t), q(t))$$

to (5.12).

Let

$$\{F, G\} = \frac{\partial F}{\partial q} \cdot \frac{\partial G}{\partial p} - \frac{\partial F}{\partial p} \cdot \frac{\partial G}{\partial q} \tag{5.14}$$

for

$$F = F(p, q), \quad G = G(p, q).$$

Then (5.12) is equivalent to

$$\dot{F} = \{F, H\}, \quad \forall F = F(q, p) \tag{5.15}$$

by

$$\dot{F} = \frac{\partial F}{\partial q} \dot{q} + \frac{\partial F}{\partial p} \dot{p},$$

which leads to the *Liouville theorem* in the following form.

Theorem 5.1 (Liouville). *The function $F = F(p, q)$ is a first integral or a conservative quantity of (5.12), that is,*

$$\frac{d}{dt} F(p, q) = 0$$

for its any solution

$$(p, q) = (p(t), q(t)),$$

if and only if

$$\{F, H\} = 0.$$

In particular, equality (5.13) holds by

$$\{H, H\} = 0.$$

The mapping $(F, G) \mapsto \{F, G\}$ in (5.14) is bilinear and skew-symmetric. It satisfies also the *Jacobi identity*

$$\{F, \{G, H\}\} + \{G, \{H, F\}\} + \{H, \{F, G\}\} = 0 \qquad (5.16)$$

and the *Leibniz identity*

$$\{FG, H\} = F\{G, H\} + G\{F, H\}. \qquad (5.17)$$

Such a mapping

$$(F, G) \mapsto \{F, G\}$$

is called a *Poisson bracket*.

If $\mathcal{M} = \mathbf{R}^{3f}$ is regarded as a manifold, then the above Hamiltonian is realized as a mapping

$$H : (p, q) \in T^*Q \mapsto H(p, q) \in \mathbf{R},$$

where $(p, q) \in T^*Q$ means $q \in T_p^*Q$. The *Hamilton vector field* is defined by

$$X = \begin{pmatrix} -\frac{\partial H}{\partial q} \\ \frac{\partial H}{\partial p} \end{pmatrix}$$

and then (5.12) is equivalent to

$$\dot{z} = X(z), \quad z = \begin{pmatrix} p \\ q \end{pmatrix}. \qquad (5.18)$$

5.3 Symplectic and Poisson Manifolds

Recall that $\mathcal{D}^1(M)$ denotes the set of 1-forms on M. Each element in $\mathcal{D}^1(M)$ induces a 2-form by *wedge product* $\wedge$ and *outer derivative* d. Henceforth the set of two forms on M is denoted by $\Lambda^2 T^*M$.

By the local chart each $\omega \in \Lambda^2 T^*M$ is represented as a smooth function valued at skew-symmetric matrices. Hence it induces a bilinear, skew-symmetric mapping

$$\omega : \mathcal{X}(M) \times \mathcal{X}(M) \to \mathbf{R}.$$

If $M = \mathbf{R}^2$ and $\omega = dx^1 \wedge dx^2$, for example, it follows that

$$\langle (u, v), dx^1 \wedge dx^2 \rangle = u^1 v^2 - u^2 v^1$$

for

$$u = u^1 \frac{\partial}{\partial x_1} + u^2 \frac{\partial}{\partial x_2}, \quad v = v^1 \frac{\partial}{\partial x^1} + v^2 \frac{\partial}{\partial x^2}.$$

The 2-form ω is said to be *non-degenerate* if the associated skew-symmetric matrix represented by local chart is non-degenerate. It is said to be *closed* if $d\omega = 0$.

A *symplectic manifold* is a manifold M provided with a non-degenerate, closed 2-form ω called the *symplectic form*. Then *Darboux's theorem* ensures that this (M, ω) is locally equivalent to the *canonical symplectic form* $(\mathbf{R}^{2n}, \omega_0)$, where

$$\omega_0 = dx \wedge dy \equiv dx^1 \wedge dy^1 + \cdots + dx^n \wedge dy^n, \quad (x, y) \in \mathbf{R}^n \times \mathbf{R}^n.$$

Then, *Liouville's 1-form*,

$$\theta = q_i dp^i,$$

is introduced in accordance with analytic mechanics, which results in the *symplectic form*

$$\omega = -d\theta, \tag{5.19}$$

where

$$p^i = x^i, \ q^i = y^i, \quad 1 \le i \le n.$$

The following theorem ensures the existence of a Poisson bracket on each symplectic manifold. Recall that $\mathcal{X}(M)$ denotes the set of vector fields on M.

Theorem 5.2. *Let (M, ω) be a symplectic manifold and $H \in C^\infty(M)$. Define $X_H \in \mathcal{X}(M)$, called the Hamilton vector field, by*

$$\omega(X_H, \cdot) = dH \tag{5.20}$$

in $\mathcal{D}^1(M) = \mathcal{X}^*$. Then

$$\{F, G\} = \omega(X_F, X_G), \quad F, G \in C^\infty(M)$$

is a skew-symmetric bilinear form provided with the Jacobi and Leibniz identities, and hence is a Poisson bracket.

Let M be a manifold, and $X \in \mathcal{X}(M)$, $f \in C^\infty(M)$ be given. Then the *Lie derivative* $\mathcal{L}_X f$ is defined by

$$\mathcal{L}_X f(x) = \frac{d}{dt} f \circ \Phi_t(x) \Big|_{t=0}, \quad x \in M,$$

where $\{\Phi_t\}$ is the local dynamical system generated by X:

$$\frac{d}{dt} \Phi_t(x) = X(\Phi_t(x)), \quad \Phi_t(x)|_{t=0} = x. \tag{5.21}$$

If $\mathcal{M} = \mathbf{R}^n$, the Lie derivative is the directional derivative toward X as in

$$\mathcal{L}_X = X \cdot \nabla, \quad \nabla = (\frac{\partial}{\partial x_1}, \cdots, \frac{\partial}{\partial x_n})^T$$

for $x = (x_i) \in \mathcal{M}$.

A manifold P provided with a Poisson bracket

$$\{,\} : C^\infty(M) \times C^\infty(M) \to C^\infty(M)$$

is called a *Poisson manifold*. A symplectic manifold is a Poisson manifold, under the bracket induced in Theorem 5.2.

Let

$$(P, \{,\})$$

be a Poisson manifold and $H \in C^\infty(P)$. We define the Hamilton vector field

$$X = X_H \in \mathcal{X}(M)$$

from this Poisson structure, that is,

$$\mathcal{L}_X F = \{F, H\}, \quad \forall F \in C^\infty(P). \tag{5.22}$$

Then we generalize (5.18) as in

$$\dot{x} = X(x) \tag{5.23}$$

and call it a Hamilton system on P. We define $\Phi_t(x)$ by (5.21) for $X = X_H$, put

$$x(t) = \Phi_t(x),$$

and obtain

$$\frac{d}{dt} H(x(t)) = \mathcal{L}_X(x(t)) = \{H, H\}(x(t)) = 0.$$

Hence $H(x(t))$ is the first integral of (5.23).

These notations are justified by the following theorem ([Holm–Schmah–Stoica (2009)]).

Theorem 5.3. *Let (M, ω) be a simpletic manifold and $H \in C^\infty(M)$. Then the Hamilton vector field X_H defined in Theorem 5.2 coinicides with that defined above, regarding (M, ω) as a Poisson manifold using the Poisson bracket $\{,\}$ defined there.*

An example of the Poisson manifold of odd dimension is

$$P = \mathbf{R}^3$$

and

$$\{F, G\} = -\nabla C \cdot \nabla F \times \nabla G$$

for fixed $C \in C^\infty(P, \mathbf{R}^3)$. Then the Hamilton system induced by the Hamiltonian $H \in C^\infty(P)$ becomes

$$\dot{x} = \nabla C \times \nabla H. \tag{5.24}$$

In fact, equation (5.22) means

$$X_H \cdot \nabla F = -\nabla C \cdot \nabla F \times \nabla H, \quad \forall F$$

in this case, and therefore,

$$X_H = \nabla C \times \nabla H.$$

This structure is observed in a Lotka Volterra system in §5.5.

5.4   Hamilton Jacobi Theory

Here we describe the theory of complete integrability of the Hamilton system. For simplicity we restrict ourselves to the autonomous system (5.12) in §5.2 with two variables, that is, momentum q and positon p. The Poisson bracket $\{,\}$ defined by (5.14) is skew-symmetric, bi-linear, and satisfies the Jacobi identity (5.16) and the Leibniz identity (5.17). Symplectic form, finally, is defined by (5.19) in §5.3, that is,

$$\omega = -d\theta = dp \wedge dq.$$

The transformation

$$z = \begin{pmatrix} p \\ q \end{pmatrix} \mapsto \zeta = \begin{pmatrix} \xi \\ \eta \end{pmatrix} \tag{5.25}$$

is called a *symplectic transformation* if it preserves the symplectic form as in

$$d\xi \wedge d\eta = dp \wedge dq. \tag{5.26}$$

This condition is quivalent to

$$\xi_p \eta_q - \xi_q \eta_p = 1 \tag{5.27}$$

by

$$\begin{aligned} d\xi \wedge d\eta &= (\xi_p dp + \xi_q dq) \wedge (\eta_p dp + \eta_q dq) \\ &= (\xi_p \eta_q - \xi_q \eta_p) dp \wedge dq. \end{aligned}$$

Then we obtain the following theorem.

Theorem 5.4. *Let*
$$(p, q) = (p(t), q(t))$$
be the solution to the Hamilton system (5.12), and (5.25) be a symplectic transformation. Define
$$\zeta(t) = (\xi(t), \eta(t))$$
by
$$\zeta(t) = \zeta(z(t)), \quad z(t) = (p(t), q(t)).$$
Then it holds that
$$\dot{\eta} = \tilde{H}_\xi, \quad \dot{\xi} = -\tilde{H}_\eta, \tag{5.28}$$
where
$$\tilde{H}(\zeta) = H(z). \tag{5.29}$$

Proof. In fact, we obtain
$$\dot{\eta} = \frac{\partial \eta}{\partial p}\dot{p} + \frac{\partial \eta}{\partial q}\dot{q}$$
$$= -\eta_p H_q + \eta_q H_p$$
$$= -\eta_p(\tilde{H}_\xi \xi_q + \tilde{H}_\eta \eta_q) + \eta_q(\tilde{H}_\xi \xi_p + \tilde{H}_\eta \eta_p)$$
$$= (\xi_p \eta_q - \xi_q \eta_p)\tilde{H}_\xi = \tilde{H}_\xi,$$
and similarly,
$$\dot{\xi} = \frac{\partial \xi}{\partial p}\dot{p} + \frac{\partial \xi}{\partial q}\dot{q}$$
$$= -\xi_p H_q + \xi_q H_p$$
$$= -\xi_p(\tilde{H}_\xi \xi_q + \tilde{H}_\eta \eta_q) + \xi_q(\tilde{H}_\xi \xi_p + \tilde{H}_\eta \eta_p)$$
$$= -(\xi_p \eta_q - \xi_q \eta_p)\tilde{H}_\eta = -\tilde{H}_\eta$$
by (5.27). Hence it follows that (5.28). □

Remark 5.1. Equality (5.27) means
$$\det \frac{\partial(\xi, \eta)}{\partial(p, q)} = \begin{vmatrix} \frac{\partial \xi}{\partial p} & \frac{\partial \xi}{\partial q} \\ \frac{\partial \eta}{\partial p} & \frac{\partial \eta}{\partial q} \end{vmatrix} = 1,$$
and therefore, if the domain D in pq-space is mapped onto the domain Ω in $\xi\eta$-space by the symplectic transformation
$$(p, q) \mapsto (\xi, \eta)$$
it hols that
$$\iint_D dp\,dq = \iint_\Omega d\xi\,d\eta. \tag{5.30}$$
Equality (5.30) is a form of the other *Liouville's theorem*.

Let (5.25) be a symplectic transformation and define the 1-form $\theta = \theta(q, \xi)$ of (q, ξ) by

$$\theta = pdq + \eta d\xi.$$

Then equality (5.26) implies

$$d\theta = dp \wedge dq + d\eta \wedge d\xi = 0$$

and hence there is a 0-form denoted by W such that

$$\theta = dW, \quad W = W(q, \xi)$$

by Poincaré's lemma. Since

$$dW = W_\xi d\xi + W_q dq$$

we obtain

$$W_\xi = \eta, \quad W_q = p. \tag{5.31}$$

On the other hand we introduce the following notion.

Definition 5.1. The Hamilton system (5.12) is said to be *completely integrable* if there is a symplectic transformation (5.25) such that

$$\frac{\partial \tilde{H}}{\partial \eta} = 0 \tag{5.32}$$

for $\tilde{H}(\xi, \eta)$ defined by (5.29).

In this case, there arises that

$$\tilde{H} = \tilde{H}(\xi)$$

and the Hamilton system (5.28) is reduced to

$$\dot{\eta} = \tilde{H}_\xi, \quad \dot{\xi} = 0.$$

This $\xi = \xi(t)$ is a constant, denoted by

$$\xi(t) = \alpha \in \mathbf{R},$$

which implies

$$\eta(t) = \eta(0) + \tilde{E}t,$$

where

$$\tilde{E} = \tilde{H}_\xi(\alpha).$$

We obtain also

$$H(p, q) = E, \quad (p, q) = (p(t), q(t))$$

with a constant E, and therefore,

$$H(W_q, q) = E,$$

by (5.31). Hence it follows that

$$\frac{\partial S}{\partial t} + H\left(\frac{\partial S}{\partial q}, q\right) = 0 \qquad (5.33)$$

for

$$S = S(\alpha, q, t) \equiv W - Et. \qquad (5.34)$$

Equation (5.33) called the *Hamilton Jacobi equation*.

If we find, conversely, a solution

$$S = S(\alpha, q, t)$$

to (5.33) in the form of (5.34) containing the parameter $\alpha \in \mathbf{R}$, we define

$$(p, \xi)$$

by (5.31) as in

$$p = p(\xi, q), \quad \eta = \eta(\xi, q)$$

for $\xi = \alpha$. If the transformation (5.25) is well-defined by this process, it is a symplectic transformation satisfying (5.32). Then the Hamilton system

$$\dot{q} = H_p, \quad \dot{p} = -H_q \qquad (5.35)$$

is transformed into

$$\dot{\eta} = \tilde{E}, \quad \dot{\xi} = 0.$$

which results in

$$\xi(t) = \alpha, \quad \eta(t) = \eta(0) + \tilde{E}t.$$

Then we obtain an explicit representation of

$$(p, q) = (p(t), q(t))$$

for the solution to (5.35). In this case (5.35) is said to be *completely integrable*.

Remark 5.2. If the Hamiltonian depends on t,

$$H = H(p, q, t),$$

the Hamilton Jacobi equation arises as in

$$\frac{\partial S}{\partial t} + H\left(\frac{\partial S}{\partial q}, q, t\right) = 0. \qquad (5.36)$$

There, the solution does not necessarily take the form (5.34), but (5.35) can be completely integrable if (5.36) admits a family of solutions involving the parameter $\alpha \in \mathbf{R}$. These arguments are valid to general $(p, q) \in \mathbf{R}^{2n}$ sysmtems.

5.5 Symplectic and Poisson Structures of Biological Models

Systems of reaction equations are used to describe time evolution of the population of multi-species in theoretical biology. Some of them take the form of the Hamilton system (5.12), while the others are described by (5.15) in §5.2 on a Poisson manifold of odd dimension. Then we reach a different type of the complete integrability from the Hamilton Jacobi theory in §5.4, formulated by the differential form (Theorem 5.13 in §5.10).

The reaction system of two species is formulated by

$$\frac{du}{dt} = f(u,v), \ \tau\frac{dv}{dt} = g(u,v) \tag{5.37}$$

with

$$(u,v)|_{t=0} = (u_0, v_0) > 0, \tag{5.38}$$

where $\tau > 0$ is a constant,

$$f, g \to \overline{\mathbf{R}}_+^2 \to \mathbf{R}, \quad \overline{\mathbf{R}}_+ = [0, +\infty),$$

are local Lipschitz continuous functions satisfying the *quasi-positivity*,

$$(f(0,v), g(u,0)) \geq 0, \ \forall (u,v) \geq 0.$$

Local-in-time well-posedness of (5.37)–(5.38) is valid, and if $T = T_{\max} \in (0, +\infty]$ denotes its maximal existence time, there arises that

$$(u(t), v(t)) > 0, \quad 0 \leq t < T.$$

Equation (5.37) forms the *prey predator system* if

$$f(u,v) = u(a - bv), \ g(u,v) = v(-c + du),$$

and $a, b, c, d > 0$ are constants:

$$\frac{du}{dt} = u(a - bv), \ \tau\frac{dv}{dt} = v(-c + du). \tag{5.39}$$

It admits the unique stationary solution

$$(u_*, v_*) = (c/d, a/b),$$

and otherwise, the orbit becomes periodic in time. This fact follows because (5.39) has the first integral

$$a \log v - bv + c \log u - du = \text{constant}, \tag{5.40}$$

and (5.40) determines a Jordan curve in the first quadrant in uv plane except for the stationary solution (u_*, v_*).

To derive (5.40), we use the growth rates

$$\xi = \log u, \quad \eta = \log v \tag{5.41}$$

to write (5.39) as

$$\frac{d\xi}{dt} = a - be^{\eta}, \quad \frac{d\eta}{dt} = -\tau^{-1}c + \tau^{-1}de^{\xi} \tag{5.42}$$

([Latos–Suzuki–Yamada (2012)]). The right-hand side of (5.42) does not depend on the variables in the left-hand side explcitly, and therefore, it takes the form of a Hamilton system

$$\xi_t = -H_{\eta}, \quad \eta_t = H_{\xi} \tag{5.43}$$

with the Hamiltonian

$$H(\xi, \eta) = -a\eta + be^{\eta} - \tau^{-1}c\xi + \tau^{-1}de^{\xi}. \tag{5.44}$$

Then we obtain (5.40) by

$$\frac{dH}{dt} = 0, \tag{5.45}$$

and hence (5.40).

The spatially homogeneous part of the Gierer Meinhard system in morphogenesis is described by

$$\frac{du}{dt} = -u + \frac{u^p}{v^q}, \quad \tau\frac{dv}{dt} = -v + \frac{u^r}{v^s}. \tag{5.46}$$

It has the same property as that of the prey predator system (5.39), if

$$\tau = \frac{s+1}{p-1}. \tag{5.47}$$

Thus there is a unique stationary solution and the other orbit is periodic-in-time in the first quadrant in the uv plane. There is actually a first integral of (5.46) for (5.47). Here we derive this fact via the Hamilton formalism as in (5.43).

For this purpose we rewrite (5.46) as

$$u^{-p}(u_t + u) = v^{-q}, \quad v^s(v_t + \tau^{-1}v) = \tau^{-1}u^r. \tag{5.48}$$

Then, regarding the first terms of the left-hand side of (5.48), we put

$$\xi = \frac{u^{-p+1}}{p-1}, \quad \eta = \frac{v^{s+1}}{s+1},$$

to obtain

$$\xi_t = u^{-p+1} - v^{-q} = (p-1)\xi - \{(s+1)\eta\}^{-\frac{q}{s+1}}$$
$$\eta_t = -\frac{v^{s+1}}{\tau} + \frac{u^r}{\tau} = -\tau^{-1}(s+1)\eta + \tau^{-1}\{(p-1)\xi\}^{-\frac{r}{p-1}}. \tag{5.49}$$

The first terms of the right-hand side of (5.49) are linear in ξ and η, which are contained in the Hamiltonian under the assumption of (5.47). More precisely, system (5.49) is reduced to the Hamilton system

$$\frac{d\xi}{dt} = H_\eta, \quad \frac{d\eta}{dt} = -H_\xi$$

for

$$H(\xi, \eta) = (p-1)\xi\eta + \left(\frac{r}{p-1} - 1\right)^{-1} A(\xi) + \left(\frac{q}{s+1} - 1\right)^{-1} B(\eta)$$

and

$$A(\xi) = \tau^{-1}(p-1)^{-\frac{r}{p-1}}\xi^{1-\frac{r}{p-1}}, \quad B(\eta) = (s+1)^{-\frac{q}{s+1}}\eta^{1-\frac{q}{s+1}},$$

if (5.47) holds. Then we obtain (5.45) ([Karali–Suzuki–Yamada (2013)]).

There is a Lotka Volterra system with Poisson structure, differently from (5.39), that is, the three system

$$\tau_1 \frac{du_1}{dt} = (u_2 - u_3)u_1$$

$$\tau_2 \frac{du_2}{dt} = (u_3 - u_1)u_2$$

$$\tau_3 \frac{du_3}{dt} = (u_1 - u_2)u_3, \qquad (5.50)$$

reduced to

$$\tau_1 \frac{d\xi_1}{dt} = e^{\xi_2} - e^{\xi_3}$$

$$\tau_2 \frac{d\xi_2}{dt} = e^{\xi_3} - e^{\xi_1}$$

$$\tau_3 \frac{d\xi_3}{dt} = e^{\xi_1} - e^{\xi_2}, \qquad (5.51)$$

in the variables

$$\xi_i = \log u_i, \quad i = 1, 2, 3.$$

If $\times$ denotes the outer product in $\mathbf{R}^3$, this (5.51) means

$$\mathcal{M}\frac{d\xi}{dt} = H(\xi) \times a, \qquad (5.52)$$

where

$$\xi = \begin{pmatrix} \xi_1 \\ \xi_2 \\ \xi_3 \end{pmatrix}, \quad a = \begin{pmatrix} 1 \\ 1 \\ 1 \end{pmatrix}, \quad H(\xi) = \begin{pmatrix} e^{\xi_1} \\ e^{\xi_2} \\ e^{\xi_3} \end{pmatrix} \qquad (5.53)$$

and

$$\mathcal{M} = \text{diag}(\tau_1, \tau_2, \tau_3)$$

([Suzuki–Yamada (2013)]).

Equation (5.52) takes the form of the Hamilton system (5.24), and this Poisson structure induces two conservation laws of (5.52),

$$0 = \frac{d}{dt}\mathcal{M}\xi \cdot a = \frac{d}{dt}(\tau_1\xi_1 + \tau_2\xi_2 + \tau_3\xi_3)$$

$$0 = \frac{d}{dt}\mathcal{M}\xi \cdot H(\xi) = \frac{d}{dt}(\tau_1 e^{\xi_1} + \tau_2 e^{\xi_2} + \tau_3 e^{\xi_3}), \tag{5.54}$$

where

$$\mathcal{M}\xi \cdot a, \quad \mathcal{M}\xi \cdot H(\xi)$$

are regarded as mass and entropy of this system.

To conclude this section, we confirm fundamental facts on the reaction system with N-components,

$$\tau_j \frac{dv_j}{dt} = f_j(v), \quad v_j|_{t=0} = v_{j0} > 0, \quad 1 \le j \le N, \tag{5.55}$$

where

$$\tau = (\tau_j) \in \mathbf{R}_+^N, \quad \mathbf{R}_+ = (0, +\infty).$$

We assume that each

$$f_j : \mathbf{R}_+^N \to \mathbf{R}, \quad 1 \le j \le N$$

is locally Lipschitz continuous, provided with the *quasi-positivity*

$$f_j(u_1, \cdots, u_{j-1}, 0, u_{j+1}, \cdots, u_N) \ge 0, \quad u = (u_j) \ge 0, \ 1 \le j \le N. \tag{5.56}$$

Then, there is a solution

$$v = (v_j(t))$$

to (5.55) local in time, satisfying

$$v_j(t) > 0, \ 1 \le j \le N, \quad 0 \le t < T,$$

where $T = T_{\max} \in (0, +\infty]$ denote the maximal existence time by the theory of *invariant region* [Nagumo (1942)]. Hence each coordinate plane in $\mathbf{R}^N$ denoted by

$$v_j = 0 \quad 1 \le \exists j \le N$$

is an invariant set of (5.66). If $T < +\infty$, furthermore, it holds that

$$\lim_{t \uparrow T} |v(t)| = +\infty, \quad |v| = \sqrt{v_1^2 + \cdots + v_N^2}.$$

5.6   Reaction Diffusion Systems

Symplectic and Poisson structures of reaction systems in the previous section are useful for the study of reaction diffusion equation, for example,

$$u_t = \varepsilon^2 \Delta u + f(u,v), \quad \tau v_t = D\Delta v + g(u,v) \quad \text{in } \Omega \times (0,T)$$

$$\frac{\partial}{\partial \nu}(u,v)\Big|_{\partial\Omega} = 0, \quad (u,v)|_{t=0} = (u_0(x), v_0(x)) > 0 \qquad (5.57)$$

in the case of two components, where $\Omega \subset \mathbf{R}^n$ is a bounded domain with smooth boundary.

In the case of the prey predator system

$$f(u,v) = u(a - bv), \quad g(u,v) = v(-c + du),$$

the ODE part is reduced to (5.43) with (5.44), for (ξ, η) defined by (5.41). Then it follows that

$$\xi_t = \varepsilon^2 e^{-\xi}\Delta\xi - H_\eta, \quad \eta_t = \tau^{-1}De^{-\eta}\Delta e^\eta + H_\xi$$

$$\frac{\partial}{\partial \nu}(\xi,\eta)\Big|_{\partial\Omega} = 0, \quad (\xi,\eta)|_{t=0} = (\log u_0, \log v_0)$$

from (5.57), which implies

$$\frac{d}{dt}\int_\Omega H(\xi,\eta)\,dx = \int_\Omega \xi_t H_\xi + \eta_t H_\eta\,dx$$

$$= \int_\Omega (-\tau^{-1}c + \tau^{-1}de^\xi)\varepsilon^2 e^{-\xi}\Delta e^\xi + (-a + be^\eta)\tau^{-1}De^{-\eta}\Delta e^\eta dx$$

$$= -\tau^{-1}\int_\Omega c\varepsilon^2 e^{-\xi}\Delta e^\xi + aDe^{-\eta}\Delta e^\eta dx$$

by

$$\int_\Omega \Delta e^\xi dx = \int_{\partial\Omega}\frac{\partial e^\xi}{\partial \nu}ds = 0, \quad \int_\Omega \Delta e^\eta dx = \int_{\partial\Omega}\frac{\partial e^\eta}{\partial \nu}\,ds = 0.$$

We thus obtain

$$\frac{d}{dt}\int_\Omega H(\xi,\eta)\,dx = -\tau^{-1}\int_\Omega c\varepsilon^2|\nabla\xi|^2 + aD|\nabla\eta|^2\,dx \le 0. \qquad (5.58)$$

Inequality (5.58) induces a homogenization asymptotic-in-time of the solution to (5.57) ([Latos–Suzuki–Yamada (2012)]), and such a profile has been observed in the other systems of Gierer Meinhardt, (5.46) in §5.5, three systems (5.50) in §5.5, and public goods game model ([Karali–Suzuki–Yamada (2013); Suzuki–Yamada (2013); Fellner–Latos–Suzuki (2019)]).

General reaction diffusion system takes the form

$$\tau_j \frac{\partial u_j}{\partial t} - d_j \Delta u_j = f_j(u) \quad \text{in } \Omega \times (0, T)$$

$$\left. \frac{\partial u_j}{\partial \nu} \right|_{\partial\Omega} = 0, \quad u_j|_{t=0} = u_{j0}(x) > 0, \tag{5.59}$$

where $\Omega \subset \mathbf{R}^n$ is a bounded domain with smooth boundary $\partial\Omega$,

$$\tau = (\tau_j) > 0, \ d = (d_j) > 0, \quad 1 \leq j \leq N,$$

are constants, and

$$u_{j0} = u_{j0}(x) \in C(\overline{\Omega}).$$

Let

$$f_j = f_j(u) : \mathbf{R}^N \to \mathbf{R}, \quad u = (u_j),$$

be locally Lipschitz continuous, to guarantee local-in-time well-posedness of the system. Let $T \in (0, +\infty]$ be the maximal existence time of the classical solution. Assuming quasi-positivity, (5.56), we reach

$$u = (u_j(x, t)) > 0$$

by the theory of invariant region and the strong maximum principle [Chueh–Conley–Smoller (1977)].

Henceforth, we assume these conditions together with mass dissipation

$$\sum_j f_j(u) \leq 0, \quad u = (u_j) \geq 0. \tag{5.60}$$

Inequality (5.60) ensures the total mass control

$$\|\tau \cdot u(\cdot, t)\|_1 \leq \|\tau \cdot u_0\|_1, \quad 0 \leq t < T.$$

As in the Lotka Volterra system, fundamental reactions are mostly modeled by quadratic nonlinearity using the law of mass action. Under the quadratic growth,

$$|\nabla f_j(u)| \leq C(1 + |u|), \ 1 \leq j \leq N, \tag{5.61}$$

Similarly to the Smoluchowski Poisson equation, then, the critical dimension for the uniform boundedness of the solution

$$T = +\infty, \ \|u(\cdot, t)\|_\infty \leq C \tag{5.62}$$

is thought to be $n = 2$. We have, actually, (5.62) for $n = 1$ by [Masuda–Takahashi (1994)].

If $n = 2$, we have actually ε-regularity as in Lemma 3.1 in §3.6, concerning the initial total mass

$$M = \|\tau \cdot u_0\|_1$$

for (5.62). The principal nonlinearity of the Lotka Volterra system, furthermore, is provided with the scaling invariance:

$$u_j^\mu(x,t) = \mu^2 u_j(\mu x, \mu^2 t), \ 1 \le j \le N.$$

Using these properties, we obtain (5.62) for $n = 2$ by [Suzuki–Yamada (2015)]. Monotonicity formula is also used there, to assure the formation of collapse, which results in a contradiction under the existence of the solution blowup in finite or infinite time.

Several arguments based on the pointwise estimate, however, are valid to (5.59) other than the ones used for the Smoluchowski Poisson equation. Among them is the L^2 estimate in space and time derived from a duality argument [Pierre (2010)].

Lemma 5.1. *If* $0 \le u = (u_j(x,t))$ *is smooth in* $\overline{\Omega} \times [0, T)$ *and satisfies*

$$\frac{\partial}{\partial t}(\tau \cdot u) - \Delta(d \cdot u) \le 0, \ in \ \Omega \times (0, T), \qquad \left.\frac{\partial}{\partial \nu}(d \cdot u)\right|_{\partial \Omega} \le 0, \qquad (5.63)$$

then it follows that

$$\|u\|_{L^2(Q_T)} \le CT^{1/2}\|u_0\|_2, \qquad u|_{t=0} = u_0. \qquad (5.64)$$

Proof. Let $u_0 = u|_{t=0}$. By (5.63) we have

$$\tau \cdot u(\cdot, t) - \tau \cdot u_0 \le \int_0^t \Delta(d \cdot u(\cdot, s)) \ ds,$$

and hence

$$(\tau \cdot u(\cdot, t), d \cdot u(\cdot, t)) - (\tau \cdot u_0, d \cdot u(\cdot, t))$$

$$\le -(\nabla d \cdot u(\cdot, t), \nabla \int_0^t d \cdot u(\cdot, s) \ ds)$$

$$= -\frac{1}{2}\frac{d}{dt}\|\nabla \int_0^t d \cdot u(\cdot, s) \ ds\|_2^2, \qquad (5.65)$$

where (,) denotes the L^2-inner product. Integration of (5.65) over $(0, T)$ implies

$$\int_0^T (\tau \cdot u(\cdot, t), d \cdot u(\cdot, t)) \ dt \le \|\tau \cdot u_0\|_2 \cdot \int_0^T \|d \cdot u(\cdot, t)\|_2 \ dt$$

$$\le T^{1/2}\|\tau \cdot u_0\|_2 \cdot \left(\int_0^T \|d \cdot u(\cdot, t)\|_2^2 \ dt\right)^{1/2},$$

and hence (5.64) holds by $u = (u_j(\cdot, t)) \ge 0$. $\qquad \square$

In [Suzuki–Yamada (2015)] it is shown that inequalities (5.61) and (5.62) imply a slightly stronger estimate on the growth rate of $f = (f_j(u))$,

$$\sum_j f_j(u) \log u_j \leq C(1 + |u|^2),$$

which implies

$$\frac{d}{dt} \int_\Omega u_j (\log u_j - 1) \, dx \leq C \int_\Omega 1 + |u|^2 \, dx, \ 1 \leq j \leq N,$$

and hence $T = +\infty$ for $n = 2$ by (5.64), as in the Smoluchowski Poisson equation.

The other application of Lemma 5.1 is the following theorem.

Theorem 5.5 ([Pierre–Suzuki–Yamada (2019)]). *If (5.56), (5.60), and (5.61) hold, there exists a weak solution global in time to (5.59) with the orbit*

$$\mathcal{O} = \{u(t)\}$$

pre-compact in $L^1(\Omega)$ for any n.

Several estimates on the parabolic regularity, particularly, regularity interpolation of [Kanel (1990)] refines Theorem 5.5 as follows, of which proof is not described here.

Theorem 5.6 ([Fellner–Morgan–Tang (2021)]). *Under the assumption of the previous theorem, system (5.59) has a unique uniformly bounded global-in-time classical solution for any n.*

5.7 Lotka Volterra Systems with Skew-Symmetry

The Lotka Volterra system is formulated by

$$\tau_j \frac{dv_j}{dt} = (-e_j + \sum_{k=1}^N a_{jk} v_k) v_j, \quad v_j|_{t=0} = v_{j0} > 0, \quad 1 \leq j \leq N, \quad (5.66)$$

where $v_j = v_j(t) > 0$ stands for the population of the j-th species, $A = (a_{jk})$ is a square matrix of order N, and $\tau_j > 0$, $e_j \in \mathbf{R}$ are constants. Since Lotka [Lotka (1925)] and Volterra [Volterra (1926)], this model has been used in several areas including ecology, economics, and chemistry, and has been studied extensively.

Any solution $v = (v_j(t))$ to (5.66) remains in the positive cone

$$\mathbf{R}_+^N \equiv \{(v_j) \in \mathbf{R}^N \mid v_j > 0, 1 \leq \forall j \leq N\}$$

as far as it exists, because of the quasi-positivity of the model. One of major questions in population dynamics, then, is whether some species go to extinction or not. Here, extinction means approaching zero as $t \to \infty$. Finding criteria for the long term coexistence of multiple species, referred to as *permanence* or uniform persistence, is an important issue.

Finding periodic-in-time orbits is one approach to the quenstion of permanence. The prey predator system in the previous section is a typical case that any orbit is periodic in time except for the unique equilibrium. This structure leads to asymptotic spatial homogenization corresponding to a partial differential equation provided with a diffusion term [Pierre–Suzuki–Yamada (2019)].

In §5.7–§5.9 we formulate a class of system (5.66) provided with such a property via conserved quantities. This class is explicitly given by a set of algebraic conditions on $A = (a_{jk})$ and $e = (e_j)$. If the system takes N components, we have $2N - 3$ and $2N - 1$ degrees of freedom without and with linear terms, respectively.

We say that the Lotka Volterra system (5.66) is provided with the *property* (P) if it satisfies the following conditions:

$(P1)$ The set of stationary solutions (equilibria),

$$E = \{v = (v_j) \in \mathbf{R}_+^N \mid \sum_{k=1}^N a_{jk} v_k = e_j, \ 1 \leq j \leq N\},$$

is the intersection of an affine space of co-dimension 2 denoted by L and the positive cone $\mathbf{R}_+^N$,

$$E = L \cap \mathbf{R}_+^N.$$

$(P2)$ Any non-stationary solution is periodic in time, with the orbit

$$\mathcal{O} \cong S^1$$

contractible to an equilibrium, that is, a point belonging to E, in $\mathbf{R}_+^N \setminus E$.

$(P3)$ Any two distinct orbits $\mathcal{O}_1, \mathcal{O}_2 \cong S^1$ do not link in $\mathbf{R}_+^N$.

Henceforth, we assume the following conditions for the matrix $A = (a_{jk})$:

$(a1)$ A is irreducible.

(*a2*) A is skew-symmetric,

$$A^T + A = 0. \tag{5.67}$$

(*a3*) A has both positive and negative components in any row.

We recall that the square matrix A is called reducible if there is a permutation matrix P such that

$$P^T A P = \begin{pmatrix} A_{11} & A_{12} \\ 0 & A_{22} \end{pmatrix}, \tag{5.68}$$

where A_{11} and A_{22} are non-trivial square matrices. Since A is skew symmetric, it holds that $A_{11}^T = -A_{11}$, $A_{22}^T = A_{22}$, and $A_{12} = 0$ in (5.68). If A is not reducible, it is called *irreducible*, which means that system (5.66) does not have any non-trivial proper sub-systems.

Here we distinguish the cases of $e = 0$ and $e \neq 0$, where $e = (e_j)$.

Theorem 5.7 ([Kobayashi–Suzuki–Yamada (2019)]). *Let* $A = (a_{jk})$ *satisfy (a1)–(a3). Let* $e = (e_j) = 0$, $N \geq 3$, *and*

$$a_{ij}a_{kl} + a_{il}a_{jk} - a_{ik}a_{jl} = 0, \quad \forall i, j, k, l \in \{1, \ldots, N\}. \tag{5.69}$$

Then, system (5.66) is provided with property (P).

Condition (5.69) is void if $N = 3$. Therefore, when $N = 3$, $e_1 = e_2 = e_3 = 0$, and

$$A = \begin{pmatrix} 0 & c_3 & -c_2 \\ -c_3 & 0 & c_1 \\ c_2 & -c_1 & 0 \end{pmatrix}, \quad c_1, c_2, c_3 \text{ have the same sign}, \tag{5.70}$$

any non-stationary solution to (5.66) is periodic in time. The set of equilibria E is a half line in $\mathbf{R}_+^3$, any non-stationary orbit $\mathcal{O}$ is homeomorphic to S^1, and any distinct two non-stationary orbits $\mathcal{O}_1$ and $\mathcal{O}_2$ do not link in $\mathbf{R}_+^3$.

Let $N \geq 4$ and $M_N(\mathbf{R})$ be the set of (N, N) matrices of real components. The set of skew-symmetric matrices,

$$X = \{A \in M_N(\mathbf{R}) \mid A^T = -A\}, \tag{5.71}$$

is identified with $\mathbf{R}^{N(N-1)/2}$. Second, if i, j, k, l are not distinct, equation (5.69) for $A = (a_{ij}) \in X$ is obvious. Third, condition (5.69) is invariant under the change of order on i, j, k, l for $A = (a_{jk}) \in X$. Therefore, system (5.69) for $N \geq 4$ is reduced to that of $_N C_4$ relations

$$a_{ij}a_{kl} + a_{il}a_{jk} - a_{ik}a_{jl} = 0, \quad 1 \leq i < j < k < l \leq N. \tag{5.72}$$

The set of irreducible skew-symmetric matrices satisfying (5.72), denoted by Y, however, is identified with a dense set in $\mathbf{R}^{2N-3}$.

To see this property, let

$$a_{12} \neq 0$$

without loss of generality. Then the entries $a_{34}, \ldots, a_{N-1N}$ of $A = (a_{jk})$ are represented by $a_{12}, \ldots, a_{2N}$ from the first $(N-2)(N-3)/2$ relations of (5.72), that is,

$$a_{kl} = \frac{a_{1k}a_{2l} - a_{1l}a_{2k}}{a_{12}}, \quad 3 \leq k < l \leq N. \tag{5.73}$$

Then the rest of relations in (5.72) are satisfied for a_{kl} defined above. In other words, (5.72) for $i = 1$, $j = 2$, $3 \leq k < l \leq N$, assures all the other relations of (5.72), provided that $a_{12} \neq 0$. Since irreducibility is a generic property of matrices, the above Y is identified with a dense set in $\mathbf{R}^{2N-3}$.

We also note that if

$$\operatorname{sign}(a_{ij}) = (-1)^{i+j}$$

is the case for $i = 1, 2$ and $i < j \leq N$, the matrix $A = (a_{jk}) \in Y$ satisfies the sign condition, $(a3)$. Thus, there are $(2N - 3)$ degrees of freedom for system (5.66) with $e = (e_j) = 0$, provided with property (P).

The last condition, $(a3)$, is associated with the persistence of (5.66). If property (P) does not arise to (5.66) in spite of

$$A = (a_{jk}) \in Y, \quad e = (e_j) = 0, \tag{5.74}$$

any non-stationary solution $v = (v_j(t))$ is global in time but satisfies

$$\lim_{t \uparrow +\infty} v_j(t) = 0, \quad 1 \leq \exists j \leq N. \tag{5.75}$$

In particular, any $v_* \in E$ is unstable (Remark 5.3 in §5.8). A simple example of this case is $N = 3$ of (5.70), with the sign condition of c_1, c_2, c_3 violated.

If

$$(Au, u) \leq 0, \ u = (u_j) \geq 0, \quad e = (e_j) \geq 0$$

holds in (5.66) for $A = (a_{ij})$, the associated reaction diffusion system

$$\tau_j u_{jt} = d_j \Delta u_j + \left(-e_j + \sum_{k=1}^{N} a_{jk} u_k\right) u_j \text{ in } \Omega \times (0, T)$$

$$\left.\frac{\partial u_j}{\partial \nu}\right|_{\partial \Omega} = 0, \ u_j|_{t=0} = u_{j0}(x) > 0, \quad 1 \leq j \leq N \tag{5.76}$$

has a uniformly bounded classical solution global in time by Theorem 5.6, where $\Omega \subset \mathbf{R}^n$ a bounded domain with smooth boundary and ν is the unit outer normal vector. Under the assumption of Theorem 5.7, there is

$$b \in \mathbf{R}^N \setminus \{0\}$$

such that (5.83) in §5.8. Then, Theorem 1.4 of [Suzuki–Yamada (2015)] gurantees that the ω-limit set of the solution is contained in $\overline{\mathbf{R}}_+^N$. Hence it is homeomorphic to S^1 or a singleton in this case of $A = (a_{ij})$ and

$$e = (e_j) = 0.$$

We turn to the case $e = (e_j) \neq 0$, which includes the prey predator system, that is, $N = 2$, and

$$e_1 \cdot e_2 < 0, \quad a_{12} \cdot a_{21} < 0.$$

This system is provided with property (P). More precisely, there is a unique equilibrium $v_* \in \mathbf{R}_+^2$, and the total set of orbits $\mathcal{F}$, which constitutes of a foliation, is composed of curves homeomorphic to S^1 and the equilibrium v_*. A general form of (5.66) provided with this property is given by the following theorem.

Theorem 5.8 ([Kobayashi–Suzuki–Yamada (2019)]). *Let $N \geq 3$, $A = (a_{jk})$ satisfy (a1)–(a3), and $e = (e_j)$ have both positive and negative components. Assume, furthermore,*

$$a_{jk}e_i - a_{ik}e_j + a_{ij}e_k = 0, \quad \forall i, j, k \in \{1, \ldots, N\}. \tag{5.77}$$

Then, system (5.66) is provided with property (P).

Condition (5.77) concerning $A = (a_{jk}) \in X$ and $e = (e_j) \in \mathbf{R}^N$ has the following reduction similar to (5.69). First, this condition is obvious if i, j, k are not distinct. Second, this relation is invariant under the change of orders of i, j, k. Hence (5.77) is reduced to the system of $_NC_3$ relations

$$a_{jk}e_i - a_{ik}e_j + a_{ij}e_k = 0, \quad 1 \leq i < j < k \leq N. \tag{5.78}$$

Finally, the degree of freedom of $\{a_{ij}, e_k\}$ satisfying (5.78) is $(2N - 1)$.

To see the last fact, we assume $a_{1N} \cdot e_1 \neq 0$ by (1.38) without loss of generality, and define $a_{23}, \ldots, a_{2N}$ by $a_{12}, \ldots, a_{1N}$ and $e_1, \ldots, e_N$, using the first $(N - 1)(N - 2)/2$ relations of (5.78),

$$a_{jk} = \frac{a_{1k}e_j - a_{1j}e_k}{e_1}, \quad 2 \leq j < k \leq N. \tag{5.79}$$

Then the other relations of (5.78) are satisfied automatically.

Let $\hat{X}$ be the set of systems (5.66) with (5.67). This time, both

$$A = (a_{ij}), \quad e = (e_k)$$

are parameters, and the set $\hat{X}$ is identified with $\mathbf{R}^{N(N+1)/2}$. Let $\hat{Y}$ be the set of $\{a_{ij}, e_k\}$ satisfying $A = (a_{ij}) \in X$ and (5.77). By the above reduction, this $\hat{Y}$ is identified with $\mathbf{R}^{2N-1}$. The irreducibility of A is, again, a generic property in $\hat{Y}$.

To detect a class of

$$\{a_{ij}, e_k\} \in \hat{Y}$$

with $A = (a_{ij})$ satisfying (*a3*), let

$$\mathrm{sign}(a_{1j}) = (-1)^j, \ 2 \leq j \leq N, \quad \mathrm{sign}(e_k) = (-1)^k, \ 3 \leq k \leq N,$$

and $e_1 > 0 > e_2$. Then it holds that

$$\mathrm{sign}(a_{2j}) = (-1)^{j+1}, \quad 3 \leq j \leq N,$$

and therefore, $A = (a_{ij})$ actually satisfies (*a3*). Thus, there are $(2N-1)$ degrees of freedom of $\{a_{ij}, e_k\}$ such that (5.66) is provided with property (P).

The proof of Theorems 5.7 and 5.8 relies on the facts on the sign of the entries of the skew-symmetric matrix A.

Lemma 5.2. *If $N \geq 4$, one of the following cases arises for $A = (a_{jk})$ satisfying (a2) and (a3):*

Case 1. *There exist distinct $1 \leq i, j, k \leq N$ such that $a_{ij} > 0$, $a_{ik} < 0$, $a_{jk} > 0$.*

Case 2. *There exist distinct $1 \leq i, j, k, l \leq N$ such that $a_{ij} > 0$, $a_{ik} < 0$, $a_{il} = 0$, $a_{jk} = 0$, $a_{jl} > 0$, $a_{kl} < 0$.*

Lemma 5.3. *Under the assumptions of Theorem 5.8, either the first or the second case of Lemma 5.2 arises with $e_i \neq 0$:*

Case 1. *There exist distinct $1 \leq i, j, k \leq N$ such that $a_{ij} > 0$, $a_{ik} < 0$, $a_{jk} > 0$, and $e_i \neq 0$.*

Case 2. *There exist distinct $1 \leq i, j, k, l \leq N$ such that $a_{ij} > 0$, $a_{ik} < 0$, $a_{il} = 0$, $a_{jk} = 0$, $a_{jl} > 0$, $a_{kl} < 0$, and $e_i \neq 0$.*

If the matrix A is given, we can actually check the above sign conditions of its entries in Lemma 5.2 and Lemma 5.3. The point here is that these conditions are the consequence of the assumption of Theorem 5.7 or Theorem 5.8. So far, an elementary but complicated proof of these lemmas is given [Kobayashi–Suzuki–Yamada (2019)].

5.8 The Case without Linear Term

To show Theorem 5.7 in §5.7, let $N \geq 3$ and $e = (e_j) = 0$ in (5.66):

$$\tau_j \frac{dv_j}{dt} = \sum_{k=1}^{N} a_{jk} v_k v_j, \quad v_j|_{t=0} = v_{j0} > 0, \quad 1 \leq j \leq N. \tag{5.80}$$

First, equality (5.67) implies the total mass conservation

$$\frac{dM}{dt} = 0, \tag{5.81}$$

where $M = \tau \cdot v$. Second, by $v = (v_j) > 0$ and (5.81) we obtain $T = +\infty$ with

$$\sup_{t \geq 0} \max_{1 \leq j \leq N} v_j(t) < +\infty. \tag{5.82}$$

Here and henceforth, $v = (v_j) > 0$ indicates

$$v_j > 0, \quad 1 \leq \forall j \leq N.$$

Let

$$\mathcal{O} = \{v(t)\} \subset \mathbf{R}_+^N$$

be the corresponding orbit.

Boltzmann's H-function or *entropy*, on the other hand, is defined under the assumption that

$$\exists b \in \mathbf{R}^N \setminus \{0\}, \quad \tilde{A}^T b = 0, \quad \tilde{A} = (\tau_j^{-1} a_{jk}), \tag{5.83}$$

that is,

$$H = b \cdot \log v = \sum_{j=1}^{N} b_j \log v_j,$$

where $b = (b_j)$ and

$$\log v = (\log v_j), \quad v = (v_j) \in \mathbf{R}_+^N.$$

Thus we obtain

$$\frac{dH}{dt} = \sum_{j=1}^{N} b_j v_j^{-1} \frac{dv_j}{dt} = \sum_{j,k=1}^{N} b_j \tau_j^{-1} a_{jk} v_k = 0$$

by (5.83) for the solution $v = (v_j(t))$ to (5.66). This H actually induces an increasing quantity for the reaction diffusion system (5.76).

We thus obtain the following lemma.

Lemma 5.4. *Assume (5.67) and the existence of linearly independent* r *vectors*

$$\vec{b}_i \in Ker\, \tilde{A}^T, \quad 1 \le i \le r.$$

Put $v_0 = (v_{j0}) > 0$ *and*

$$\beta_0 = \tau \cdot v_0, \quad \beta_i = \vec{b}_i \cdot \log v_0, \ 1 \le i \le r.$$

Then it holds that $\mathcal{O} \subset \mathcal{O}_*$, *where*

$$\mathcal{O}_* = \{v \in \mathbf{R}_+^N \,|\, | \ \tau \cdot v = \beta_0, \quad \vec{b}_i \cdot \log v = \beta_i, \ 1 \le i \le r\}. \qquad (5.84)$$

Now we show the following lemma.

Lemma 5.5. *If* $r = N - 2$ *and*

$$\vec{b}_i > 0, \quad 1 \le \exists i \le r$$

in Lemma 5.4, the set $\mathcal{O}_*$ *defined by (5.84) is a Jordan curve or singleton.*

Proof. Under the transformation $\xi = \log v \in \mathbf{R}^N$, the set $\mathcal{O}_*$ is homeomorphic to

$$\tilde{\mathcal{O}}_* = \{\xi \in \mathbf{R}^N \mid \tau \cdot e^\xi = \beta_0, \quad \vec{b}_i \cdot \xi = \beta_i, \ 1 \le i \le N - 2\}. \qquad (5.85)$$

Let

$$M = \{\xi \in \mathbf{R}^N \mid \tau \cdot e^\xi \le \beta_0\}.$$

This set is strictly convex, and $\tilde{\mathcal{O}}_*$ is the intersection of its boundary

$$\partial M = \{\xi \in \mathbf{R}^N \mid \tau \cdot e^\xi = \beta_0\}$$

and linearly independent $(N - 2)$ hyperplanes

$$P_i = \{\xi \in \mathbf{R}^N \mid \vec{b}_i \cdot \xi = \beta_i\}, \quad 1 \le i \le N - 2.$$

Hence $\tilde{\mathcal{O}}_* \subset \mathbf{R}^N$ is a non-intersecting curve or singleton.

Now we show the boundedness of $\tilde{\mathcal{O}}_*$. First, there arises

$$\sup_{1 \le j \le N} \xi_j \le C, \quad \forall \xi = (\xi_j) \in M.$$

Second, since $b^i > 0$ for some $1 \le i \le N - 2$ we have

$$\inf_{1 \le j \le N} \xi_j \ge -C, \quad \forall \xi = (\xi_j) \in M \cap P_i.$$

Therefore, $\tilde{\mathcal{O}}_* \subset \mathbf{R}^N$ is bounded. We see that $\mathcal{O}_*$ is a Jordan curve or singleton by the inverse transformation

$$v_j = e^{\xi_j}, \quad 1 \le j \le N,$$

and then the proof is complete. $\qquad \square$

Remark 5.3. If $\tilde{\mathcal{O}}_*$ is not bounded, the set $\mathcal{O}_*$ is an open curve in $\mathbf{R}_+^N$. Then any non-stationary solution $v = (v_j(t)) > 0$ satisfies (5.75) by (5.82), because any stationary solution does not lie on $\mathcal{O}_*$ by Lemma 5.7 below.

Now we show the following lemma, using Lemma 5.2.

Lemma 5.6. *The requirements of Lemma 5.5 are fulfilled under the assumption of Theorem 5.7.*

Proof. If $N = 3$ the assumption implies (5.70). Since the relation

$$b = \begin{pmatrix} b_1 \\ b_2 \\ b_3 \end{pmatrix} \in \text{Ker } \tilde{A}^T, \quad b_1 b_2 b_3 \neq 0$$

means

$$\frac{b_3}{b_2} = \frac{c_3}{c_2} \cdot \frac{\tau_3}{\tau_2}, \quad \frac{b_1}{b_3} = \frac{c_1}{c_3} \cdot \frac{\tau_1}{\tau_3}, \quad \frac{b_2}{b_1} = \frac{c_2}{c_1} \cdot \frac{\tau_2}{\tau_1},$$

we can take

$$0 < b = (b_1, b_2, b_3)^T \in \text{Ker } \tilde{A}^T.$$

Let $N \geq 4$. First, we examine the condition for the existence of linearly independent $(N-2)$ vectors in Ker $\tilde{A}^T$, denoted by

$$\vec{b_i} = (b_i^j) \in \text{Ker } \tilde{A}^T, \quad 1 \leq i \leq N - 2.$$

This condition is equivalent to the linear independence of

$$\vec{B_i} = (B_i^j), \ 1 \leq i \leq N - 2, \quad B_i^j = \tau_j^{-1} b_i^j,$$

which follows from rank $A = 2$.

To confirm this property under the assumption of Theorem 5.7, let $A = [\vec{a_1} \cdots \vec{a_N}]$. First, it holds that

$$\vec{a_k} = (a_{1k}, \cdots, a_{Nk})^T \neq 0, \quad 1 \leq k \leq N$$

because A is irreducible. Second, since A is skew-symmetric, the vectors

$$\{\vec{a_k}, \vec{a_l}\}, \quad k \neq l,$$

are linearly independent if

$$a_{kl} = -a_{lk} \neq 0.$$

To see this property, we assume

$$c_1 \vec{a_k} + c_2 \vec{a_l} = 0, \quad c_1, c_2 \in \mathbf{R},$$

to deduce

$$c_1 a_{jk} + c_2 a_{jl} = 0, \quad 1 \le j \le N.$$

Then putting $j = l$ and $j = k$, it follows that $c_2 = c_1 = 0$.

For such k, l, which actually exists, condition (5.69) implies

$$\vec{a}_i - \frac{a_{il}}{a_{kl}} \vec{a}_k - \frac{a_{ik}}{a_{lk}} \vec{a}_l = 0, \quad \forall i \ne k, l, \tag{5.86}$$

and hence rank $A = 2$. Thus we have actually linearly indepent $(N - 2)$ vectors in Ker $\tilde{A}^T$. Now we shall show the existence of $0 < b \in \mathrm{Ker}\ \tilde{A}^T$, or equivalently, that of $0 < B \in \mathrm{Ker}\ A^T = \mathrm{Ker}\ A$.

In fact, in the first case of Lemma 5.2 we have $1 \le i, j, k \le N$ such that

$$a_{ij} > 0, \ a_{ik} < 0, \ a_{jk} > 0.$$

Here, the condition $\vec{B} = (B^1, \cdots, B^N)^T \in \mathrm{Ker}\ A^T$ means

$$\sum_{n=1}^{N} a_{ln} B^n = 0, \quad 1 \le l \le N, \tag{5.87}$$

and hence

$$B^k = -\frac{1}{a_{jk}} \sum_{n \ne j,k} a_{jn} B^n,$$

or

$$B^j = \frac{1}{a_{jk}} \sum_{n \ne j,k} a_{kn} B^n.$$

Therefore, the space Ker A^T is generated by $\vec{B}_l = (B_l^n)$, $l \ne j, k$, where

$$B_l^n = \begin{cases} 1, & n = l \\ 0, & n \ne l, j, k \\ \frac{1}{a_{jk}} a_{kl}, & n = j \\ -\frac{1}{a_{jk}} a_{jl}, & n = k \end{cases}$$

Consequently, we obtain Ker $A^T = \langle \vec{c}_l \mid l \ne j, k \rangle$ for

$$\vec{c}_l = (c_l^1, \cdots, c_l^N)^T \in \mathbf{R}^N$$

defined by

$$c_l^n = \begin{cases} a_{jk}, & n = l \\ a_{kl}, & n = j \\ -a_{jl}, & n = k \\ 0, & n \ne l, j, k. \end{cases} \tag{5.88}$$

In particular, $\vec{c}_i$ is a vector of which i-th, j-th, and k-th components are positive and the others are zero.

We obtain, on the other hand,

$$\sum_{l \neq i,j,k} \vec{c}_l = (\tilde{c}^1, \cdots, \tilde{c}^N)^T$$

with

$$\tilde{c}^n = \sum_{l \neq i,j,k} c_l^n = a_{jk} > 0, \quad n \neq i,j,k$$

by (5.88), and hence

$$0 < \vec{C} \equiv \sum_{l \neq i,j,k} \vec{c}_l + s\vec{c}_i \in \mathrm{Ker}\, A^T, \quad s \gg 1.$$

We turn to the second case of Lemma 5.2. First, such $1 \leq i,j,k,l \leq N$ are distinct. Second, the condition $\vec{B} = (B^1, \cdots, B^N) \in \mathrm{Ker}\, A^T$ means

$$B^j = -\frac{1}{a_{ij}} \sum_{n \neq i,j} a_{in} B^n$$

or

$$B^i = \frac{1}{a_{ij}} \sum_{n \neq i,j} a_{jn} B_n.$$

by (5.87). Then there arises that

$$\mathrm{Ker}\, A^T = \langle \vec{c_m} \mid m \neq i,j \rangle$$

for $\vec{c_m} = (c_m^1, \cdots, c_m^N)^T \in \mathbf{R}^N$ defined by

$$c_m^n = \begin{cases} a_{ij}, & n = m \\ a_{jm}, & n = i \\ -a_{im}, & n = j \\ 0, & n \neq i,j,m \end{cases} \tag{5.89}$$

We observe that $\vec{c_k}$ is a vector of which the j-th and k-th components are positive and the others are zero. Also, $\vec{c_l}$ is a vector of which the i-th and l-th components are positive and the others are zero.

Since $\tilde{c}_m = a_{ij} > 0$, $m \neq k,l$, follows for

$$\sum_{m \neq i,j,k,l} \vec{c_m} = (\tilde{c}_1, \cdots, \tilde{c}_N)^T$$

from (5.89), we obtain

$$0 < \vec{C} \equiv \sum_{m \neq i,j,k,l} \vec{c_m} + s(\vec{c_k} + \vec{c_l}) \in \mathrm{Ker}\, A^T, \quad s \gg 1,$$

and the proof is complete. $\qquad\qquad\qquad\qquad\qquad\qquad\qquad\qquad \square$

Remark 5.4. In (5.86), the pair $(c_1, c_2) = (a_{il}/a_{kl}, a_{ik}/a_{lk})$ is the unique choice for

$$\vec{a_i} - c_1 \vec{a_k} - c_2 \vec{a_l} = 0$$

to hold, regarding the l-th and the k-th components on the left-hand side. Consequently, the condition

$$a_{ij} a_{kl} + a_{il} a_{jk} - a_{ik} a_{jl} = 0, \ 1 \le \forall i, j, k, l \le N, \quad a_{kl} \ne 0, \qquad (5.90)$$

slightly weaker than (5.69), is sufficient for rank $A = 2$ to hold, if

$$A = [\vec{a_1} \cdots \vec{a_N}], \quad \vec{a_\ell} \ne 0, \ 1 \le l \le N,$$

is skew-symmetric. Condition (5.90) is also necessary for rank $A = 2$, if, furthermore, $\{\vec{a_k}, \vec{a_l}\}$ are linearly dependent whenever $a_{kl} = 0$. By this reason it is natural that the solution set to (5.69) has $(2N - 3)$ degrees of freedom, because two vectors

$$\vec{a_k}, \vec{a_l}, \quad k \ne l,$$

can be free in $A = [\vec{a_1} \cdots \vec{a_N}]$ if rank $A = 2$.

The following lemma excludes both hetero-clinic and homo-clinic orbits of (5.66) in §5.7 with $e = (e_j) = 0$, when $A = (a_{ij})$ satisfies $(a1)$, $(a2)$, and (5.69). Recall the set of skew-symmetric matirices X defined by (5.71).

Lemma 5.7. *Let $A \in X$ satisfy rank $A = 2$ and*

$$\vec{b_i} \in Ker \ \tilde{A}^T, \quad 1 \le i \le N - 2,$$

be linearly independent $(N - 2)$ vectors. Then it holds that

$$\mathcal{O}_* \cap Ker \ A \ne \emptyset \quad \Rightarrow \quad \sharp \mathcal{O}_* = 1,$$

where $\mathcal{O}_*$ is the set defined by (5.84) with $r = N - 2$.

Proof. Recall $\tilde{\mathcal{O}}_*$, M, and P_i used in the proof of Lemma 5.5. Let $\mathcal{M} = \partial M$ and

$$\tau e^\xi = (\tau_j e^{\xi_j}), \quad \tau = (\tau_j), \quad \xi = (\xi_j).$$

First, from

$$\mathcal{M} = \{\xi \in \mathbf{R}^N \mid \tau \cdot e^\xi = \beta_0\},$$

it follows that

$$T_\xi \mathcal{M} = \langle \tau e^\xi \rangle^\perp \equiv \{\eta \in \mathbf{R}^N \mid \tau e^\xi \cdot \eta = 0\}, \quad \xi \in \mathcal{M}. \qquad (5.91)$$

Second, it holds that

$$\text{Ker } A = \text{Ker } A^T = \langle \tau^{-1}\vec{b_i} \mid 1 \leq i \leq N-2 \rangle,$$

where

$$\tau^{-1}\vec{b_i} = (\tau_j^{-1}b_i^j), \quad \vec{b_i} = (b_i^j).$$

Each $v \in \mathcal{O}_* \cap \text{Ker } A$ is associated with $\xi \in \tilde{\mathcal{O}}_*$ by

$$v = e^\xi.$$

Then, there arises that

$$\tau e^\xi \in \langle \vec{b_i} \mid 1 \leq i \leq N-2 \rangle. \tag{5.92}$$

We have, on the other hand,

$$P \equiv \{\zeta \in \mathbf{R}^N \mid \vec{b_i} \cdot \zeta = \beta_i, \ 1 \leq i \leq N-2\}$$
$$= \{\xi\} + \bigcap_{i=1}^{N-2} \{\vec{b_i}\}^\perp = \{\xi\} + \langle \vec{b_i} \mid 1 \leq i \leq N-2 \rangle^\perp \tag{5.93}$$

by

$$\vec{b_i} \cdot \log v = \beta_i, \quad 1 \leq i \leq N-2,$$

and therefore,

$$(T_\xi \mathcal{M})^\perp = \langle \tau e^\xi \rangle \subset \langle \vec{b_i} \mid 1 \leq i \leq N-2 \rangle = (P - \{\xi\})^\perp$$

by (5.91), (5.92), and (5.93). We thus end up with

$$P \subset T_\xi \mathcal{M} + \{\xi\} \tag{5.94}$$

The inclusion (5.94) implies

$$\tilde{\mathcal{O}}_* = P \cap \mathcal{M} \subset T_\xi \mathcal{M} + \{\xi\}. \tag{5.95}$$

Equation (5.95) implies $\tilde{\mathcal{O}}_* = \{\xi\}$ because $\mathcal{M}$ is the boundary of the strictly convex set

$$M = \{\xi \in \mathbf{R}^N \mid \tau \cdot e^\xi \leq \beta_0\}.$$

$\square$

Now we give the following proof.

Proof of Theorem 5.7. By Lemma 5.4, Lemma 5.5, and Lemma 5.6, it holds that $\mathcal{O} \subset \mathcal{O}_*$, where $\mathcal{O}_*$ is a Jordan curve or singleton. If $\mathcal{O}_*$ is a Jordan curve, we have $\mathcal{O} = \mathcal{O}_*$ by Lemma 5.7, which means that the solution $v = v(t)$ is periodic in time. If $\mathcal{O}_*$ is singleton, on the other hand, the orbit $\mathcal{O}$ is composed of an equilibrium. Hence the solution is stationary.

The set of equilibria to (5.80) is given by

$$E = \mathbf{R}_+^N \cap \operatorname{Ker} A.$$

This set is non-empty because of the existence of $0 < \vec{B} \in \operatorname{Ker} A$. Since rank $A = 2$, it is the intersection of the vector space $\operatorname{Ker} A$ of co-dimension two and the positive cone $\mathbf{R}_+^N$.

We recall that $\mathcal{M}$ is the boundary of the strictly convex space

$$M = \{\xi \in \mathbf{R}^N \mid \tau \cdot e^\xi \le \beta_0\}$$

determined by β_0, that is, $\mathcal{M} = \partial M$. As $\beta_0 > 0$ varies, $\mathcal{F} = \{\mathcal{M}\}$ forms a nested family of hypersurfaces of co-dimension one, covering $\mathbf{R}^N \setminus \{0\}$. Furthermore, each orbit $\mathcal{O}$ of (5.80) is homeomorphic to the intersection of some $\mathcal{M} \in \mathcal{F}$ and an affine space P of dimension two defined by (5.93). Since M with $\partial M = \mathcal{M}$ is strictly convex and $\mathcal{O}$ is either a Jordan curve or singleton composed of an equbirium, it is contractible in $\mathbf{R}_+^N \setminus E$ to an equibirium point as P moves in $\mathbf{R}^N$.

Finally, if two distinct Jordan orbits $\mathcal{O}_1, \mathcal{O}_2 \subset \mathbf{R}_+^N$ are realized on the same hypersurface $\mathcal{M} = \partial M \in \mathcal{F}$, they do not link because of the strict convexity of M. Therefore, the same property arises to any distinct two non-stationary orbits in $\mathbf{R}_+^N$. $\square$

5.9   The Case with Linear Term

Here we study the case $N \ge 3$ of (5.66) and show Theorem 5.8 in §5.7. To begin with, we have

$$\frac{d}{dt}\tau \cdot v = -e \cdot v \le a\tau \cdot v, \quad a = \frac{\max_j(-e_j)}{\min_j \tau_j} \tag{5.96}$$

for $\tau = (\tau_j) > 0$ and $e = (e_j)$, which implies

$$\tau \cdot v \le (\tau \cdot v_0)e^{at}$$

for $v_0 = (v_{0j})$. Then it holds that $T = +\infty$ and

$$0 < v_j(t) \le \tau_j^{-1}(\tau \cdot v_0)e^{at}.$$

We transform (5.66) to

$$\tau_j \frac{d\xi_j}{dt} = -e_j + \sum_{k=1}^{N} a_{jk} e^{\xi_k}, \quad \xi_j\big|_{t=0} = \xi_{j0} \in \mathbf{R}, \quad 1 \le j \le N, \tag{5.97}$$

using $\xi_j = \log v_j$. Let $\xi = (\xi_j)$, and $\xi_0 = (\xi_{j0})$. Without loss of generality we assume

$$(i, j, k) = (1, j, N), \quad 2 \le j \le N - 1$$

and

$$(i, j, k, l) = (1, j, N, l), \quad 2 \le j \ne l \le N - 1,$$

in the first and the second cases of Lemma 5.3, respectively. Thus we have $e_1 \ne 0$ and either

$$a_{1j} > 0, \ a_{1N} < 0, \ a_{jN} > 0, \quad 2 \le \exists j \le N - 2 \tag{5.98}$$

or

$$a_{1j} > 0, \ a_{1l} = 0, \ a_{1N} < 0, \ a_{jN} = 0, \ a_{jl} > 0, \ a_{N\ell} < 0$$
$$2 \le \exists j \ne \exists l \le N - 2. \tag{5.99}$$

In both cases it holds that $a_{1N} < 0$.

Lemma 5.8. *Under the above assumptions, define* $V_j(\xi)$, $1 \le j \le N - 1$, *by*

$$V_1(\xi) = a_{1N} \sum_{k=1}^{N} \tau_k e^{\xi_k} + e_N \tau_1 \xi_1 - e_1 \tau_N \xi_N$$
$$V_j(\xi) = a_{jN} \tau_1 \xi_1 - a_{1N} \tau_j \xi_j + a_{1j} \tau_N \xi_N, \quad 2 \le j \le N - 1. \tag{5.100}$$

Then $V_j(\xi)$, $1 \le j \le N - 1$, *are invariant with respect to (5.97). Hence it holds that*

$$\tilde{\mathcal{O}} \equiv \{\xi(t) \mid t \ge 0\} \subset \tilde{\mathcal{O}}_*,$$

where

$$\tilde{\mathcal{O}}_* \equiv \{\xi \in \mathbf{R}^N \mid V_j(\xi) = c_j, \ 1 \le j \le N - 1\}, \quad c_j = V_j(\xi_0). \tag{5.101}$$

Moreover, $\tilde{\mathcal{O}}_*$ *is a Jordan curve or singleton.*

Proof. We shall confirm

$$\frac{d}{dt}V_j(\xi(t)) = 0, \quad 1 \le j \le N-1 \tag{5.102}$$

for the solution $\xi = (\xi_j(t))$ to (5.97). First, (5.67) implies

$$\frac{d}{dt}\sum_{k=1}^{N}\tau_k e^{\xi_k} = -\sum_{k=1}^{N}e_k e^{\xi_k},$$

while

$$e_N\tau_1\frac{d\xi_1}{dt} - e_1\tau_N\frac{d\xi_N}{dt} = \sum_{k=1}^{N}(e_N a_{1k} - e_1 a_{Nk})e^{\xi_k}$$

is obvious. Then, (5.102) for $j = 1$ is a consequence of (5.77) with $(i,j,k) = (1,N,k)$:

$$\frac{dV_1}{dt} = \sum_{k=1}^{N}(-a_{1N}e_k + e_N a_{1k} - e_1 a_{Nk})e^{\xi_k} = 0.$$

Here we use (5.77) for $(i,j,k) = (1,j,N)$, $(i,j,k) = (1,j,k)$, and $(i,j,k) = (1,N,k)$ to obtain

$$a_{jN}e_1 - a_{1N}e_j + a_{1j}e_N = 0$$
$$a_{jk}e_1 - a_{1k}e_j + a_{1j}e_k = 0$$
$$a_{Nk}e_1 - a_{1k}e_N + a_{1N}e_k = 0.$$

Then it follows that

$$(a_{jN}a_{1k} + a_{1j}a_{Nk} - a_{1N}a_{jk})e_1 = 0$$

and hence (5.69) for $(i,j,k,l) = (1,j,k,N)$, that is,

$$a_{jN}a_{1k} + a_{1j}a_{Nk} - a_{1N}a_{jk} = 0, \quad 2 \le j \ne k \le N-1. \tag{5.103}$$

From (5.100) and (5.103) we obtain

$$\frac{dV_j}{dt} = \sum_{k=1}^{N}(a_{jN}a_{1k} - a_{1N}a_{jk} + a_{1j}a_{Nk})e^{\xi_k} + (-a_{jN}e_1 + a_{1N}e_j - a_{1j}e_N)$$
$$= -a_{jN}e_1 + a_{1N}e_j - a_{1j}e_N, \quad 2 \le j \le N-1, \tag{5.104}$$

which implies (5.102) for $2 \le j \le N-1$ by (5.77) with $(i,j,k) = (1,j,N)$.

We show that $\tilde{O}_*$ is a Jordan curve or singleton. In fact, it is the intersection of the boundary of

$$M = \{\xi \in \mathbf{R}^N \mid V_1(\xi) \ge c_1\},$$

which is strictly convex by $a_{1N} < 0$, and the $N - 2$ linearly independent hyperplanes

$$P_j = \{\xi \in \mathbf{R}^N \mid V_j(\xi) = c_j\}, \quad 2 \leq j \leq N - 1.$$

Thus we have only to show that $\tilde{\mathcal{O}}_*$ defined by (5.101) is bounded. In fact, first, it hols that

$$c_1 = V_1(\xi) \leq a_{1N}(\tau_N e^{\xi_N} + \tau_1 e^{\xi_1}) + e_N \tau_1 \xi_1 - e_N \tau_N \xi_N,$$

by $a_{1N} < 0$, and therefore, the components ξ_1, ξ_N are bounded above on $\tilde{\mathcal{O}}_*$. Second, from these bounds the first equality of (5.100) implies

$$-C \leq a_{1N} \sum_{k=2}^{N-1} \tau_k e^{\xi_k}$$

with a constant $C > 0$, Therefore, the components $\xi_2, \cdots, \xi_{N-1}$ are also bounded above on $\tilde{\mathcal{O}}_*$ by $a_{1N} < 0$.

Thus we have only to derive the boundedness below of $\xi_1, \cdots, \xi_N$ for $\xi = (\xi_1, \cdots, \xi_N) \in \tilde{\mathcal{O}}_*$. For this purpose, we distinguish two cases of Lemma 5.3.

Case 1. It holds that $a_{1j} > 0$, $a_{1N} < 0$, $a_{jN} > 0$, and $e_1 \neq 0$ for some $2 \leq j \leq N - 1$. Then, V_j is a linear function of (ξ_1, ξ_j, ξ_N), and their coefficients are positive. Since ξ_1, ξ_j, ξ_N are bounded above on $\tilde{\mathcal{O}}_*$, any of them is bounded below there. From this property and $V_i(\xi) = c_i$, so is ξ_i for $i \neq 1, j, N$. Hence $\tilde{\mathcal{O}}_*$ is bounded below.

Case 2. There are $j \neq l$ satisfying $2 \leq j, l \leq N - 1$, $a_{1j} > 0$, $a_{1N} < 0$, $a_{1l} = 0$, $a_{jN} = 0$, $a_{jl} > 0$, $a_{Nl} < 0$, and $e_1 \neq 0$. Then, V_j and V_l are linear functions of (ξ_j, ξ_N) and (ξ_1, ξ_l), respectively, and all of their coefficients are positive. Hence the coordinates $\xi_1, \xi_j, \xi_N, \xi_l$ are bounded below on $\tilde{\mathcal{O}}_*$ because they are bounded above. From this property and $V_i(\xi) = c_i$, so is ξ_i for $i \neq 1, j, N, l$. Hence $\tilde{\mathcal{O}}_*$ is bounded below. $\qquad \square$

Lemma 5.9. *Under the assumptions of Theorem 5.8, the set of stationary solutions to (5.66) is given by*

$$E = \mathbf{R}_+^N \cap E_*$$

for

$$E_* = \{v_*\} + \langle \vec{b_j} \mid 2 \leq j \leq N - 1 \rangle,$$

where $v_* = \frac{1}{a_{1N}}(-e_N, 0, \cdots, 0, e_1)^T$ and

$$\vec{b_j} = (b_j^n), \quad b_j^n = \begin{cases} a_{jN}, & n = 1 \\ a_{1j}, & n = N \\ -a_{1N}, & n = j \\ 0, & n \neq 1, j, N, \end{cases} \quad 2 \leq j \leq N - 1.$$

Proof. We have $v = (v_k) \in E$ if and only if $v > 0$ and

$$I_j \equiv -e_j + \sum_{k=1}^{N} a_{jk}v_k = 0, \quad 1 \leq j \leq N. \tag{5.105}$$

By this definition and (5.103) we obtain

$$a_{1j}I_N - a_{Nj}I_1 = a_{1N}I_j, \quad 2 \leq j \leq N - 1,$$

recalling (5.77) with $(i, j, k) = (1, j, N)$.

Since $a_{1N} \neq 0$, therefore, system of equations (5.105) is reduced to $I_1 = I_N = 0$, that is,

$$v_1 = \frac{1}{a_{1N}}\left(-e_N - \sum_{k=2}^{N-1} a_{kN}v_k\right), \quad v_N = \frac{1}{a_{1N}}\left(e_1 - \sum_{k=2}^{N-1} a_{1k}v_k\right),$$
$$\tag{5.106}$$

where v_k, $2 \leq k \leq N - 2$, are arbitrary. Then the result follows. $\quad\square$

Let

$$\tilde{E} = \{\xi \in \mathbf{R}^N \mid \xi = \log v, \ v \in E\}.$$

Lemma 5.10. *Under the above assumptions, the property $\tilde{\mathcal{O}}_* \cap \tilde{E} \neq \emptyset$ arises only if $\sharp \tilde{\mathcal{O}}_* = 1$.*

Proof. The hypersurface

$$\mathcal{M} = \{\xi \in \mathbf{R}^N \mid V_1(\xi) = c_1\}$$

is the boundary of a strictly convex set $M = \{\xi \in \mathbf{R}^N \mid V_1(\xi) \geq c_1\}$, and

$$P = \{\xi \in \mathbf{R}^N \mid V_j(\xi) = c_j, \ 2 \leq j \leq N - 1\}$$

is a two-dimensional affine space.

Given $\xi \in \mathcal{M}$, we have

$$T_\xi\mathcal{M} = \{(\eta_j) \in \mathbf{R}^N \mid (a_{1N}\tau_1 e^{\xi_1} + e_N\tau_1)\eta_1$$
$$+ \sum_{i=2}^{N-1} a_{1N}\tau_i e^{\xi_i}\eta_i + (a_{1N}\tau_N e^{\xi_N} - e_1\tau_N)\eta_N = 0\}.$$

In addition, for $\xi \in \tilde{\mathcal{O}}_* = \mathcal{M} \cap P$, the plane P is paralle to $T_\xi \mathcal{M}$ if and only if

$$a_{1N}\tau_1 e^{\xi_1} + e_N \tau_1 = \sum_{i=2}^{N-1} k_i a_{iN} \tau_1$$

$$a_{1N}\tau_j e^{\xi_j} = -k_j a_{1N} \tau_j, \quad 2 \le j \le N-1$$

$$a_{1N}\tau_N e^{\xi_N} - e_1 \tau_N = \sum_{i=2}^{N-1} k_i a_{1i} \tau_N$$

for some $k_2, \cdots, k_{N-1} \in \mathbf{R}$.

This condition is equivalent to

$$(e^{\xi_1}, e^{\xi_2}, e^{\xi_3}, \cdots, e^{\xi_{N-1}}, e^{\xi_N})^T \in E,$$

or $\xi \in \tilde{E}$ by Lemma 5.10. Therefore, if $\xi \in \tilde{\mathcal{O}}_* \cap \tilde{E}$, the plane P is parallel to $T_\xi \mathcal{M}$. Then $\tilde{\mathcal{O}}_* = \{\xi\}$ follows because M is strictly convex. $\square$

Proof of Theorem 5.8. To begin with, we show the existence of $v = (v_k) > 0$ satisfying (5.106).

In fact, in the case of (5.98) we have $2 \le j \le N-1$ such that $a_{1j} > 0$, $a_{jN} > 0$. Then the requirement $v = (v_k) > 0$ holds in (5.106) for $v_j > 0$ sufficiently large by $a_{1N} < 0$. In the other case of (5.99), on the other hand, we have $2 \le j \le N-1$ such that $a_{1j} > 0 = a_{jN}$. Here we find $2 \le i \le N-1$ with $a_{iN} > 0$, using (a1), (a3), and $a_{1N} < 0$. Then $v = (v_k) > 0$ holds in (5.106) for sufficiently large $v_j > 0$ and $v_i > 0$.

Since (5.105) is equivalent to (5.106), we have $E \ne \emptyset$. The rest of the proof is similar to that of Theorem 5.7, using Lemmas 5.8, 5.9, and 5.10. $\square$

Persistence of (5.66) means that any solution $v = (v_j(t))$ exists global in time and it holds that

$$\liminf_{t\uparrow+\infty} v_j(t) > 0, \quad 1 \le j \le N.$$

From the above method we can show the following theorem.

Theorem 5.9. *System (5.66), $N \ge 3$, exibits persistence if*

$$A^T + A \le 0 \tag{5.107}$$

and there exists $1 \le i, j, l \le N$ such that $a_{jl} > 0$, $a_{ij} > 0$, $a_{il} < 0$, and

$$a_{ij}a_{kl} + a_{il}a_{jk} - a_{ik}a_{jl} \le 0, \quad a_{jk}e_i - a_{ik}e_j + a_{ij}e_k \le 0 \tag{5.108}$$

for $1 \le k \le N$.

Proof. By (5.107) we have

$$\frac{d}{dt}\tau \cdot v \leq -e \cdot v$$

for (5.96). Then it follows that $T = +\infty$ from $v = (v_k(t)) > 0$.

We assume $i = 1$, $l = N$ without loss of generality, and take $V = V_j(\xi)$ for $2 \leq j \leq N - 1$ in (5.100), recalling $\xi = (\log v_j)$. By the first equality of (5.104) and the assumption (5.108), it holds that

$$\frac{dV}{dt} \geq 0.$$

Here, the set

$$\hat{\mathcal{O}}_* = \{\xi \in \mathbf{R}^N \mid V(\xi) \geq V(\xi_0)\}$$

is bounded below for $\xi_0 = (\log v_{j0})$, because each coefficient of the linear function $V_j(\xi)$ is positive from the assumption of $a_{jl}, a_{ij} > 0 > a_{il}$ with $i = 1$, $l = N$. Then the result follows.   □

5.10   Integrable Systems and Differential Forms

Several models in §5.5, §5.8, and §5.9 are provided with multiple first integrals. They are written as

$$\frac{d\xi}{dt} = f(\xi) \tag{5.109}$$

in $\xi \in \mathbf{R}^N$-variable, where $f = (f_j(\xi)) : \mathbf{R}^N \to \mathbf{R}^N$ is locally Lipschitz continuous.

Here we regard $\xi(t) = (\xi_j(t)) \in \mathbf{R}^N$ as the 1-form

$$\xi(t) = \sum_{j=1}^{N} \xi_j(t) d\xi_j, \tag{5.110}$$

and identify the left-hand side of (5.109) with

$$\frac{d\xi}{dt} = \sum_{j=1}^{N} \dot{\xi}_j(t)\, d\xi_j. \tag{5.111}$$

Then (5.109) is reduced to

$$\frac{d\xi}{dt} = \omega, \quad \omega = \sum_{j=1}^{N} f_j(\xi) d\xi_j.$$

We extend the wedge operator $\wedge$ in §1.5 to N-variables as in

$$d\xi_i \wedge d\xi_j = -d\xi_j \wedge d\xi_i, \quad 1 \leq i, j \leq N.$$

Given $1 \le p \le N$, we call

$$\theta = \sum_{1 \le j_1 < \cdots < j_p \le N} \theta_{j_1, \cdots, j_p} d\xi_{j_1} \wedge \cdots \wedge d\xi_{jp} \qquad (5.112)$$

a *p-form* on $\mathbf{R}^N$, where

$$\theta_{j_1, \cdots, j_p} \in C^\infty(\mathbf{R}^n).$$

The set of p-forms is denoted by Λ^p. Then the *outer derivative*

$$d : \Lambda^p \to \Lambda^{p+1}$$

is defined for $1 \le p \le N - 1$, satisfying

$$d^2 = 0,$$

where

$$\Lambda^0 = C^\infty(\mathbf{R}^N), \quad \Lambda^{N+1} = \{0\}.$$

We put also

$$*(dx_{j_1} \wedge \cdots \wedge dx_{j_p}) = \operatorname{sgn} \sigma \cdot dx_{j_{p+1}} \wedge \cdots \wedge dx_{j_N}$$

for

$$\sigma : (1, \ldots, N) \mapsto (j_1, \cdots, j_N), \quad \sigma \in \mathbf{S}_N,$$

to define the *Hodge operator*, where $\mathbf{S}_N$ denotes the N-th *alternative group*:

$$* : \Lambda^p \to \Lambda^{N-p}, \quad 0 \le p \le N.$$

Then (5.111) means

$$*\frac{d\xi}{dt} = \sum_{j=1}^{N} (-1)^{j+1} \frac{d\xi_j}{dt} d\xi_1 \wedge \cdots \wedge \widehat{d\xi_j} \wedge \cdots \wedge d\xi_N$$

as an $(N-1)$ form.

Theorem 5.10. *If*

$$H = H(\xi), \quad h = h(\xi)$$

are 0-form and $(N-2)$-form, respectively, then the system

$$*\frac{d\xi}{dt} = dH \wedge h \qquad (5.113)$$

admits H as the first integral, that is,

$$\frac{d}{dt} H(\xi) = 0 \qquad (5.114)$$

for any solution $\xi = \xi(t)$.

Proof. Equation (5.113) implies

$$dH \wedge * \frac{d\xi}{dt} = dH \wedge dH \wedge h = 0,$$

while there holds that

$$dH \wedge * \frac{d\xi}{dt} = \sum_j \frac{\partial H}{\partial \xi_j} \, d\xi_j \wedge \sum_k \frac{d\xi_k}{dt} \, d\xi_1 \wedge \cdots \wedge \widehat{d\xi_k} \wedge \cdot \wedge d\xi_N$$

$$= \sum_j (-1)^{j+1} \frac{\partial H}{\partial \xi_j} \, d\xi_j \wedge \frac{d\xi_j}{dt} \, d\xi_1 \wedge \cdots \wedge \widehat{d\xi_j} \wedge \cdots \wedge d\xi_N$$

$$= \sum_j \frac{\partial H}{\partial \xi_j} \frac{d\xi_j}{dt} \, d\xi_1 \wedge \cdots \wedge d\xi_N = \frac{dH}{dt} \, d\xi_1 \wedge \cdots \wedge d\xi_N,$$

and hence (5.114). □

The following result is a direct consequence.

Corollary 5.1. *Let* $1 \leq r \leq N - 1$, *and*

$$H^j = H^i(\xi), \ 1 \leq i \leq r, \quad h = h(\xi)$$

be 0-forms and $(N - r - 1)$-*form, respectively. Then the system*

$$* \frac{d\xi}{dt} = dH^1 \wedge \cdots \wedge dH^r \wedge h \tag{5.115}$$

admits

$$H^i, \quad 1 \leq i \leq r$$

as the first integrals. Hence it holds that

$$\frac{d}{dt} H^i(\xi) = 0, \quad 1 \leq i \leq r$$

for any solution $\xi = \xi(t)$ *to (5.115).*

Definition 5.2. We call (5.115) an integrable system of order r.

By the definition, if system (5.115) is order r, $2 \leq r \leq N - 1$, it is $r - 1$.

Example 5.1. The Hamilton system with $N = 2$,

$$\frac{d\xi_1}{dt} = \frac{\partial H}{\partial \xi_2}, \quad \frac{d\xi_2}{dt} = -\frac{\partial H}{\partial \xi_1}$$

is an integrable system of order 1.

In fact we have

$$\xi = \xi_1 d\xi_1 + \xi_2 d\xi_2$$

and hence

$$*\xi = -\xi_2 d\xi_1 + \xi_2 d\xi_2.$$

Then there arises that

$$* \frac{d\xi}{dt} = \frac{\partial H}{\partial \xi_1} d\xi_1 + \frac{\partial H}{\partial \xi_2} d\xi_2 = dH.$$

5.11 Integrable Systems of Order 1

The integrable system of order 1 is classified as follows.

Theorem 5.11. *Equation (5.109) in §5.10 is an integrable system of order 1 if and only if it is*

$$\frac{d\xi}{dt} = h\nabla H, \tag{5.116}$$

where

$$h = h(\xi)$$

is an $N \times N$ skew-symmetric matrix,

$$H = H(\xi)$$

is a 0-form, and

$$\nabla H = (\frac{\partial H}{\partial \xi_j}), \quad \xi = (\xi_j).$$

Proof. Equation (5.109) is integrable with order 1, if and only if

$$*\frac{d\xi}{dt} = dH \wedge h_2, \tag{5.117}$$

where ξ is the 1-form defined by (5.110), h is an $(N-2)$-form, and H is the 0-form to be the first integral. Here we obtain

$$h_2 = * \sum_{j,k} \tilde{h}^2_{jk} \, d\xi_j \wedge d\xi_k, \quad \exists h^2_{jk} = -h^2_{kj},$$

and hence

$$dH \wedge h_2 = \sum_i \frac{\partial H}{\partial \xi_i} \, d\xi_i \wedge \sum_{j,k} \tilde{h}^2_{jk} \, d\xi_1 \wedge \cdots \wedge \widehat{d\xi_j} \wedge \cdot \wedge \widehat{d\xi_k} \wedge \cdots \wedge d\xi_N$$

$$= \sum_{i,j,k} \frac{\partial H}{\partial \xi_i} \tilde{h}^2_{jk} \, d\xi_1 \wedge (d\xi_1 \wedge \cdots \wedge \widehat{d\xi_j} \wedge \cdots \wedge \widehat{d\xi_k} \wedge \cdots \wedge d\xi_N)$$

$$= \sum_{i,k} \frac{\partial H}{\partial \xi_i} \tilde{h}^2_{ik} (-1)^{i+1} \, d\xi_1 \wedge \cdots \wedge \widehat{d\xi_k} \wedge \cdots \wedge d\xi_N.$$

Equation (5.117) is thus equivalent to

$$\frac{d\xi_k}{dt} = \sum_{i=1}^{N} \frac{\partial H}{\partial \xi_i} \tilde{h}^2_{ik} (-1)^{i+k}, \quad 1 \le k \le N,$$

or

$$\frac{d\xi}{dt} = h\nabla H, \quad h = (\tilde{h}^2_{ki}(-1)^{i+k}).$$

Then the result follows. □

Remark 5.5. We have

$$\frac{dH}{dt} = (h\nabla H, \nabla H) = 0$$

directly from (5.116) by

$$h^T = -h. \tag{5.118}$$

Remark 5.6. If (5.118) holds, equation (5.116) is equivalent to

$$\dot{F} = \{F, H\}, \quad \forall F,$$

where

$$\{F, G\} = (\nabla F, h\nabla G). \tag{5.119}$$

Then Liouville's theorem in §5.2 guarantees that

$$\frac{dF}{dt} = 0$$

if and only if

$$\{F, H\} = (\nabla F, h\nabla H) = 0.$$

Remark 5.7. The above $\{\ ,\ \}$ in (5.119) is the Poisson bracket. In fact, first, the skew-symmetry

$$\{F, G\} = -\{G, F\}$$

is obvious. Second, Leibniz's identity (5.17) in §5.2 holds as

$$\begin{aligned}
\{FG, H\} &= (\nabla(FG), h\nabla H) \\
&= F(\nabla G, h\nabla H) + G(\nabla F, h\nabla H) \\
&= F\{G, H\} + G\{F, H\} \\
&= F\{G, H\} - G\{H, F\}.
\end{aligned}$$

To confirm Jacobi's identity, finally, we note

$$\{F, G\} = (\nabla F, h\nabla G) = (\nabla F)^T h\nabla G.$$

Writing

$$a_i = \frac{\partial a}{\partial x_i}, \quad 1 \leq i \leq N,$$

then we obtain

$$\begin{aligned}
\{F, G\}_i &= (\nabla F_i)^T h\nabla G + (\nabla F)^T h\nabla G_i \\
&= (\nabla F_i)^T h\nabla G + (h\nabla G_i)^T \nabla F \\
&= (h^T \nabla F_i)^T \nabla G + (h\nabla G_i)^T \nabla F \\
&= -(h\nabla F_i)^T \nabla G + (h\nabla G_i)^T \nabla F,
\end{aligned}$$

which implies

$$\nabla\{F, G\} = -(h\nabla^2 F)^T \nabla G + (h\nabla^2 G)^T \nabla F.$$

It thus holds that

$$\begin{aligned}
\{H, \{F, G\}\} &= (h^T \nabla H, \nabla\{F, G\}) \\
&= -(h^T \nabla H, (h\nabla^2 F)^T \nabla G) + (h^T \nabla H, (h\nabla^2 G)^T \nabla F) \\
&= -(h(\nabla^2 F)h^T \nabla H, \nabla G) + (h(\nabla^2 G)h^T \nabla H, \nabla F) \\
&= -(h(\nabla^2 F)h^T)[\nabla H, \nabla G] + (h(\nabla^2 G)h^T)[\nabla F, \nabla H],
\end{aligned}$$

where

$$\nabla^2 H = (H_{ij})_{1 \le i, j \le N}$$

and

$$A[x, y] = x^T A y, \quad x, y \in \mathbf{R}^N$$

are the Hesse matrix of H and the quadratic form associated with the real symmetrix matrix

$$A = A^T,$$

respectively. There arises that

$$\begin{aligned}
\{H, \{F, G\}\} &= -(h(\nabla^2 F)h^T)[\nabla H, \nabla G] + (h(\nabla^2 G)h^T)[\nabla H, \nabla F] \\
\{G, \{H, F\}\} &= -(h(\nabla^2 G)h^T)[\nabla F, \nabla H] + (h(\nabla^2 H)h^T)[\nabla F, \nabla G] \\
\{H, \{F, G\}\} &= -(h(\nabla^2 H)h^T)[\nabla G, \nabla F] + (h(\nabla^2 F)h^T)[\nabla G, \nabla h]
\end{aligned}$$

and hence (5.16) in §5.2.

Remark 5.8. In the Poisson structure induced by (5.119) with (5.118) on $\mathbf{R}^N$, let

$$H = H(\xi), \quad \xi \in \mathbf{R}^N$$

be the Hamiltonian and X be the Hamilton vector field generated by H:

$$\mathcal{L}_X F = \{F, H\}, \quad \forall F.$$

Then it holds that

$$\mathcal{L}_X F = (X \cdot \nabla)F$$

(§5.3), and hence

$$X = h\nabla H.$$

The Hamilton system associated with H on $P = (\mathbf{R}^N, \{\ ,\ \})$, is thus equivalent to

$$\frac{d\xi}{dt} = h\nabla H(\xi), \tag{5.120}$$

or (5.116).

Remark 5.9. Let $x = Q\xi$ and define $\tilde{H}(x)$ by

$$\tilde{H}(x) = H(\xi),$$

where Q is a non-singular $N \times N$ matrix. It then follows that

$$\frac{\partial(x_1, \cdots, x_n)}{\partial(\xi_1, \cdots, \xi_n)} = \begin{pmatrix} \frac{\partial x_1}{\partial \xi_1} & \cdots & \frac{\partial x_1}{\partial \xi_n} \\ \cdots & \vdots & \vdots \\ \frac{\partial x_n}{\partial \xi_1} & \cdots & \frac{\partial x_n}{\partial \xi_n} \end{pmatrix} = Q$$

and hence

$$\nabla_x H = Q^T \nabla_\xi \tilde{H}.$$

It also holds that

$$\frac{dx}{dt} = Q \frac{d\xi}{dt},$$

and therefore, equation (5.120) is reduced to

$$\frac{dx}{dt} = \tilde{h}\nabla\tilde{H}, \quad \tilde{h} = QhQ^T \tag{5.121}$$

with $\tilde{h}^T = -\tilde{h}$. By this transformation of variables the integrable system of order 1 is reduced to the Hamilton system of even number of components.

Example 5.2. Assuming $\tau_1 = \tau_2 = \tau_3 = 1$ in the three system (5.51) in §5.5 for simplicity, we obtain (5.120) for

$$h = \begin{pmatrix} 0 & 1 & -1 \\ -1 & 0 & 1 \\ 1 & -1 & 0 \end{pmatrix}. \tag{5.122}$$

It holds that

$$x \equiv h\xi = \omega \times \xi, \quad \omega = \begin{pmatrix} 1 \\ 1 \\ 1 \end{pmatrix}, \quad \xi = \begin{pmatrix} \xi_1 \\ \xi_2 \\ \xi_3 \end{pmatrix},$$

and therefore, $y \mapsto hy$ is a combination of the rotation of y with respect to ω by $\pi/2$ counter-clockwise and the dilation by $\sqrt{3}$. Hence there is an orthogonal matrix Q,

$$\exists Q^{-1} = Q^T,$$

such that

$$x = \tilde{h}\xi = \tilde{\omega} \times \xi, \quad \tilde{\omega} = \sqrt{3} \begin{pmatrix} 0 \\ 0 \\ 1 \end{pmatrix}$$

for $\tilde{h} = QhQ^T$. It holds that

$$\tilde{h} = \sqrt{3} \begin{pmatrix} 0 & -1 & 0 \\ 1 & 0 & 0 \\ 0 & 0 & 0 \end{pmatrix},$$

and therefore, equation (5.120) for (5.122) is reduced to the Hamilton system

$$\frac{dx_2}{dt} = \hat{H}_{x_1}, \quad \frac{dx_1}{dt} = -\hat{H}_{x_2},$$

where $\hat{H} = \sqrt{3}\tilde{H}(x)$ and $x_3 = $ constant.

Example 5.3. We write the Lotka Volterra system without linear term, (5.80) in §5.8, like

$$\frac{d\xi_i}{dt} = \sum_{j=1}^{N} \tau_i^{-1} a_{ij} e^{\xi_j}, \quad 1 \le i \le N, \tag{5.123}$$

using $\xi_i = \log v_i$. Equation (5.123) takes the form (5.116) for

$$H = \tau \cdot e^\xi \equiv \sum_{i=1}^{N} \tau_i e^{\xi_i}, \quad h = (\tau_i^{-1} \tau_j^{-1} a_{ij}). \tag{5.124}$$

Then (5.118) is equivalent to

$$A^T + A = 0, \quad A = (a_{ij}). \tag{5.125}$$

For the general system with linear term, (5.66) in §5.7,

$$\frac{d\xi_i}{dt} = \sum_{j=1}^{N} \tau_i^{-1} a_{ij} e^{\xi_j} - \tau_i^{-1} e_i, \quad 1 \le i \le N \tag{5.126}$$

with (5.125), if

$$\tau^{-1} e = (\tau_i^{-1} e_i) \in R(h)$$

holds for h in (5.124), equation (5.116) arises with

$$H = -\vec{b} \cdot \xi + \tau \cdot e^\xi, \quad h\vec{b} = \tau^{-1} e$$

for some $\vec{b} \in \mathbf{R}^N$. This case includes the prey predator system described in §5.5.

5.12   Integrable Systems of Order 2

We begin with the following example, dealing with general order.

Example 5.4. Three system (5.51) in §5.5 is an integrable system of order 2, which takes the form

$$* \frac{d\xi}{dt} = dH^1 \wedge dH^2$$

for

$$\xi = \xi_1 d\xi_1 + \xi_2 d\xi_2 + \xi_3 d\xi_3$$
$$H^1 = \tau_1 \xi_1 + \tau_2 \xi_2 + \tau_3 \xi_3$$
$$H^2 = -(\tau_1 \tau_2 \tau_3)^{-1}(\tau_1 e^{\xi_1} + \tau_2 e^{\xi_2} + \tau_3 e^{\xi_3}).$$

Example 5.4 is generalized as follows. The proof is omitted.

Theorem 5.12. *The integrable system of order* r, $1 \le r \le N-1$, *takes the form*

$$\frac{d\xi_i}{dt} = \sum_{1 \le i_1 \cdots i_r \le N} \frac{\partial H^1}{\partial \xi_{i_1}} \cdots \frac{\partial H^r}{\partial \xi_{i_r}} h_{ii_1 \cdots i_r}, \quad 1 \le i \le N, \tag{5.127}$$

where

$$H^j, \ 1 \le j \le r,$$

are 0-forms and

$$h_{ii_1 \cdots i_r} = h_{ii_1 \cdots i_r}(\xi)$$

are tensors fields satisfying

$$h_{\sigma(i)\sigma(i_1)\cdots\sigma(i_r)} = sgn\ \sigma \cdot h_{ii_1 \cdots i_r}, \quad \forall \sigma \in \mathbf{S}_{r+1}.$$

Hence it follows that

$$\frac{dH^j}{dt} = 0, \quad 1 \le j \le r, \tag{5.128}$$

for

$$H^j = H^j(\xi)$$

if $\xi = \xi(t)$ *is a solution to* (5.127).

Remark 5.10. Under the presence of Jacobi's identity, (5.16) in §5.2, there arises that

$$\{F_1, H\} = \{F_2, H\} = 0 \;\Rightarrow\; \{F_3, H\} = 0,$$

where $F_3 = \{F_1, F_2\}$ (Poisson's theorem). This fact is sometime used to detect the other first integrals by Liouville's theorem, Theorem 5.1. If $r = 3$ occurs to (5.115) in §5.10, however, there arises actually that

$$\{H_1, H\} = \{H_2, H\} = 0$$

for $H = H_3$, but it holds also that

$$\{H_1, H_2\} = 0$$

by Theorem 5.12. Hence this method is invalid.

Remark 5.11. Similarly to Remark 5.5, equality (5.128) can be derived directly from (5.127). In fact, we obtain

$$\frac{dH^j}{dt} = \sum_i \frac{\partial H^j}{\partial \xi_i} \frac{d\xi_i}{dt}$$

$$= \sum_{i i_1 \cdots i_r} \frac{\partial H^j}{\partial \xi_i} \frac{\partial H^1}{\partial \xi_{i_1}} \cdots \frac{\partial H^r}{\partial \xi_{i_r}} h_{i i_1 \cdots i_r}$$

$$= \sum_{i, i_1, \cdots, i_r} \frac{\partial H^1}{\partial \xi_{i_1}} \cdot \frac{\partial H^{j-1}}{\partial \xi_{i_{j-1}}} \left(\frac{\partial H^j}{\partial \xi_i} \cdot \frac{\partial H^j}{\partial \xi_{i_j}} \right) \frac{\partial H^{j+1}}{\partial \xi_{i_{j+1}}} \cdots \frac{\partial H^r}{\partial \xi_{i_r}}$$

$$\cdot h_{i i_1 \cdots i_{j-1} i_j i_{j+1} \cdots i_r}$$

with

$$h_{i i_1 \cdots i_{j-1} i_j i_{j+1}, \cdots, i_r} = -h_{i_j i_1 \cdots i_{j-1} i i_{j+1} \cdots i_r}$$

and hence (5.128).

Remark 5.12. The system of order 2,

$$\frac{d\xi_i}{dt} = \sum_{jk} \frac{\partial H}{\partial \xi_j} \frac{\partial L}{\partial \xi_k} h_{ijk}, \quad 1 \le i \le N \tag{5.129}$$

with

$$h_{\sigma(i)\sigma(j)\sigma(k)} = \operatorname{sgn} \sigma \cdot h_{ijk}, \quad \forall \sigma \in \mathbf{S}_3 \tag{5.130}$$

is reduced to that of order 1,

$$\frac{d\xi}{dt} = \tilde{h} \nabla H, \quad \tilde{h} = (\tilde{h}_{ij}), \quad \tilde{h}_{ij} = \sum_k \frac{\partial L}{\partial \xi_k} h_{ijk}. \tag{5.131}$$

It holds that

$$\tilde{h}_{ij} = -\tilde{h}_{ji}$$

and the Poisson bracket is induced by

$$\{F, G\} = (\nabla F, \tilde{h} \nabla G).$$

Then L in (5.129) is actually the first integral of equation (5.131) and Liouville's theorem actually holds as

$$\{L, H\} = (\nabla L, \tilde{h} \nabla H)$$
$$= \sum_{ij} \frac{\partial L}{\partial \xi_i} \tilde{h}_{ij} \frac{\partial H}{\partial \xi_j}$$
$$= \sum_{ijk} \frac{\partial L}{\partial \xi_i} \frac{\partial L}{\partial \xi_k} \frac{\partial H}{\partial \xi_j} h_{ijk}$$
$$= \sum_j \frac{\partial H}{\partial \xi_j} \sum_{ik} \frac{\partial L}{\partial \xi_i} \frac{\partial L}{\partial \xi_k} h_{ijk} = 0$$

by

$$h_{ijk} = -h_{kji}.$$

Example 5.5. The Lotka Volterra system (5.126) is an integrable system of order 2, if

$$a_{ij} = \tau_i \tau_j \sum_{k=1}^{N} c_k h_{ijk}, \quad e_i = \sum_{j=1}^{N} b_j a_{ij},$$

for h_{ijk} satisfying (5.130). It holds that

$$\frac{dH^1}{dt} = \frac{dH^2}{dt} = 0$$

for

$$H^1 = \tau \cdot e^\xi - (\tau \vec{b}) \cdot \xi, \quad H^2 = \vec{c} \cdot \xi,$$

where

$$\xi = (\xi_i), \quad \tau = (\tau_i), \quad e^\xi = (e^{\xi_i}), \quad \tau \vec{b} = (\tau_i b_i), \quad \vec{c} = (c_i).$$

If $\vec{c} > 0$, this system is provided with the permanence,

$$\liminf_{t \to \infty} \xi_j(t) > -\infty, \quad 1 \leq j \leq N.$$

This case is reduced to the prey pretator system (5.39) if $N = 2$, which is not included in Theorem 5.8 in §5.7.

5.13 Integrable Systems of Order $(N-1)$

Property (P) in §5.7 is induced from the following theorem.

Theorem 5.13. *The integrable system of order* $(N-1)$,

$$*\frac{d\xi}{dt} = dH^1 \wedge \cdots \wedge dH^{N-1}, \tag{5.132}$$

takes the form

$$(-1)^{i+1}\frac{d\xi_i}{dt} = \det A_i, \quad 1 \le i \le N, \tag{5.133}$$

where

$$A_i = (\vec{h}_1, \cdots, \vec{h}_{i-1}, \vec{h}_{i+1}, \cdots, \vec{h}_N)$$

and

$$h_i = (h_i^1, \cdots, h_i^{N-1})^T, \quad h_i^j = \frac{\partial H^j}{\partial \xi_i}, \; 1 \le j \le N-1 \tag{5.134}$$

for $1 \le i \le N$.

Proof. Writing $h^j = H^j$, $1 \le j, N-1$, we have

$$dH^1 \wedge \cdots \wedge dH^{N-1} = \sum_{i_1} h_{i_1}^1 d\xi_{i_1} \wedge \sum_{i_2} h_{i_2}^2 d\xi_{i_2} \wedge \cdots \wedge \sum_{i_{N-1}} h_{i_{N-1}}^{N-1} d\xi_{i_{N-1}}$$

$$= \sum_{i_1, \cdots, i_{N-1}} h_{i_1}^1 \cdots h_{i_{N-1}}^{N-1} d\xi_{i_1} \wedge \cdots \wedge d\xi_{i_{N-1}}.$$

Put

$$dH^1 \wedge \cdots \wedge dH^{N-1} = \sum_i g_i d\xi_1 \wedge \wedge \widehat{d\xi_i} \wedge \cdots \wedge d\xi_N,$$

fix i, and let

$$\mathbf{S}_{N-1}^i$$

be the alternative group of

$$\{1, \cdots, i-1, i+1, \cdots, N\}.$$

Then we obtain

$$g_i d\xi_1 \wedge \cdots \wedge \widehat{d\xi_i} \wedge d\xi_N$$
$$= \sum_{\sigma \in \mathbf{S}_{N-1}^i} (\text{sgn } \sigma) h_{\sigma(1)}^1 \cdots h_{\sigma(N)}^{N-1} d\xi_1 \wedge \widehat{d\xi_i} \wedge \cdots \wedge d\xi_N,$$

and hence (5.132) is equivalent to

$$(-1)^{i+1}\frac{d\xi_i}{dt} = \sum_{\sigma \in \mathbf{S}_{N-1}^i} (\mathrm{sgn}\ \sigma) \cdot h^1_{\sigma(1)} \cdots h^{N-1}_{\sigma(N-1)}, \quad 1 \le i \le N.$$

We thus obtain the result by

$$\sum_{\sigma \in \mathbf{S}_{N-1}^i} (\mathrm{sgn}\ \sigma) \cdot h^1_{\sigma(1)} \cdots h^{N-1}_{\sigma(N)} = \det A_i.$$

$\square$

Example 5.6. The Lotka Volterra system in the form of

$$\frac{d\xi_i}{dt} = (-1)^{i+1} \det(\vec{h}_1, \cdots, \vec{h}_{i-1}, \vec{h}_{i+1}, \cdots \vec{h}_N), \quad 1 \le i \le N \qquad (5.135)$$

has the first integrals

$$H^1 = \tau \cdot e^\xi - (\tau \vec{b}) \cdot \xi, \quad H^\ell = \vec{c}_{\ell-1} \cdot \xi,\ 2 \le \ell \le N-1,$$

where

$$\vec{b} = (b^j), \quad \vec{c}_1 = (c^j_1), \quad \cdots, \quad \vec{c}_{N-2} = (c^j_{N-2}),$$

are arbitrary,

$$\tau \vec{b} = (\tau_j b^j), \quad \tau = (\tau_j),$$

and

$$\vec{h}_i = \begin{pmatrix} \tau_i(e^{\xi_i} - b_i) \\ c^i_1 \\ \vdots \\ c^i_{N-2} \end{pmatrix}, \quad 1 \le i \le N. \qquad (5.136)$$

Hence this system admits the property (P) if $\tau = (\tau_i) > 0$,

$$\vec{c}_\ell > 0, \quad 1 \le \exists \ell \le N-2,$$

and $\vec{c}_\ell$, $1 \le \ell \le N-2$, are linearly independent. The mechanism for (P) to arise in (5.135) with (5.136) is the same. It may not be a real extension of the Lotka Volterra (LV) system studied in §5.8–§5.9. System (5.133), however, includes other systems than the LV system which is provided with the property (P).

Chapter 6

Interface Vanishing

There is a class of partial differential equations which admits the property of interface vanishing. In these equations if outer forces are piecewise regular, the singularity on the interface of some components of the solution vanishes. They are actually formulated as $d - \delta$ system using differential forms. A typical example is Maxwell's equation associated with Minkowski metric, where some components of electric magnetic fields do not suffer the effect of interface and their singularity propagates under the light speed across there. In this chapter, first, we describe this property for the Ampère equation, using layer potentials in the context of magnet-encephalography (§6.1), and then summarize the result on non-stationary Maxwell equation (§6.2). Then we take preliminaries on differential forms (§6.3), to develop a general theory of two forms on Riemann spaces (§6.4) and Minkowski spaces (§6.5). Then we turn to the result on one forms (§6.6), and finally, extends them to general forms (§6.7).

6.1   Geselowitz Equation

Magnet-encephalography is a technology to detect the activity of brains by measuring the magnetic field on the head using a super conducting quantum interface device (SQUID). The *Geselowitz equations* [Geselowitz (1967, 1970)]

$$\frac{\sigma_I + \sigma_O}{2} V(\xi) = - \int_\Omega \nabla \cdot J^p(y) \Gamma(\xi - y) dy$$

$$- (\sigma_I - \sigma_O) \int_{\partial\Omega} V(\eta) \frac{\partial \Gamma}{\partial \nu_\eta} \Gamma(\xi - \eta) \, dS_\eta, \quad \xi \in \partial\Omega \qquad (6.1)$$

and

$$\mu_0^{-1}B = -\int_\Omega J^p(y) \times \nabla\Gamma(x-y)dy$$

$$+(\sigma_I - \sigma_O)\int_{\partial\Omega} V(\eta)\nu_\eta \times \nabla\Gamma(x-\eta)\,dS_\eta, \quad x \notin \partial\Omega \qquad (6.2)$$

are used to relate the magnetic field B measured at the head and the *primary current* J^p caused by the activity of neurons in the brain, where

$$\Gamma(x) = \frac{1}{4\pi|x|}$$

stands for the *Newton potential* and $\mu_0 > 0$ is the permeability. Hence it holds that

$$J^p(x) = 0, \quad x \notin \overline{\Omega}. \qquad (6.3)$$

The *secondary current* $-\sigma(x)\nabla V$ is taken into account with the piecewise constnat conductivity

$$\sigma(x) = \begin{cases} \sigma_I, & x \in \Omega \\ \sigma_O, & x \notin \overline{\Omega}, \end{cases} \qquad (6.4)$$

where $\Omega \subset \mathbf{R}^3$ is a bounded domain indicating the brain and ν denotes the outer unit normal vector on $\partial\Omega$. Equation (6.2) is derived from the *Viot Savard law* applied to the *Ampère equation*

$$\nabla \cdot B = 0, \ \mu_0^{-1}\nabla \times B = J \quad \text{in } \mathbf{R}^3, \qquad (6.5)$$

where

$$J = J^p - \sigma(x)\nabla V \qquad (6.6)$$

denotes the *total current density*.

Theory of compact operators is applicable to confirm the well-posedness of (6.1)–(6.2) and the spectrum of layer potentials is studied in the context of inverse problems [Miyanishi–Suzuki (2017)]. In fact, if $\sigma_O = 0$ in (6.1), the unknown $V(\xi)$ is determined up to an additive constant by the fundamental property of the *second layer potential*,

$$\int_{\partial\Omega} \frac{\partial}{\partial\nu_\eta}\Gamma(x-\eta)\,dS_\eta = \begin{cases} -1, & x \in \Omega \\ -\frac{1}{2}, & x \in \partial\Omega \\ 0, & x \notin \overline{\Omega} \end{cases} \qquad (6.7)$$

(Chapter 10 of [Suzuki (2022a)]). This ambiguity, however, is cancelled in (6.2) by

$$F \equiv \int_{\partial\Omega} V(\eta)\nu_\eta \times \nabla\Gamma(x-\eta)dS_\eta$$

$$= \int_{\partial\Omega} \nu_y \times \nabla V(\eta)\Gamma(x-\eta)\,dS_\eta, \quad x \notin \partial\Omega$$

derived from the *Stokes formula*, because $\nu_\eta \times \nabla$ is a tangential derivative on $\partial\Omega$ (Chapter 4 of [Suzuki (2022a)]).

The Stokes formula implies also

$$-\int_\Omega J^p(y) \times \nabla\Gamma(x-y)dy = \int_\Omega \nabla \times J^p(y)\Gamma(x-y)dy + H \qquad (6.8)$$

for

$$H = -\int_{\partial\Omega} \nu_y \times J^p(y)\Gamma(x-y)dS_y$$

in the second equation of (6.2). Since the first term on the right-hand side of (6.8) is smooth in x if so is $J^p(x)$, we obtain

$$\mu_0^{-1}B \sim H + (\sigma_I - \sigma_0)F \equiv G, \qquad (6.9)$$

where $A \sim B$ means that $A - B$ is smooth in $\mathbf{R}^3$.

We define $A_\pm$ on $\partial\Omega$ by

$$A_\pm(\xi) = \lim_{x\in\Omega_\pm \to \xi\in\partial\Omega} A(x), \quad \Omega_+ = \overline{\Omega}^c, \quad \Omega_- = \Omega,$$

to put

$$[A]_-^+ = A_+ - A_-.$$

Then it follows that

$$[\nu \times \nabla \times G]_-^+ = [\nu \times \nabla \times \mu_0^{-1}B]_-^+ = [\nu \times J]_-^+$$
$$= -\nu \times J^p + (\sigma_I - \sigma_O)\nu \times \nabla V \qquad (6.10)$$

from (6.3), (6.4), (6.5), (6.6), and (6.9), because ν is the outer normal vector on $\partial\Omega_-$. Since H and F are the *first layer potential*, there arises that

$$[H]_-^+ = [F]_-^+ = 0, \quad \left[\frac{\partial H}{\partial\nu}\right]_-^+ = \nu \times J^p, \quad \left[\frac{\partial F}{\partial\nu}\right]_-^+ = -\nu \times \nabla V$$

(Chapter 10 of [Suzuki (2022a)]), which implies

$$[G]_-^+ = 0, \quad [\nu \times \nabla \times G]_-^+ = \left[-\frac{\partial G}{\partial\nu}\right]_-^+ \qquad (6.11)$$

by (6.10).

Here we use the identity

$$(\nu \cdot \nabla)G + \nu \times \nabla \times G = \nabla(\nu \cdot G) - (\nabla \otimes \nu) \cdot G,$$

to get

$$[\nabla(\nu \cdot G)]_-^+ = [(\nu \cdot \nabla)G + \nu \times \nabla \times G]_-^+ = \left[(\nu \cdot \nabla)G - \frac{\partial G}{\partial \nu}\right]_-^+ = 0 \quad (6.12)$$

by (6.11). Then

$$[\nabla(\nu \cdot B)]_-^+ = 0 \tag{6.13}$$

follows from (6.9).

Equation (6.13) assures that the first derivatives of the normal component of the magnetic field B is continuous accross the interface $\partial\Omega$ in spite of the discontinuity of the conductivity $\sigma(x)$ in the Ampère equation (6.5) for (6.6).

6.2   Maxwell Equation

Maxwell's equation is concerned on the electric field E, the magnetic field B, the current density J, and the electric charge density ρ, depending on the space-time variables (x, t). Using the gradient operator

$$\nabla = \nabla_x \equiv \begin{pmatrix} \frac{\partial}{\partial x_1} \\ \frac{\partial}{\partial x_2} \\ \frac{\partial}{\partial x_3} \end{pmatrix}, \quad x = (x_1, x_2, x_3),$$

and the outer and the inner products in $\mathbf{R}^3$ denoted by $\times$ and $\cdot$, respectively, it is given by

$$\nabla \times H - \frac{\partial D}{\partial t} = J, \quad \nabla \cdot D = \rho, \quad \nabla \times E + \frac{\partial B}{\partial t} = 0, \quad \nabla \cdot B = 0$$

with

$$D = \varepsilon E, \quad B = \mu H, \quad J = \sigma E,$$

where ε, μ, and σ denote permittivity, permeability, and conductivity, respectively.

Assuming that the first two physical constants ε and μ are independent of the media as in magneto-encephalograpy, we put them to be 1, to reach

$$\nabla \times B - \frac{\partial E}{\partial t} = J, \quad \nabla \cdot E = \rho$$

$$\nabla \times E + \frac{\partial B}{\partial t} = 0, \quad \nabla \cdot B = 0 \qquad \text{in } \Omega, \tag{6.14}$$

with the normalized light speed $c = \mu\varepsilon = 1$, where $\Omega \subset \mathbf{R}^4$ is a domain of space-time variables.

We take the case that this Ω is composed of two media, with the interface $\mathcal{M}$ forming a smooth hyper-surface, and that the terms J and ρ are piecewise regular. Then, our purpose is to clarify the components of B and E of which singularities propagate accross the interface $\mathcal{M}$ without suffering any effects from it. This case occurs actually in magneto-encephalography studied in the previous section, where (6.14) is reduced to (6.5) by $B_t = 0$.

To provide a unified description to both stationary and nonstationary problems, we formulate the above described geometric profile of Ω in a general setting as follows.

Definition 6.1. The bounded Lipschitz domain $\Omega \subset \mathbf{R}^n$ is said to have interface, denoted by $\mathcal{M}$, if this $\mathcal{M}$ is a non-compact $C^{2,1}$ hyper-surface in $\mathbf{R}^n$ without boundary such that $\Gamma \equiv \mathcal{M} \cap \Omega \neq \emptyset$ is connected.

In this case the domain Ω is divided into two domains by $\mathcal{M}$, denoted by $\Omega_{\pm}$, and then Γ is recognized as $\Gamma_{\pm}$, regarded as subsets of $\partial\Omega_{\pm}$:

$$\Omega = \Omega_+ \cup \Gamma \cup \Omega_-, \quad \Gamma_{\pm} = \partial\Omega_{\pm} \setminus \partial\Omega(=\Gamma). \tag{6.15}$$

Then, ν denotes the outer unit normal vector on Γ_-, extended on Ω as a $C^{1,1}$ vector field.

Our assumption is the piecewise regularity of $J \in L^2(\Omega)$ and $\rho \in L^2(\Omega)$ in (6.14), precisely,

$$\nabla \times J \in L^2(\Omega_{\pm})^3, \quad \frac{\partial J}{\partial t} + \nabla\rho \in L^2(\Omega_{\pm})^3, \tag{6.16}$$

where the differentiations are taken in the sense of distributions in $\Omega_{\pm}$.

Theorem 6.1 ([Suzuki (2021)]). *Let $\Omega \subset \mathbf{R}^4$ be a domain with interface. Let $B, E \in H^1(\Omega)^3$ be the solution to (6.14), for $J \in L^2(\Omega)^3$ and $\rho \in L^2(\Omega)$ satisfying (6.16). Then it holds that*

$$\Box(\nu^0 B + \tilde{\nu} \times E) \in L^2(\Omega)^3, \quad \Box(\tilde{\nu} \cdot B) \in L^2(\Omega), \tag{6.17}$$

where

$$\nu = \begin{pmatrix} \nu^1 \\ \nu^2 \\ \nu^3 \\ \nu^0 \end{pmatrix}, \quad \tilde{\nu} = \begin{pmatrix} \nu^1 \\ \nu^2 \\ \nu^3 \end{pmatrix}, \quad \Box = -\frac{\partial^2}{\partial t^2} + \Delta_x.$$

Remark 6.1. Since equations (6.14) imply

$$-\Delta B - \frac{\partial}{\partial t}\nabla \times E = \nabla \times J, \quad \nabla\rho - \Delta E + \frac{\partial}{\partial t}\nabla \times B = 0,$$

there arises that

$$\Box B = -\nabla \times J, \quad \Box E = \nabla \rho + \frac{\partial J}{\partial t}.$$

The assumption (6.16) thus implies

$$\Box B, \ \Box E \in L^2(\Omega_\pm)^3,$$

which, however, does not mean $\Box B, \ \Box E \in L^2(\Omega)^3$. The conclusion (6.17) of the above theorem, therefore, assures that H^2-singularities of $\nu^0 B + \tilde\nu \times E$ and $\tilde\nu \cdot B$ propagate through the interface $\mathcal{M}$ under the light speed $c = 1$, without suffering any effects from it.

In the stationary state, $E_t = B_t = 0$, the system (6.14) sprits into

$$\nabla \times B = J, \quad \nabla \cdot B = 0 \qquad \text{in } \Omega \tag{6.18}$$

and

$$\nabla \times E = 0, \quad \nabla \cdot E = \rho \qquad \text{in } \Omega \tag{6.19}$$

where $\Omega \subset \mathbf{R}^3$ is a three-dimensional domain. By Theorem 6.1, we obtain the following facts because $\nu_0 = 0$ under the notation there. The first result is actually shown in the previous section using layer potentials.

(1) If $B \in H^1(\Omega)^3$ solves (6.18) with $J \in L^2(\Omega)$ satisfying

$$\nabla \times J \in L^2(\Omega_\pm)^3,$$

it follows that

$$\Delta(\nu \cdot B) \in L^2(\Omega),$$

and hence $\nu \cdot B \in H^2_{loc}(\Omega)$ ([Kobayashi–Suzuki–Watanabe (2003)]).
(2) If $E \in H^1(\Omega)^3$ solves (6.19) with $\rho \in L^2(\Omega)$ satisfying

$$\nabla \rho \in L^2(\Omega_\pm)^3,$$

it follows that

$$\Delta(\nu \times E) \in L^2(\Omega),$$

and hence $\nu \times E \in H^2_{loc}(\Omega)^3$ ([Kanou–Sato–Watanabe (2013)]).

Later, differential forms used in §1.5 and §5.10 are applied for the proof of the above results. They are actually extended to the results on 1-forms on Riemann manifolds of any dimension [Kanou–Sato–Watanabe (2016)]. Here we use several notions in the theory of harmonic integration [Morita (2001)], confirming that $\Omega \subset \mathbf{R}^n$ is the domain with interface.

First, $\Lambda^p = \Lambda^p(\Omega)$ denots the set of p-forms on Ω. Second, the *outer derivative*

$$d : \Lambda^p \to \Lambda^{p+1}$$

and the *Hodge operator*

$$* : \Lambda^p(\Omega) \to \Lambda^{n-p}(\Omega)$$

induces the *co-derivative*

$$\delta : \Lambda^p \to \Lambda^{p-1}.$$

Then it holds that

$$-\Delta = \delta d + d\delta : \Lambda^p \to \Lambda^p,$$

where Δ denotes the *Laplacian*. Analogous result is valid piecewisely to $\Lambda^p_\pm = \Lambda^p(\Omega_\pm)$, the set of p-forms on $\Omega_\pm$.

A p-form, denoted by ω, is said to belong to $H^q(\Omega)$ if any of its coefficients belongs to $H^q(\Omega)$ for $q = 0, 1, 2$. Let

$$B = B^1 dx_1 + \cdots + B^n dx_n \in L^2(\Omega), \tag{6.20}$$

be 1-form, and put

$$B^\nu = (B, \nu)\nu, \quad B^\tau = B - B^\nu, \tag{6.21}$$

where (,) denotes the $\mathbf{R}^n$ inner product. Here, the vector field $\nu = (\nu^i)$ is identified with the 1-form

$$\nu = \nu^1 dx_1 + \cdots + \nu^n dx_n, \tag{6.22}$$

and hence

$$(B, \nu) = \sum_{i=1}^n B^i \nu^i.$$

Theorem 6.2 ([Kanou–Sato–Watanabe (2013)]). *Let $B \in H^1(\Omega)$ be 1-form.*

(1) Let $J \in L^2(\Omega)$ be 2-form satisfying

$$dB = J, \quad \delta B = 0 \quad in \ \Omega,$$

and assume $\delta J \in L^2(\Omega_\pm)$ in the sense of distributions in $\Omega_\pm$. Then it holds that

$$\Delta B^\nu \in L^2(\Omega)$$

and hence $B^\nu \in H^2_{loc}(\Omega)$.

(2) Let $g \in L^2(\Omega)$ be 0-form satisfying

$$dB = 0, \ \delta B = g \quad in \ \Omega,$$

and assume $dg \in L^2(\Omega_\pm)$ in the sense of distributions in $\Omega_\pm$. Then it holds that

$$\Delta B^\tau \in L^2(\Omega)$$

and hence $B^\tau \in H^2_{loc}(\Omega)$.

Remark 6.2. In both cases of the above theorem, the 1-form $B \in H^1(\Omega)$ actually satisfies $\Delta B \in L^2(\Omega_\pm)$ by

$$-\Delta = d\delta + \delta d.$$

This property, however, implies only $B \in H^2_{loc}(\Omega_\pm)$, which does not mean $B \in H^2_{loc}(\Omega)$.

Henceforth, X' denotes the dual space of the Banach space X over $\mathbf{R}$, and $\langle \ , \ \rangle$ denotes the paring between X and X'.

Remark 6.3. The trace of the component of $B \in H^1(\Omega)$ to $\Gamma_\pm$ is defined as an element in $H_0^{1/2}(\Gamma_\pm)$. We use the paring between $H_0^{1/2}(\Gamma_\pm)$ and $H^{-1/2}(\Gamma_\pm) = H_0^{1/2}(\Gamma_\pm)'$ in the proof of Theorem 6.2 (§6.6).

We recall that the nonstationary Maxwell equation (6.14) is formulated by a 2-form in Minkowski space. To show Theorem 6.1, first, we extend Theorem 6.2 on 1-form B to that on 2-form ω. Then we transform this result to the 2-form on the Minkowski space.

The essential difficulty in this process is that the 2-form ω does not take any natural normal or tangential components, in contrast with that of the 1-form B, as in B^ν and B^τ defined by (6.21), respectively. We have to formulate, therefore, the component of 2-form, provided with analogous properties to Theorem 6.2 on 1-forms.

To this end, given 2-form on Ω, denoted by

$$\omega = \sum_{i<j} \omega^{ij} dx_i \wedge dx_j,$$

we define 1-forms $\hat{\omega}^i$ by

$$\hat{\omega}^i = \sum_\ell \tilde{\omega}^{\ell i} dx_\ell, \quad 1 \le i \le n,$$

where

$$\tilde{\omega}^{ij} = \begin{cases} \omega^{ij}, & i < j \\ 0, & i = j \\ -\omega^{ji}, & i > j. \end{cases} \tag{6.23}$$

Recall that the vector field $\nu = (\nu^i)$ is identified with the 1-form as in (6.22). Then the result on 2-forms, comparable to 1-forms on Theorem 6.2, is given as follows.

Theorem 6.3 ([Suzuki (2021)]). *Let $\Omega \subset \mathbf{R}^n$ be a domain with interface and $\omega \in H^1(\Omega)$ be a 2-form. Assume*

$$d\omega = \theta, \quad \delta\omega = 0 \quad in \ \Omega,$$

where $\theta \in L^2(\Omega)$ is a 3-form satisfying $\delta\theta \in L^2(\Omega_\pm)$ in the sense of distributions in $\Omega_\pm$. Then it holds that

$$\Delta(\nu, \hat{\omega}^i) \in L^2(\Omega), \quad 1 \le i \le n.$$

Once Theorem 6.3 is obtained, Theorem 6.1 is proven by replacing the Euclidean metric to the Minkowski metric (§6.5).

6.3 Differential Forms in Higher Dimension

Henceforth $D \subset \mathbf{R}^n$ denotes Lipschitz domain, standing for $\Omega_\pm$ or Ω in the previous section (§7.1). In this case, the trace operator

$$\gamma : H^1(D) \to H^{1/2}(\partial D)$$

is well-defined. It holds that

$$H^{1/2}(\partial D) \cong H^1(D)/H_0^1(D)$$

by this homomorphism and $C^\infty(\overline{D})$ is dense in $H^1(D)$ ([Girault–Raviart (1986); Kufner–John–Fučik (1977); Nečas (1967)]). Then we write

$$\varphi|_{\partial D} = \gamma\varphi$$

for $\varphi \in H^1(D)$.

We confirm several fundamental facts on differential forms on the Euclidean space with smooth coefficients ([Flanders (1963); Morita (2001)]). First, the Euclidean inner product of the 1-forms

$$\alpha = \sum_\ell \alpha^\ell dx_\ell, \quad \beta = \sum_\ell \beta^\ell dx_\ell$$

is given by

$$(\alpha, \beta) = \sum_\ell \alpha^\ell \beta^\ell.$$

Second, if $\lambda = \alpha_1 \wedge \cdots \wedge \alpha_p$ and $\mu = \beta_1 \wedge \cdots \wedge \beta_p$ are p-forms made by 1-forms α_i and β_i for $1 \le i \le p$, we put

$$(\lambda, \mu) = \det \left((\alpha_i, \beta_j) \right)_{i,j}. \tag{6.24}$$

Let $\Lambda^p(D)$ be the set of p-forms on D. The outer derivative and the wedge product of these differential forms are denoted by

$$d : \Lambda^p(D) \to \Lambda^{p+1}(D) \tag{6.25}$$

and $\wedge$, respectively. The Hodge operator $* : \Lambda^p(D) \to \Lambda^{n-p}(D)$ is defined by

$$\omega \wedge \tau = (*\omega, \tau) \, dx_1 \wedge \cdots \wedge dx_n \tag{6.26}$$

for $\omega \in \Lambda^p(D)$ and $\tau \in \Lambda^{n-p}(D)$, and then it follows that

$$*(dx_{j_1} \wedge \cdots \wedge dx_{j_p}) = \operatorname{sgn} \sigma \cdot dx_{j_{p+1}} \wedge \cdots \wedge dx_{j_n}, \tag{6.27}$$

where $\sigma : (1, \cdots, n) \mapsto (j_1, \cdots, j_n)$, $\sigma \in \mathbf{S}_n$. Then we define the co-derivative

$$\delta = (-1)^p *^{-1} d* : \Lambda^p(D) \to \Lambda^{p-1}(D), \tag{6.28}$$

to obtain

$$\delta d + d\delta = -\Delta : \Lambda^p(D) \to \Lambda^p(D).$$

By this definition, if

$$B = \sum_i B^i dx_i$$

and

$$\omega = \sum_{i<j} \omega^{ij} dx_i \wedge dx_j$$

are 1-form and 2-form, respectively, it follows that

$$\delta B = -\sum_i B^i_i \tag{6.29}$$

and

$$\delta \omega = -\sum_{i,\ell} \tilde{\omega}^{\ell i}_\ell dx_i, \tag{6.30}$$

for $\tilde{\omega}^{ij}$ defined by (6.23). Here and henceforth, we put

$$B_i^i = \frac{\partial B^i}{\partial x_i}, \quad \tilde{\omega}_\ell^{\ell i} = \frac{\partial \tilde{\omega}^{\ell i}}{\partial x_\ell}.$$

Given

$$\omega \in \Lambda^p(D), \quad \theta \in \Lambda^{p-1}(D),$$

we have

$$(d\theta, \omega) \, dx_1 \wedge \cdots \wedge dx_n = d\theta \wedge *\omega$$
$$= d(\theta \wedge *\omega) + (-1)^p \theta \wedge d * \omega$$

and hence

$$\int_D (d\theta, \omega) \, dx_1 \wedge \cdots \wedge dx_n = \int_{\partial D} \theta \wedge *\omega$$
$$+ \int_D (\theta, \delta\omega) \, dx_1 \wedge \cdots \wedge dx_n \tag{6.31}$$

if the coefficients of θ and ω are in $H^1(D)$.

The area element ds on ∂D, on the other hand, is defined by

$$ds = \sum_i \nu^i * dx_i, \tag{6.32}$$

which results in the *vector area element*

$$\nu ds = (*dx_1, \cdots, *dx_n)^T,$$

or

$$\nu^i ds = *dx_i, \quad 1 \le i \le n$$

as in (1.27) in §1.5.

Lemma 6.1. *It holds that*

$$*B = (B, \nu) \, ds, \quad B \wedge *C = (\nu \wedge B, C) \, ds$$

if B and C are 1-form and 2-form, respectively.

Proof. Let

$$B = \sum_i B^i dx_i, \quad C = \sum_{i<j} \omega^{ij} dx_i \wedge dx_j, \quad \nu = \sum_i \nu^i dx_i.$$

First, we have

$$(B, \nu) \, ds = \sum_i B^i \nu^i ds = \sum_i B^i * dx_i = * \sum_i B^i dx_i = *B.$$

Second, it holds that

$$\nu \wedge B = \sum_{i,j} (\nu^i dx_i) \wedge (B^j dx_j) = \sum_{i<j} (\nu^i B^j - \nu^j B^i) dx_i \wedge dx_j$$

and hence

$$(\nu \wedge B, C)\, ds = \sum_{i<j} (\nu^i B^j - \nu^j B^i) C^{ij}\, ds$$

$$= \sum_{i<j} C^{ij} (B^j * dx_i - B^i * dx_j).$$

Here, we have

$$*C = \sum_{i<j} C^{ij} * (dx_i \wedge dx_j),$$

which implies

$$B \wedge *C = \sum_{k} \sum_{i<j} B^k C^{ij} dx_k \wedge *(dx_i \wedge dx_j)$$

$$= \sum_{i<j} C^{ij} (B^j * dx_i - B^i * dx_j).$$

Then it follows that

$$B \wedge *C = (\nu \wedge B, C)\, ds$$

and hence the result. □

Henceforth, we write

$$\int_D \cdots dx_1 \wedge \cdots \wedge dx_n = \int_D \cdots$$

and

$$\int_{\partial D} \cdots ds = \int_{\partial D} \cdots,$$

in short. Confirm that if $\theta \in H^1(D)$ is p-form, its trace $\theta|_{\partial D}$ belongs to $H^{1/2}(\partial D)$.

We show the following lemma of Gauss and Stokes formulate, noticing their validity to differential forms on Minkowski spaces.

Lemma 6.2. *If φ, B, and J are 0-form, 1-form, and 2-form belonging to $H^1(D)$, respectively, it holds that*

$$\int_D (\delta B, \varphi) = \int_D (B, d\varphi) - \int_{\partial D} (B, \nu)\varphi \tag{6.33}$$

$$\int_D (dB, J) = \int_D (B, \delta J) + \int_{\partial D} (\nu \wedge B, J). \tag{6.34}$$

Proof. Having

$$\varphi|_{\partial D}, B|_{\partial D}, J|_{\partial D} \in H^{1/2}(\partial D),$$

we apply (6.31) for $\omega = B$ and $\theta = \varphi$. Since

$$\theta \wedge *\omega = \varphi \cdot *B = \varphi \cdot (B, \nu)\, ds \quad \text{on } \partial D,$$

it holds that (6.33). For (6.34) we put

$$\omega = J, \quad \theta = B$$

in (6.31). There arises

$$\theta \wedge *\omega = B \wedge *J = (\nu \wedge B, J)\, ds \quad \text{on } \partial D,$$

and hence the conclusion. $\qquad\square$

We use $H^{-1/2}(\partial D) = H^{1/2}(\partial D)'$ and the paring $\langle\,,\,\rangle$ between $H^{1/2}(\partial D)$ and $H^{-1/2}(\partial D)$.

Lemma 6.3. *Let p be 0-form in $H^1(D)$.*

(1) If $\Delta p \in H^1(D)'$, then

$$(dp, \nu)|_{\partial D} \in H^{-1/2}(\partial D) \tag{6.35}$$

is well-defined, and it holds that

$$\langle \varphi, (dp, \nu) \rangle = \int_D (d\varphi, dp) + \langle \varphi, \Delta p \rangle, \quad \forall \varphi \in H^1(D). \tag{6.36}$$

(2) The 2-form

$$\nu \wedge dp|_{\partial D} \in H^{-1/2}(\partial D)$$

is well-defined, and is continuous in $p \in H^1(D)$. It holds that

$$\langle J, \nu \wedge dp \rangle = -\int_D (\delta J, dp) \tag{6.37}$$

for any 2-form $J \in H^1(D)$.

Proof. In the first case we have $p \in H^1(D)$ and $\Delta p \in H^1(D)'$, and hence the mapping

$$\varphi \in H^1(D) \mapsto \int_D (d\varphi, dp) + \langle \varphi, \Delta p \rangle$$

is bounded linear. We show that this mapping is reduced to

$$\varphi \in H^{1/2}(\partial D) \mapsto \int_D (\varphi, dp) + \langle \varphi, \Delta p \rangle, \tag{6.38}$$

to define

$$(dp, \nu)|_{\partial D}$$

in (6.35) by (6.38).

In fact, since

$$H^{1/2}(\partial D) \cong H^1(D)/H_0^1(D) \tag{6.39}$$

holds, the well-posedness of (6.38) follows from

$$\int_D (d\varphi, dp) + \langle \varphi, \Delta p \rangle = 0, \quad \forall \varphi \in H_0^1(D).$$

This equality is reduced to

$$\int_D (d\varphi, dp) + \langle \varphi, \Delta p \rangle = 0, \quad \forall \varphi \in C_0^\infty(D),$$

or

$$\int_D (d\varphi, dp) + (\Delta \varphi, p) = 0, \quad \forall \varphi \in C_0^\infty(D), \tag{6.40}$$

which is valid to $p \in H^1(D)$ by (6.33).

If $p \in H^2(D)$, the above

$$(dp, \nu)|_{\partial D}$$

coincides with

$$(dp|_{\partial D}, \nu) \in H^{1/2}(D), \tag{6.41}$$

by (6.33) because

$$B = dp, \quad \delta d = -\Delta$$

on $\Lambda^0(D)$, and therefore, this

$$(dp, \nu)|_{\partial D}$$

in (6.35) for $p \in H^1(D)$ with $\Delta p \in H^1(D)'$ is consistent to (6.41) for $p \in H^2(D)$.

The proof of the second case is similar. First, given $p \in H^1(D)$, we regard the right-hand side of (6.36) as a bounded linear mapping of 2-forms belonging to $H^1(D)$:

$$J \in H^1(D) \mapsto -\int_D (dp, \delta J).$$

We note that this mapping is continuous in $p \in H^1(D)$ in the operator norm. Second, this mapping is regarded as an element in $H^{-1/2}(\partial D)$ by (6.39), because the right-hand side is 0 for $J \in H_0^1(D)$:

$$J \in H^{1/2}(D) \mapsto -\int_D (dp, \delta J),$$

which ensures the well-posedness of $\nu \wedge dp \in H^{-1/2}(\partial D)$ by

$$\langle J, \nu \wedge dp \rangle = -\int_D (dp, \delta J).$$

Finally, we observe that equality (6.37) for $p \in H^2(D)$ holds with

$$(\nu \wedge dp|_{\partial D}, \nu) \in H^{1/2}(\partial D)$$

by equality (6.34) applied to $B = dp$ by $d^2 = 0$. Hence the above

$$\nu \wedge dp \in H^{-1/2}(\partial D)$$

is identified with

$$\nu \wedge dp|_{\partial D} \in H^{1/2}(\partial D)$$

if $p \in H^2(\Omega)$. $\qquad\square$

Remark 6.4. Writing

$$\langle dp, \nu \rangle = \frac{\partial p}{\partial \nu} \in H^{-1/2}(\partial D)$$

in (6.35), we obtain *Green's formula* in the form of

$$\langle g, \Delta h \rangle_{H^1(D), H^1(D)'} - \langle h, \Delta g \rangle_{H^1(D), H^1(D)'}$$
$$= \left\langle g, \frac{\partial h}{\partial \nu} \right\rangle_{H^{1/2}(\partial D), H^{-1/2}(\partial D)} - \left\langle h, \frac{\partial g}{\partial \nu} \right\rangle_{H^{1/2}(\partial D), H^{-1/2}(\partial D)} \tag{6.42}$$

valid to $g, h \in H^1(D)$ satisfying

$$\Delta g, \Delta h \in H^1(D)'.$$

Remark 6.5. If $\Omega = \Omega_+ \cup \Gamma \cup \Omega_-$ is a bounded Lipschitz domain with $C^{0,1}$ interface $\mathcal{M}$ and $\Gamma = \Gamma_\pm = \partial \Omega_\pm$, any 0-form $p \in H^1(\Omega_\pm)$ admits 2-forms on $\Gamma_\pm$ as in

$$\nu \wedge dp|_{\Gamma_\pm} \in H^{-1/2}(\Gamma_\pm) = H_0^{1/2}(\Gamma_\pm)'.$$

6.4   2-Forms on Euclidean Spaces

Here we show Theorem 6.3 in §6.2. Let $\Omega \subset \mathbf{R}^n$ be a domain with interface. If $p \in H^1(\Omega)$ is 0-form, the 2-forms

$$\nu \wedge dp|_{\Gamma_\pm} \in H^{-1/2}(\Gamma_\pm))$$

are well-defined by Remark 6.5. Identifying $H^{-1/2}(\Gamma_\pm)$ with $H^{-1/2}(\Gamma)$, we define 2-form on Γ by

$$[\nu \wedge dp]_-^+ = \nu \wedge dp|_{\Gamma_+} - \nu \wedge dp|_{\Gamma_-} \in H^{-1/2}(\Gamma).$$

Lemma 6.4 ([Kobayashi–Suzuki–Watanabe (2003)]). *If*

$$p \in H^1(\Omega),$$

it holds that

$$[\nu \wedge dp]_-^+ = 0 \quad in \ H^{-1/2}(\Gamma). \tag{6.43}$$

Proof. Recall that ν is the outer unit normal vector on Γ_- extended smoothly on Ω. Given 2-form J on Ω of which coefficients are in $C_0^\infty(\Omega)$, therefore, we obtain

$$\pm \langle J, \nu \wedge dp \rangle_{H_0^{1/2}(\Gamma_\pm), H^{-1/2}(\Gamma_\pm)} = \int_{\Omega_\pm} (\delta J, dp)$$

by Lemma 6.3, which implies

$$\left[\langle J, \nu \wedge dp \rangle_{H_0^{1/2}(\Gamma), H^{-1/2}(\Gamma)} \right]_-^+ = \int_\Omega (\delta J, dp). \tag{6.44}$$

The right-hand side of (6.44) is equal to 0 if

$$p \in H^2(\Omega)$$

by (6.34) for $B = dp$ and $D = \Omega$, because the coefficients of J are in $C_0^\infty(\Omega)$:

$$\left[\langle J, \nu \wedge dp \rangle_{H_0^{1/2}(\Gamma), H^{-1/2}(\Gamma)} \right]_-^+ = 0. \tag{6.45}$$

Then it follows that (6.43) because $J \in C_0^\infty(\Omega)$ is arbitrary.

This equality (6.43) is extended to

$$p \in H^1(\Omega)$$

from the continuity of

$$p \in H^1(\Omega) \mapsto \nu \wedge dp \in H^{-1/2}(\Gamma_\pm)$$

because $C^\infty(\overline{\Omega})$ is dense in $H^1(\Omega)$. $\square$

Lemma 6.5. *If* $p \in H^1(\Omega)$ *is 0-form, it holds that*

$$[\nu^i p_j - \nu^j p_i]_-^+ = 0, \ 1 \le i, j \le n, \quad in \ H^{-1/2}(\Gamma),$$

where

$$p_i = \frac{\partial p}{\partial x_i}.$$

Proof. The result is a direct consequence of Lemma 6.4 because

$$dp = \sum_i p_i dx_i$$

and

$$\nu \wedge dp = \sum_{i<j} (\nu^i p_j - \nu^j p_i) dx_i \wedge dx_j.$$

$\square$

Given a 0-form f, let

$$(\nu, d)f = \sum_i \nu^i f_i = \frac{\partial f}{\partial \nu}, \quad f_i = \frac{\partial f}{\partial x_i}. \tag{6.46}$$

Given 2-form ω, recall also $\hat{\omega}^i$ defined in Theorem 6.3.

Lemma 6.6. *If* $\omega \in H^1(\Omega)$ *is 2-form, it holds that*

$$\left[\delta\omega + \sum_i [(\nu, d)(\nu, \hat{\omega}^i) dx_i \right]_-^+ = 0 \quad in \ H^{-1/2}(\Gamma).$$

Proof. Since (6.30) holds for

$$\omega = \sum_{i<j} \omega^{ij} dx_i \wedge dx_j,$$

it suffices to confirm

$$\left[\sum_\ell \tilde{\omega}_\ell^{\ell i} - \frac{\partial}{\partial \nu} (\nu, \hat{\omega}^i) \right]_-^+ = 0 \quad in \ H^{-1/2}(\Gamma), \ 1 \le i \le n. \tag{6.47}$$

Henceforth, we write again

$$A \sim B$$

if $A - B \in H^1(\Omega)$. It then follows that

$$[A - B]_-^+ = 0 \quad in \ H^{1/2}(\Gamma).$$

Fix i, and put $B^\ell = \tilde\omega^{\ell i}$ and

$$B = \sum_\ell B^\ell dx_\ell (= \hat\omega^i).$$

It follows that

$$\sum_\ell \tilde\omega_\ell^{\ell i} - \frac{\partial}{\partial\nu}(\nu, \hat\omega^i) = \sum_\ell \{B_\ell^\ell - \nu^\ell(\nu, B)_\ell\}$$

for

$$(\nu, B) = \sum_k \nu^k B^k, \quad (\nu, B)_\ell = \frac{\partial}{\partial x_\ell}(\nu, B).$$

We thus obtain

$$\sum_\ell \tilde\omega_\ell^{\ell i} - \frac{\partial}{\partial\nu}(\nu, \hat\omega^i) \sim \sum_\ell B_\ell^\ell - \sum_{\ell,k} \nu^\ell \nu^k B_\ell^k$$

$$= \sum_\ell B_\ell^\ell - \sum_{k,\ell} \nu^k \nu^\ell B_k^\ell$$

$$= \sum_\ell \{B_\ell^\ell - \nu^\ell(\nu, d)B^\ell\}.$$

For $p = B^\ell$ we have

$$B_\ell^\ell - \nu^\ell(\nu, d)B^\ell = p_\ell - \nu^\ell(\nu, d)p$$

$$= \sum_k \{(\nu^k)^2 p_\ell - \nu^\ell \nu^k p_k\} = \sum_k \nu^k(\nu^k p_\ell - \nu^\ell p_k)$$

Then (6.47) follows from Lemma 6.5 as

$$\left[B_\ell^\ell - \nu^\ell(\nu, d)B^\ell\right]_-^+ = \sum_k \nu^k \left[\nu^k p_\ell - \nu^\ell p_k\right]_-^+ = 0 \quad \text{in } H^{-1/2}(\Gamma).$$

$\square$

We are ready to complete the following proof.

Proof of Theorem 6.3. Since $\omega \in H^1(\Omega)$ is 2-form satisfying $\delta\omega = 0$, it holds that

$$\left[\frac{\partial}{\partial\nu}(\nu, \hat\omega^i)\right]_-^+ = 0 \quad \text{in } H^{-1/2}(\Gamma), \ 1 \le i \le n \qquad (6.48)$$

by Lemma 6.6. Using $d\omega = \theta \in L^2(\Omega)$ and $\delta\theta \in L^2(\Omega_\pm)$, we obtain

$$-\Delta\omega = (d\delta + \delta d)\omega = \delta\theta \in L^2(\Omega_\pm)$$

and hence there exists

$$-\Delta(\nu, \hat{\omega}^i) \in L^2(\Omega_\pm),$$

denoted by $h^i_\pm$ for $1 \le i \le n$:

$$-\Delta(\nu, \hat{\omega}^i) = h^i_\pm \quad \text{in } L^2(\Omega_\pm).$$

Define $h^i \in L^2(\Omega)$ by

$$h^i = h^i_\pm \quad \text{in } \Omega_\pm.$$

Equality (6.48) then implies

$$\int_\Omega h^i \varphi = \left(\int_{\Omega_+} + \int_{\Omega_-} \right) (d(\nu, \hat{\omega}^i), d\varphi) = \int_\Omega (d(\nu, \hat{\omega}^i), d\varphi)$$

$$= \int_\Omega (-\Delta\varphi) \cdot (\nu, \hat{\omega}^i), \quad \forall \varphi \in C_0^\infty(\Omega)$$

by (6.33), and hence

$$-\Delta(\nu, \hat{\omega}^i) = h^i \in L^2(\Omega), \quad 1 \le i \le n$$

in the sense of distributions in Ω. □

6.5 2-Forms on Minkowski Spaces

Turning to the proof of Theorem 6.1 in §6.2, we confirm the structure of differential forms in the *Minkowski space*

$$\mathbf{R}^4 \cong \mathbf{R}^3 \times \mathbf{R},$$

associated with the coordinates $x = (x_1, x_2, x_3) \in \mathbf{R}^3$ and $t = x_0 \in \mathbf{R}$. Let $D \subset \mathbf{R}^4$ be a domain in this space and $\Lambda^p(D)$ be the set of p-forms on D. The outer derivative and the wedge product are denoted similarly by d in (6.25) and $\wedge$, respectively.

Given 1-forms

$$\alpha = \sum_{i=1}^3 \alpha^i dx_i + \alpha^0 dx_0, \quad \beta = \sum_{i=1}^3 \beta^i dx_i + \beta^0 dx_0,$$

the Minkowski inner product is defined by

$$(\alpha, \beta) = -\sum_{i=1}^3 \alpha^i \beta^i + \alpha^0 \beta^0. \tag{6.49}$$

Then the Minkowski inner product of p-froms on D is defined similarly by (6.24).

The Hodge operator $*$ is now defined by (6.27) in §6.3, where $(\ ,\ )$ is replaced by the Minkowski inner product. It then holds that

$$*(dx_0 \wedge dx_i \wedge dx_j) = -dx_k, \quad (i, j, k) = (1, 2, 3)$$
$$*(dx_1 \wedge dx_2 \wedge dx_3) = dx_0$$

and

$$*(dx_0 \wedge dx_j) = -dx_k \wedge dx_\ell$$
$$*(dx_j \wedge dx_k) = dx_0 \wedge dx_\ell, \quad (j, k, \ell) = (1, 2, 3).$$

Putting δ as in (6.28), this time we obtain

$$d\delta + \delta d = \square \equiv -\frac{\partial^2}{\partial t^2} + \Delta_x : \Lambda^p(D) \to \Lambda^p(D), \tag{6.50}$$

where

$$\Delta_x = \frac{\partial^2}{\partial x_1^2} + \frac{\partial^2}{\partial x_2^2} + \frac{\partial^2}{\partial x_3^2}.$$

Then Maxwell's equation (6.14) is represented by differential forms on $D = \Omega$,

$$d\omega = 0, \ d * \omega = -j \quad \text{in } \Omega, \tag{6.51}$$

for

$$\omega = E^2 dx_0 \wedge dx_1 + E^2 dx_0 \wedge dx_2 + E^3 dx_0 \wedge dx_3$$
$$\quad - B^1 dx_2 \wedge dx_3 - B^2 dx_3 \wedge dx_1 - B^3 dx_1 \wedge dx_2$$
$$j = J^1 dx_0 \wedge dx_2 \wedge dx_3 + J^2 dx_0 \wedge dx_3 \wedge dx_1 + J^3 dx_0 \wedge dx_1 \wedge dx_2$$
$$\quad + \rho dx_1 \wedge dx_2 \wedge dx_3$$

and

$$E = \begin{pmatrix} E^1 \\ E^2 \\ E^3 \end{pmatrix}, \quad B = \begin{pmatrix} B^1 \\ B^2 \\ B^3 \end{pmatrix}, \quad J = \begin{pmatrix} J^1 \\ J^2 \\ J^3 \end{pmatrix}.$$

This (6.51) is equivalent to

$$d\theta = j, \ \delta\theta = 0 \quad \text{in } \Omega$$

for $\theta = -*\omega$.

Henceforth we write ω for θ, and θ for j. Thus we handle with

$$d\omega = \theta, \ \delta\omega = 0 \quad \text{in } \Omega \tag{6.52}$$

for

$$\omega = E^2 dx_0 \wedge dx_1 + E^2 dx_0 \wedge dx_2 + E^3 dx_0 \wedge dx_3$$
$$\quad + B^1 dx_2 \wedge dx_3 + B^2 dx_3 \wedge dx_1 + B^3 dx_1 \wedge dx_2$$
$$\theta = J^1 dx_0 \wedge dx_2 \wedge dx_3 + J^2 dx_0 \wedge dx_3 \wedge dx_1 + J^3 dx_0 \wedge dx_1 \wedge dx_2$$
$$\quad + \rho dx_1 \wedge dx_2 \wedge dx_3.$$

We are ready to give the following proof.

Proof of Theorem 6.1. In the above setting, the domain $D \subset \mathbf{R}^4$ is provided with interface and $\omega \in H^1(\Omega)$ is 2-form satisfying (6.52). As in Theorem 6.3, therefore, if

$$\delta\theta \in L^2(\Omega_\pm), \tag{6.53}$$

it holds that

$$(\delta d + d\delta)(\nu, \hat{\omega}^i) \in L^2(\Omega), \quad 0 \le i \le 3 \tag{6.54}$$

in accordance with the Minkowski inner product (6.49).

First, since

$$\begin{aligned}
\delta\theta = &-\left(\frac{\partial J^2}{\partial x_1} - \frac{\partial J^1}{\partial x_2}\right) dx_0 \wedge dx_3 - \left(\frac{\partial J^1}{\partial x_3} - \frac{\partial J^3}{\partial x_1}\right) dx_0 \wedge dx_2 \\
&-\left(\frac{\partial J^3}{\partial x_2} - \frac{\partial J^2}{\partial x_3}\right) dx_0 \wedge dx_1 + \left(\frac{\partial J^1}{\partial x_3} + \frac{\partial \rho}{\partial x_1}\right) dx_2 \wedge dx_3 \\
&+\left(\frac{\partial J^2}{\partial x_0} + \frac{\partial \rho}{\partial x_2}\right) dx_3 \wedge dx_1 + \left(\frac{\partial J^3}{\partial x_0} + \frac{\partial \rho}{\partial x_3}\right) dx_1 \wedge dx_2.
\end{aligned}$$

the condition (6.53) is equivalent to (6.16).

Second, we obtain

$$\hat{\omega}^0 = \tilde{\omega}^{01} dx_0 + \tilde{\omega}^{20} dx_2 + \tilde{\omega}^{30} dx_3 = -B^1 dx_1 - B^2 dx_2 - B^3 dx_3$$

and hence

$$(\nu, \hat{\omega}^0) = \nu^1 B^1 + \nu^2 B^2 + \nu^3 B^3.$$

It holds also that

$$\hat{\omega}^1 = \tilde{\omega}^{01} dx_0 + \tilde{\omega}^{21} dx_2 + \tilde{\omega}^{31} dx_3 = B^1 dx_0 - E^3 dx_2 + E^2 dx_3$$

and therefore,

$$(\nu, \hat{\omega}^1) = \nu^0 B^1 + \nu^2 E^3 - \nu^3 E^2.$$

We obtain, similarly,

$$(\nu, \hat{\omega}^2) = \nu^0 B^2 + \nu^3 E^1 - \nu^1 E^3$$
$$(\nu, \hat{\omega}^3) = \nu^0 B^3 + \nu^1 E^2 - \nu^2 E^1.$$

Hence (6.54) is equivalent to (6.17), and we obtain the result by (6.50). $\quad\square$

6.6   1-Forms

Here we prove Theorem 6.2 for 1-forms, where $\Omega \subset \mathbf{R}^n$ is a domain with interface,

$$B = \sum_i B^i dx_i \in H^1(\Omega)$$

is 1-form, and B^ν and B^τ are defined by (6.21). We recall (6.29), (6.46), and

$$\delta B = -\sum_i B^i_i, \quad \frac{\partial}{\partial \nu} = \sum_k \nu^k \frac{\partial}{\partial x_k} = (\nu, d),$$

to put

$$\delta_\nu B = -\sum_i \nu^i \frac{\partial B^i}{\partial \nu}$$

$$\delta_\tau B = \delta B - \delta_\nu B = -\sum_i \left(B^i_i - \nu^i (\nu, d) B^i \right). \tag{6.55}$$

Lemma 6.7. *If* $B \in H^1(\Omega)$ *is* 1-*form, it holds that*

$$[\delta_\tau B]^+_- = 0 \quad in \ H^{-1/2}(\Gamma).$$

Proof. In (6.55) we have

$$B^i_i - \nu^i(\nu, d)B^i = \sum_k \nu^k (\nu^k B^i_i - \nu^i B^i_k)$$

Then it holds that

$$\left[\nu^k B^i_i - \nu^i B^i_k \right]^+_- = 0 \quad in \ H^{-1/2}(\Gamma)$$

by Lemma 6.5.   □

Lemma 6.8. *If* $B \in H^1(\Omega)$ *is* 1-*form, it holds that*

$$[\delta B^\tau]^+_- = 0 \quad in \ H^{-1/2}(\Gamma). \tag{6.56}$$

Proof. We have

$$\delta B^\tau = \delta(B - \nu(\nu, B)) = \sum_i \left(-B^i_i + (\nu^i(\nu, B))_i \right).$$

As in the proof of Lemma 6.6, therefore, it holds that

$$\delta B^\tau \sim \sum_i \left(-B^i_i + \nu^i(\nu, B)_i \right)$$

$$\sim \sum_i (-B^i_i) + \sum_{i,\ell} \nu^i \nu^\ell B^\ell_i = \sum_i (-B^i_i) + \sum_{i,\ell} \nu^i \nu^\ell B^i_\ell$$

$$= -\sum_i \left(B^i_i - \nu^i(\nu, d) B^i \right) = \delta_\tau B,$$

and the result follows from the Lemma 6.7.   □

Lemma 6.9. *If $B \in H^1(\Omega)$ is 1-form, it holds that*

$$[dB^\nu]_-^+ = 0 \quad in \ H^{-1/2}(\Gamma).$$

Proof. Since

$$B^\nu = \sum_i (B, \nu)\nu^i dx_i$$

it holds that

$$dB^\nu = \sum_{i<j} \left[((B, \nu)\nu^j)_i - ((B, \nu)\nu^i)_j \right] dx_i \wedge dx_j$$

$$\sim \sum_{i<j} \left\{ (B, \nu)_i \nu^j - (B, \nu)_j \nu^i \right\} dx_i \wedge dx_j.$$

Then we obtain the result by Lemma 6.5. $\qquad\qquad\square$

Proof of the first case of Theorem 6.2. Given the 1-form $B \in H^1(\Omega)$, we have

$$\delta B = \delta B^\tau + \delta B^\nu \quad in \ \Omega,$$

where equality (6.56) is applicable for the first term on the right-hand side.
For the second term we use

$$\delta B^\nu = \delta_\nu B^\nu + \delta_\tau B^\nu,$$

where Lemma 6.7 is applicable to the second term on the right-hand side of this equality. Since

$$\delta_\nu B^\nu = -\sum_{i,k} \nu^i \nu^k ((B.\nu)\nu^i)_k \sim -\sum_{i,k} (\nu^i)^2 \nu^k (B, \nu)_k$$

$$= (\nu, d)(B, \nu) = \frac{\partial}{\partial \nu}(B, \nu)$$

we see that $\delta B = 0$ in Ω implies

$$\left[\frac{\partial}{\partial \nu}(B, \nu) \right]_-^+ = 0 \quad in \ H^{-1/2}(\Gamma).$$

Then we obtain

$$(d\delta + \delta d)B \in L^2(\Omega)$$

in the sense of distributions, similarly to the proof of Theorem 6.3. $\qquad\square$

Proof of the second case of Theorem 6.2. Since

$$0 = dB = dB^\tau + dB^\nu \quad \text{in } \Omega$$

we obtain

$$[dB^\tau]_-^+ = 0 \quad \text{in } H^{-1/2}(\Gamma) \tag{6.57}$$

by Lemma 6.9.

Given 2-form $J \in C_0^\infty(\Omega)$, we have

$$\int_\Omega (d\delta B^\tau, C) = \int_\Omega (B^\tau, d\delta C)$$

by Lemma 6.8, while equality (6.57) implies

$$\int_\Omega (\delta d B^\tau, C) = \int_\Omega (B^\tau, \delta d C).$$

Hence we obtain $-\Delta B^\tau \in L^2(\Omega)$. □

6.7 $d - \delta$ Systems

Here we describe the generalization of Theorem 6.2 and Theorem 6.3 on 1-forms and 2-forms, respectively.

Let $\Omega \subset \mathbf{R}^n$ be a domain with interface, and define $\Omega_\pm$ and $\Gamma = \Gamma_\pm$ by (6.15). Let, next, $\nu = (\nu^i)$ be the outer unit normal vector on Γ_- extended smoothly on Ω, which is identified with the 1-form

$$\nu = \nu^1 dx_1 + \cdots + \nu^n dx_n.$$

Let, finally, $\omega \in H^1(\Omega)$ be p-form for $1 \le p \le n - 1$.

Theorem 6.4. *Let* $\theta \in L^2(\Omega)$ *be* $(p+1)$-*form, and assume*

$$d\omega = \theta, \ \delta\omega = 0 \ in \ \Omega, \quad \delta\theta \in L^2(\Omega_\pm). \tag{6.58}$$

Then it holds that

$$\Delta(\nu \wedge *\omega) \in L^2(\Omega).$$

Theorem 6.5. *Let* $\theta \in L^2(\Omega)$ *be* $(p-1)$-*form, and assume*

$$d\omega = 0, \ \delta\omega = \theta \ in \ \Omega, \quad d\theta \in L^2(\Omega_\pm).$$

Then it holds that

$$\Delta(\nu \wedge \omega) \in L^2(\Omega).$$

Remark 6.6. First, Theorem 6.4 and Theorem 6.5 are valid to any

$$1 \le p \le n - 1,$$

and are equivalent each other. Second, Theorem 6.2 in §6.2 is derived from both theorems for $p = 1$. Finally, Theorem 6.3 is equivalent to Theorem 6.4 for $p = 2$ ([Suzuki–Watanabe (2023)]).

Here we describe the proof of Theorem 6.4. Let

$$\omega = \sum_{i_1 < \cdots < i_p} \omega^{i_1 \cdots i_p} dx_{i_1} \wedge \cdots \wedge dx_{i_p} \in H^1(\Omega), \tag{6.59}$$

and put

$$\tilde{\omega}^{i_1 \cdots i_p} = \operatorname{sgn} \sigma \cdot \omega^{i_1' \cdots i_p'}, \quad 1 \le i_1, \cdots, i_p \le n, \tag{6.60}$$

where

$$\sigma : (i_1, \cdots, i_p) \mapsto (i_1', \cdots, i_p'), \quad i_1' < \cdots < i_p'. \tag{6.61}$$

Then it holds that

$$\delta\omega = - \sum_{i_2 < \cdots < i_p} \sum_{\ell} \tilde{\omega}^{\ell i_2 \cdots i_p} dx_{i_2} \wedge \cdots \wedge dx_{i_p} \tag{6.62}$$

First, we extend Lemma 6.6 in §6.4 as follows, recalling

$$(\nu, d) = \sum_i \nu^i \frac{\partial}{\partial x_i}.$$

Lemma 6.10. *If $\omega \in H^1(\Omega)$ is p-form, it holds that*

$$\left[\delta\omega + \sum_{i_2 < \cdots < i_p} (\nu, d)(\nu, \hat{\omega}^{i_2 \cdots i_p}) dx_{i_2 \wedge \cdots \wedge dx_{i_p}} \right]_-^+ = 0 \quad in \ H^{-1/2}(\Gamma), \tag{6.63}$$

where

$$\hat{\omega}^{i_2 \cdots i_p} = \sum_{\ell} \tilde{\omega}^{\ell i_2 \cdots i_p} dx_\ell \tag{6.64}$$

Proof. The proof is similar. Take ω as in (6.59), and fix $i_2 < \cdots < i_p$. We put

$$B = \sum_{\ell} B^\ell dx_\ell, \quad B^\ell = \tilde{\omega}^{\ell i_2 \cdots i_p},$$

recalling (6.60). Then it holds that

$$\sum_\ell \tilde\omega^{\ell i_2\cdots i_p} - (\nu,d)(\hat\omega^{i_2\cdots i_p},\nu) = \sum_\ell \{B^\ell_\ell - \nu^\ell(B,\nu)_\ell\}$$

$$\sim \sum_\ell B^\ell_\ell - \sum_{\ell,k}\nu^k\nu^\ell B^k_\ell = \sum_\ell B^\ell_\ell - \sum_{\ell,k}\nu^k\nu^\ell B^\ell_k$$

$$= \sum_\ell \{B^\ell_\ell - \nu^\ell(\nu,d)B^\ell\}$$

because ν is extended as a $C^{0,1}$ vector field on Ω from the assumption.

Here we fix ℓ, put $p = B^\ell$, and notice

$$B^\ell_\ell - \nu^\ell(\nu,d)B^\ell = p_\ell - \nu^\ell(\nu,d)p = \sum_k \{(\nu_k)^2 p_\ell - \nu^\ell\nu^k p_k\}$$

$$= \sum_k \nu^k(\nu^k p_\ell - \nu^\ell p_k).$$

Then it follows that

$$[B^\ell_\ell - \nu^\ell(\nu,d)B^\ell]^+_- = 0 \quad \text{in } H^{-1/2}(\Gamma)$$

from Lemma 6.5 in §6.4. We thus obtain

$$\left[\sum_\ell \tilde\omega^{\ell i_2\cdots i_p}_\ell - (\nu,d)(\hat\omega^{i_2\cdots i_p},\nu)\right]^+_- = 0 \quad \text{in } H^{-1/2}(\Gamma),$$

and then (6.62) implies (6.63). □

Proof of Theorem 6.4: Since $\delta\omega = 0$ in Theorem 6.4, it follows that

$$\left[\sum_{i_2<\cdots<i_p}[(\nu,d)(\nu,\hat\omega^{i_2\cdots i_p})]dx_{i_2}\wedge\cdots\wedge dx_{i_p}\right]^+_- = 0 \quad \text{in } H^{-1/2}(\Gamma). \quad (6.65)$$

We obtain, second,

$$-\Delta\omega = (d\delta + \delta d)\omega = \delta\theta \in L^2(\Omega_\pm)$$

from the assumption of this theorem. It thus holds that

$$\Delta\omega^{i_1\cdots i_p} \in L^2(\Omega_\pm), \quad i_1 < \cdots < i_p$$

by (6.59).

Let

$$h^{i_2\cdots i_p} = \Delta(\nu,\hat\omega^{i_2\cdots i_p}) \in L^2(\Omega_\pm).$$

Then equality (6.65) implies

$$-\int_\Omega (\hat\omega^{i_2\cdots i_p}, \nu)\Delta\varphi = \left(\int_{\Omega_+} + \int_{\Omega_-}\right) h^{i_2\cdots i_p}\varphi$$

$$= \int_\Omega h^{i_2\cdots i_p}\varphi$$

for any $\varphi \in C_0^\infty(\Omega)$ by Green's formula, (6.42) in §6.3. We thus obtain

$$\Delta(\hat\omega^{i_2\cdots i_p}, \nu) \in L^2(\Omega)$$

and hence

$$\Delta\beta \in L^2(\Omega)$$

for the $(p-1)$-form β defined by

$$\beta = \sum_{i_2<\cdots<i_p} (\hat\omega^{i_2\cdots i_p}, \nu) dx_{i_2} \wedge \cdots \wedge dx_{i_p}.$$

The conclusion (6.31) is thus reduced to the following lemma. $\square$

Lemma 6.11. *It holds that*

$$\nu \wedge *\omega = *\beta \quad in \ \Lambda^{n-p+1}(\Omega), \tag{6.66}$$

where

$$\beta = \sum_{i_2<\cdots<i_p} (\nu, \hat\omega^{i_2\cdots i_p}) dx_{i_2} \wedge \cdots \wedge dx_{i_p}.$$

Proof. It suffices to show (6.66) for

$$\omega = dx_1 \wedge \cdots \wedge dx_p. \tag{6.67}$$

In this case it holds that

$$\tilde\omega^{i_1\cdots i_p} = \begin{cases} \operatorname{sgn}\sigma, & \{i_1, \cdots, i_p\} = \{1, \cdots, p\} \\ 0, & \text{otherwise,} \end{cases}$$

where $\sigma : (i_1, \cdots, i_p) \mapsto (1, \cdots, p)$. Then (6.64) implies

$$\hat\omega^{i_2\cdots i_p} = \sum_\ell \tilde\omega^{\ell i_2\cdots i_p}\, dx_\ell$$

$$= \begin{cases} \sum_\ell \operatorname{sgn}\sigma_\ell\, dx_\ell, & 1 \le i_2, \cdots, i_p \le p \\ 0, & \text{otherwise,} \end{cases}$$

where $\sigma_\ell : (\ell, i_2, \cdots, i_p) \mapsto (1, \cdots, p)$. We have, for example,

$$(\hat\omega^{2,\cdots,p}, \nu) = \nu^1,$$

and hence

$$\beta = \sum_{i_2 < \cdots < i_p} (\hat{\omega}^{i_2 \cdots i_p}, \nu) dx_{i_2} \wedge \cdots \wedge dx_{i_p}$$
$$= \nu^1 dx_2 \wedge \cdots \wedge dx_p + \nu^2 dx_3 \wedge \cdots \wedge dx_p \wedge dx_1$$
$$+ \cdots + \nu^p dx_1 \wedge \cdots \wedge dx_{p-1}.$$

It thus holds that

$$*\beta = \nu^1 dx_1 \wedge dx_{p+1} \wedge \cdots dx_n + \nu^2 dx_2 \wedge dx_{p+1} \wedge \cdots \wedge dx_n$$
$$+ \nu^p dx_p \wedge dx_{p+1} \wedge \cdots \wedge dx_n.$$

By (6.67), on the other hand, we obtain

$$*\omega = dx_{p+1} \wedge \cdots \wedge dx_n$$

and hence

$$\nu \wedge *\omega = (\nu^1 dx_1 + \cdots + \nu^n dx_n) \wedge dx_{p+1} \wedge \cdots \wedge dx_n$$
$$= \nu^1 dx_1 \wedge dx_{p+1} \wedge \cdots \wedge dx_n + \nu^2 dx_2 \wedge dx_{p+1} \wedge \cdots \wedge dx_n$$
$$+ \cdots + \nu^p dx_p \wedge dx_{p+1} \wedge \cdots \wedge dx_n = *\beta,$$

which completes the proof.   □

PART 3
Theory of Transformations

Chapter 7

Non-Standard Elliptic Regularity

As is observed in §4.3, sometimes, cancellation of the singularity is detected by using moving coordinates. Study on the transformation of variables is useful to make this process clear. The second fundamental form on the boundary takes an important role there, and non-standard elliptic regularity arises as an application. First, we confirm fundamental elliptic H^1 regularity on Lipschitz domains (§7.1–§7.2), and then introduce sectional curvatures to describe Laplacian on the boundary (§7.3) Finally, H^2 regularity on convex domains concerning the Poisson equation, in accordance with an identity on the boundary associated with its second fundamental form (§7.4).

7.1   Lipschitz Domains

We confirm several properties on the Lipschitz domain used in §6.3. Given an open set $\Omega \subset \mathbf{R}^N$, we say that its boundary $\Gamma = \partial\Omega$ is Lipschitz continuous if the following conditions are satisfied:

(1) Each $x \in \Gamma$ takes an open rectangle O with $x \in O$ and new orthogonal coordinates $y = (y', y_N)$, $y' = (y_1, \cdots, y_{N-1})$ such that

$$O = \{y \mid |y_j| < a_j, \ 1 \le j \le N\},$$

where $a_j > 0$, $1 \le j \le N$.

(2) There is a Lipschitz continuous function $\varphi : O' \to \mathbf{R}$,

$$O' = \{y' \mid |y_j| < a_j, \ 1 \le j \le N - 1\},$$

such that $|\varphi(y')| \le a_N/2$ for $y' \in O'$ and

$$\Omega \cap O = \{y \mid y_N > \varphi(y'), \ y' \in O\}, \quad \Gamma \cap O = \{y \mid y_N = \varphi(y'), \ y' \in O\}.$$

A bounded domain $\Omega \subset \mathbf{R}^N$ is called Lipschitz domain if its boundary is Lipschitz continuous. In the case of $N = 2$, such domain admits corners on the boundary except for cusps.

We study the elliptic boundary value problem, that is, the Dirichlet problem for the *Laplace equation*

$$\Delta u = 0 \text{ in } \Omega, \quad u = g \text{ on } \partial\Omega \tag{7.1}$$

and that for the *Poisson equation*

$$-\Delta u = f \text{ in } \Omega, \quad u = 0 \text{ on } \partial\Omega \tag{7.2}$$

using the Laplacian

$$\Delta = \sum_{i=1}^{N} \frac{\partial^2}{\partial x_i^2}.$$

The well-posedness of such problems is classical if $\partial\Omega$ is sufficiently smooth. Three approaches are possible to (7.1), called the methods of Perron, Fredholm, and Riemann [Suzuki (2022a)].

Perron's method ensures, given $g \in C(\partial\Omega)$, that there is a unique solution $u \in C^2(\Omega) \cap C(\overline{\Omega})$ to (7.1) if and only if any boundary point is regular in the sense of potential theory. A well-known sufficient condition for this regularity is the outscribing ball condition, which means that any $x_0 \in \partial\Omega$ takes an open ball B such that $\overline{\Omega} \cap \overline{B} = \{x_0\}$. Not all Lipschitz domains satisfy this condition. A sufficient condition of this outscribing ball condition is that $\partial\Omega$ is C^1, which means that the above $\varphi : O' \to \mathbf{R}$ is C^1. The other is the convexity of Ω.

Schauder theory on (7.2) requires for $\partial\Omega$ to be $C^{2,\theta}$, $0 < \theta < 1$, which means again that the above $\varphi : O' \to \mathbf{R}$ is provided with this regularity. If this condition is satisfied, there is a unique solution $u \in C^{2,\theta}(\overline{\Omega})$ for given $f \in C^\theta(\overline{\Omega})$. In the L^p, $1 < p < \infty$, *theory* one requires that $\partial\Omega$ is $C^{1,1}$, which is defined similarly by $\varphi : O' \to \mathbf{R}$. In this case, given $f \in L^p(\Omega)$, there is a unique solution $u \in W^{2,p}(\Omega)$ to (7.2) with the boundary value $u|_{\partial\Omega} \in W^{2-1/p,p}(\partial\Omega)$ taken in the sense of *trace*. Here, to define the Sobolev space on the boundary with the fractional exponent we use a norm associated with boundary integrals [Adams (1978); Nečas (1967, 1983)], recalling that the Sobolev space $W^{s,p}(\Omega)$, $s = m+\sigma$, $0 < \sigma < 1$, $1 \le p < \infty$, is the set of $v \in W^{m,p}(\Omega)$ satisfying

$$\iint_{\Omega \times \Omega} \frac{|\partial^\alpha u(x) - \partial^\alpha u(x')|^p}{|x - x'|^{N+\sigma p}} \, dx dx' < +\infty, \quad |\alpha| = m.$$

If $p = 2$ this space is defined equivalently using the local chart $\varphi : O' \to$ **R** and Fourier transformation [Mizohata (1973)]. The higher regularity is also available for $m = 2, 3, \cdots$. Thus if $\partial\Omega$ is $C^{m,1}$ and $f \in W^{m-1,p}(\Omega)$ then it holds that $u \in W^{m+1,p}(\Omega)$. Similarly, if $\partial\Omega$ is $C^{m+1,\theta}$, $0 < \theta < 1$, and $f \in C^{m-1,\theta}(\overline{\Omega})$ then it holds that $u \in C^{m+1,\theta}(\overline{\Omega})$.

These regularity theories are, however, invalid for general Lipschitz domains. Typical examples are observed for polygons on the plane [Brenner–Scott (2007); Fujita–Saito–Suzuki (2001)]. In fact, if $f(z)$ is holomorphic in $z = x_1 + \imath x_2$ then its real part is harmonic in $x = (x_1, x_2)$. We take the harmonic function

$$u = r^m \sin m\theta = \text{Im}\ (x_1 + \imath x_2)^m, \quad x = re^{\imath\theta}, \ m > 0, \qquad (7.3)$$

on the sector $0 < \theta < \alpha$. This function satisfies the Dirichlet boundary condition on $\theta = 0$ and $\theta = \alpha$ if and only if $m\alpha = \pi$. Then

$$u_{rr} \sim r^{m-2}, \quad r \downarrow 0$$

unless $m = 1$, and therefore, $u_{rr} \in L^2$ near $r = 0$ if and only if $m > 1$, which means $\alpha < \pi$.

This example indicates the role of the angular α in the L^p theory, $1 < p < \infty$, for (7.2). We recall that if $\partial\Omega$ is C^2 then this elliptic regularity is always satisfied. Here, we notice the role of convexity of the domain again, that is, for the H^2-regularity of the Poisson equation (7.2).

In fact, if Ω is convex and $f \in L^2(\Omega)$, there is a unique solution $u \in H^2(\Omega) \cap H_0^1(\Omega)$ to (7.2). The Neumann case is similar; if Ω is convex and $f \in L^2(\Omega)$ satisfies $\int_\Omega f\ dx = 0$, there is a unique $u \in H^2(\Omega)$ up to an additive constant such that

$$-\Delta u = f \text{ in } \Omega, \quad \frac{\partial u}{\partial \nu} = 0 \text{ on } \partial\Omega, \qquad (7.4)$$

where the normal derivative $\partial u/\partial \nu \in H^{1/2}(\partial\Omega) = W^{1/2,2}(\partial\Omega)$ of u is taken in the sense of trace. It may be useful to take

$$u = \text{Re}\ z^m = r^m \cos m\theta$$

in the sector $0 < \theta < \alpha$ this time, to understand this property of H^2-regularity in accordance with the convexity of Ω (§7.4).

The mixed problem is more delicate. This problem is formulated as

$$-\Delta u = f \text{ in } \Omega, \quad u = 0 \text{ on } \gamma_0, \quad \frac{\partial u}{\partial \nu} = 0 \text{ on } \gamma_1, \qquad (7.5)$$

using γ_i, $i = 0, 1$, closed sets with non-void interiors γ_i°, $i = 0, 1$, such that

$$\gamma_0 \cup \gamma_1 = \partial\Omega, \quad \gamma_0^\circ \cap \gamma_1^\circ = \emptyset.$$

We can use again example (7.3), $u = r^m \sin m\theta$ in the sector $0 < \theta < \alpha$. Here we take $\gamma_1 : \theta = 0$ and $\gamma_2 : \theta = \alpha$. Actually, this harmonic function satisfies the second boundary condition of (7.5) if and only if $m\alpha = \pi/2$. Hence the requirement $u \in H^2(\Omega)$ near $r = 0$ is reduced to $\theta \leq \pi/2$.

As we have described, H^2-regularity actually holds in both (7.2) and (7.4) if either $\partial\Omega$ is C^2 or Ω is convex. Since these results are localized, this H^2-regularity holds also to (7.5), provided that $\gamma_1 \cap \gamma_2 = \emptyset$, that is, the portions of the boundary where Dirichlet and Neumann boundary conditions are imposed are disconnected. Thus we obtain the H^2-regularity in (7.5) in this case, if either one of γ_i, $i = 1, 2$, is a boundary of a convex domain containing Ω and the other is C^2, or both γ_i, $i = 1, 2$, are C^2.

The elliptic regularity for corner domains has been studied in detail as in [Kondrat'ev (1967)]. Roughly speaking, the seeds of irregularity of the solution are only of finite dimension, which derives the Fredholm property of the elliptic operator [Dauge (1988)]. For a convex domain on the plane we have $W^{1,p}$-regularity for $|p-2| \ll 1$ under the Dirichlet boundary condition [Mayers (1963)]. This $W^{1,p}$-regularity of the Dirichlet problem holds for any bounded domain with C^1-boundary of arbitrary dimension [Simader (1972)]. From the above example, on the other hand, we need $m > 1 - \frac{2}{p}$ for this property on the sector $0 < \theta < \alpha$ to be valid, where $m\alpha = \pi/2$. Hence $W^{1,p}$-regularity on the planar domain for any $1 < p < \infty$ is valid only for the convex case.

7.2 H^1-Solution

Lipschitz domains are used in §6.3 to justify the Gauss and Stokes formulae in Lemma 6.3. Here we describe their properties more [Nečas (1967); Kufner–John–Fučik (1977); Grisvard (1985); Girault–Raviart (1986)], in accordance with the unique solvability of the boundary value problems for Laplace and Poisson equations, say, (7.1) and (7.2), respectively.

First, Poincaré's inequality

$$\|v\|_2 \leq C\|\nabla v\|_2, \quad v \in H_0^1(\Omega), \tag{7.6}$$

guarantees that $-\Delta : H_0^1(\Omega) \to H^{-1}(\Omega) = H_0^1(\Omega)'$ is an isomorphism for any bounded domain Ω. Hence each $f \in H^{-1}(\Omega)$ takes a unique $u \in H_0^1(\Omega)$ such that

$$(\nabla u, \nabla v) = \langle v, f \rangle, \quad \forall v \in H_0^1(\Omega).$$

Here and henceforth, (,) and $\langle$, $\rangle$ denote the standard L^2 norm and the duality pairing between $H_0^1(\Omega)$ and $H^{-1}(\Omega) = H_0^1(\Omega)'$, respectively. As far as the existence of the H^1-solution is concerned, therefore, the Dirichlet problem for the Laplace equation (7.1) is reduced to that of the Poisson equation (7.2) if the boundary value g is represented as $h|_{\partial\Omega}$ in the sense of trace for some $h \in H^1(\Omega)$. The following theorem actually guarantees this procedure.

Theorem 7.1. *If $\Omega \subset \mathbf{R}^N$ is a Lipschitz domain, the mapping*

$$v \in H^1(\Omega)/H^1_0(\Omega) \;\mapsto\; v|_{\partial\Omega} \in H^{1/2}(\partial\Omega)$$

is well-defined and an isomorphism. The integration by parts

$$(v, \frac{\partial w}{\partial x_i}) = -(\frac{\partial v}{\partial x_i}, w) + \int_{\partial\Omega} w\nu^i v \, dS, \quad 1 \le i \le N \tag{7.7}$$

is valid for $v, w \in H^1(\Omega)$, where dS and

$$\nu = (\nu^i) \in L^\infty(\partial\Omega, \mathbf{R}^N)$$

denote the area element and the outer unit normal vector on $\partial\Omega$, respectively.

The following property also assures several fundamental properties of $H^1(\Omega)$ for the Lipschitz domain Ω such as the Sobolev embedding

$$H^1(\Omega) \hookrightarrow L^{2N/(N-2)}(\Omega)$$

for $N \ge 3$.

Theorem 7.2. *If $\Omega \subset \mathbf{R}^N$ is a Lipschitz domain then $C^\infty(\overline{\Omega})$ is dense in $H^1(\Omega)$ where*

$$C^\infty(\overline{\Omega}) = \{v : \overline{\Omega} \to \mathbf{R} \mid \exists \tilde{v} \in C^\infty_0(\mathbf{R}^N), \ \tilde{v}|_{\overline{\Omega}} = v\}.$$

The following theorem is equivalent to the first case of Lemma 6.3 in §6.3 represented by 1-form, where the trace $\varphi|_{\partial\Omega} \in H^{1/2}(\partial\Omega)$ is taken for $\varphi \in H^1(\Omega)$, and furthermore, Δv of $v \in H^1(\Omega)$ is taken in the sense of distributions. Here we recall that $H^1_0(\Omega) \subset H^1(\Omega)$ implies

$$H^1(\Omega)' \subset H^{-1}(\Omega) = H^1_0(\Omega)'.$$

Aslo, the inclusion

$$H^{1/2}(\partial\Omega) \hookrightarrow L^2(\partial\Omega)$$

is dense if $\partial\Omega$ is Lipschitz continuous, which realizes the Gel'fand triple (Chapter 7 of [Suzuki (2022b)])

$$H^{1/2}(\partial\Omega) \hookrightarrow L^2(\partial\Omega) \cong L^2(\partial\Omega)' \hookrightarrow H^{-1/2}(\partial\Omega) = H^{1/2}(\partial\Omega)'.$$

Theorem 7.3. *If $\Omega \subset \mathbf{R}^N$ is a Lipschitz domain and $v \in H^1(\Omega)$ admits $\Delta v \in H^1(\Omega)'$ then there is a unique element in $H^{-1/2}(\partial\Omega)$ denoted by $\partial v/\partial\nu$, satisfying*

$$(\nabla v, \nabla\varphi) = \langle \varphi, -\Delta v \rangle + \left\langle \varphi, \frac{\partial v}{\partial \nu} \right\rangle, \quad \forall \varphi \in H^1(\Omega). \tag{7.8}$$

Proof. The proof is the same. We show that the element

$$\partial v/\partial\nu \in H^{-1/2}(\partial\Omega)$$

is defined by

$$\langle \varphi, \frac{\partial v}{\partial \nu} \rangle = (\nabla v, \nabla\varphi) + \langle \varphi, \Delta v \rangle, \quad \varphi \in H^1(\Omega),$$

that is, the right-hand side is independent of the choice of the extension of $\varphi \in H^{1/2}(\partial\Omega)$ to $H^1(\Omega)$. This property means

$$(\nabla v, \nabla\varphi) = \langle \varphi, -\Delta v \rangle, \quad \forall \varphi \in H_0^1(\Omega), \tag{7.9}$$

which follows from the definition of Δv, that is,

$$(\nabla v, \nabla\varphi) = \langle \varphi, -\Delta v \rangle, \quad \forall \varphi \in C_0^\infty(\Omega). \tag{7.10}$$

In fact (7.9) follows from (7.10) because $H_0^1(\Omega)$ is the closure of $C_0^\infty(\Omega)$ in $H^1(\Omega)$. □

The above $\partial v/\partial\nu \in H^{-1/2}(\partial\Omega)$ is not regarded as a function on $\partial\Omega$, which, however, is identified with the normal derivative of v in the Neumann boundary value problem such as (7.4), if it is Lipschitz continuous on $\overline{\Omega}$. Actually, the weak form of (7.4) is formulated as finding $u \in H^1(\Omega)$ such that

$$(\nabla u, \nabla\varphi) = \langle \varphi, f \rangle, \quad \forall \varphi \in H^1(\Omega), \tag{7.11}$$

using this notion of $\partial u/\partial\nu$. Then, given

$$f \in V = H^1(\Omega)', \quad \langle 1, f \rangle = 0, \tag{7.12}$$

there is a unique solution $u \in H^1(\Omega)$ to (7.11) satisfying

$$\int_\Omega u \, dx = 0.$$

To confirm this fact, first, we note the Poincaré-Wirtinger inequality,

$$\|v\|_2 \le C\|\nabla v\|_2, \quad v \in V_0 \tag{7.13}$$

valid for

$$V_0 = \{v \in H^1(\Omega) \mid \int_\Omega v \, dx = 0\}.$$

Given $f \in H^1(\Omega)'$, next, we regard $\varphi \in V_0 \mapsto \langle \varphi, f \rangle$ as a bounded linear functional. Then the representation theorem of Riesz guarantees the existence of a unique $u \in V_0$ satisfying

$$(\nabla u, \nabla \varphi) = \langle \varphi, f \rangle, \quad \varphi \in V_0.$$

Finally, this u satisfies (7.11), if the second relation of (7.12) holds.

We have confirmed that if Ω is a Lipschitz domain and $v \in H^1(\Omega)$, then $v \in H^{1/2}(\partial\Omega)$ holds in the sense of trace. If $\Delta v \in H^1(\Omega)'$, furthermore, then

$$\frac{\partial v}{\partial \nu} \in H^{-1/2}(\partial\Omega) = H^{1/2}(\partial\Omega)'$$

is defined by (7.8). This mapping is continuous in the sense that

$$\left\| \frac{\partial v}{\partial \nu} \right\|_{H^{-1/2}(\partial\Omega)} \le C(\|\nabla v\|_{L^2(\Omega)} + \|\Delta v\|_{H^1(\Omega)'}).$$

We cannot, however, define $v \in H^{-1/2}(\partial\Omega)$ for general $v \in L^2(\Omega)$, nor $\frac{\partial v}{\partial \nu} \in H^{-1/2}(\partial\Omega)$ for general $v \in H^1(\Omega)$.

If $v \in H^2(\Omega)$, on the contrary, we have $\nabla v \in H^1(\Omega, \mathbf{R}^N)$. Using $\nu \in L^\infty(\partial\Omega, \mathbf{R}^N)$, we can thus define

$$\nu \cdot \nabla v \in L^2(\partial\Omega).$$

We have also $\Delta v \in L^2(\Omega) \hookrightarrow H^1(\Omega)'$ and hence $\partial v/\partial \nu \in H^{-1/2}(\partial\Omega)$ is defined by (7.8). These two terms are actually consistent.

Theorem 7.4. *Let $\Omega \subset \mathbf{R}^N$ be a Lipschitz domain and $v \in H^2(\Omega)$. Then the above $\nu \cdot \nabla v \in L^2(\partial\Omega)$ is identified with $\partial v/\partial \nu \in H^{-1/2}(\partial\Omega)$ in Theorem 7.3 as a distribution on $\partial\Omega$.*

Proof. Let $\omega \subset \mathbf{R}^N$ be an open set containing $\partial\Omega$. Take $\varphi \in C_0^\infty(\omega)$ and its zero extension which is regarded as an element in $H^1(\Omega)$.

Given $v \in H^2(\Omega)$, then, we have

$$\left\langle \varphi, \frac{\partial v}{\partial \nu} \right\rangle = (\nabla v, \nabla \varphi) + (\varphi, \Delta v)$$

by (7.8). The right-hand side is now equal to $\langle \nabla v, \nu\varphi \rangle$ by (7.7). Hence we obtain the conclusion by Theorem 7.3. $\qquad\square$

7.3 Sectional Curvatures

If $\Omega \subset \mathbf{R}^N$ is a Lipschitz domain, we can define the standard measure on $\Gamma = \partial\Omega$ using the local chart. Any Lipschitz continuous function on Γ admits a Lipschitz continuous extension to its neighbourhood.

For this purpose we use the flow $\{T_t\}$ generated by a Lipschitz continuous vector field v in a neighbourhood of Γ as in Example 8.1 in §8.1. Putting $G = \{T_t x \mid |t| < \varepsilon,\ x \in \Gamma\}$, $0 < \varepsilon \ll 1$, we thus define $\tilde{d} \in C^{0,1}(\Gamma)$ for given $d \in C^{0,1}(\Gamma)$ by

$$\tilde{d}(x) = d(T_{-t}x), \quad T_t x \in G.$$

If Ω is $C^{1,1}$, the outer normal unit vector ν is Lipschitz continuous on Γ. Then the above vector field v can be perpendicular to Γ everywhere. Extending ν to G by the flow $\{T_t\}$ generated by this v, then we obtain

$$\frac{\partial \nu}{\partial \nu} = 0 \quad \text{on } \Gamma. \tag{7.14}$$

where $\frac{\partial}{\partial \nu}$ denotes the directional derivative to ν. Henceforth, property (7.14) is always assumed, which implies

$$\frac{\partial s_i}{\partial \nu} \cdot \nu = 0 \quad \text{on } \Gamma,\ 1 \le i \le N-1 \tag{7.15}$$

by $s_i \cdot \nu = 0$. The frame satisfying (7.15) is called a *Fermi coordinate*.

The tangential space $T_\xi(\Gamma)$, next, is defined for almost every $\xi \in \Gamma$. Using an orthonomal basis $\{s_i \mid i = 1, \cdots, N-1\}$ of $T_\xi(\Gamma)$, we thus obtain the frame

$$\{s_1, \cdots, s_{N-1}, \nu\} \tag{7.16}$$

almost everywhere on Γ.

If $\Omega \subset \mathbf{R}^N$ is a $C^{1,1}$ domain, the frame (7.16) on Γ is Lipschitz continuous. Then we cut Γ by the plane made by $\{s_i, \nu\}$, $1 \le i \le N-1$, denoted by Π, to obtain a Lipschitz curve denoted by $\gamma = \Pi \cap \Gamma$. Taking the parametrization $x = x(s) \in \gamma$ with the arc-length s as in §1.2, we define the unit tangential vector on Π by

$$s_i = \dot{x} \equiv \frac{\partial x}{\partial s}, \quad |s_i| = 1.$$

It holds that

$$\dot{s}_i \cdot s_i = 0.$$

Since $\dot{s}_i \in \Pi$, we define the *sectional curvature* κ_i by

$$\dot{s}_i = -\kappa_i \nu, \tag{7.17}$$

recalling that the outer unit normal vector ν on Γ is in the outer direction of Ω. We have, on the other hand,

$$\nu \cdot s_i = 0, \quad |\nu| = 1,$$

which results in $\dot{\nu} \cdot s_i = \kappa_i$ and $\dot{\nu} \cdot \nu = 0$ along this parametrization. Hence it holds that

$$\dot{\nu} = \kappa_i s_i \tag{7.18}$$

by $\dot{\nu} \in \Pi$.

Concerning s_j for $j \neq i$, we use

$$s_j \cdot s_j = s_j \cdot \nu = 0, \quad |s_j| = 1.$$

Since $\dot{s}_j$ defined on γ belongs to the space generated by s_i, s_j, ν, it follows that

$$\dot{s}_j = 0 \tag{7.19}$$

similarly. We write the relations (7.17) and (7.19) as

$$\frac{\partial s_i}{\partial s_i} = -\kappa_i \nu, \quad \frac{\partial s_j}{\partial s_i} = 0, \ j \neq i, \tag{7.20}$$

parametrizing γ by s_i. Hence the vector s_j on γ is parametrized by the line elements s_i, $1 \leq i \leq N - 1$, in (7.20).

We use, next, a tubular neighborhood $\mathcal{O}$ and extend the above frame (7.16) to a neighborhood of Γ. Hence we rewrite γ locally as a graph of $z = f(y)$ satisfying $f(0) = f'(y) = 0$ to take

$$\mathcal{O} = \{ (y, f(y) + t) \mid |y|, \ |t| < \varepsilon \}, \quad 0 < \varepsilon \ll 1.$$

We thus have a fiber of γ on Π with the length parameter t, which results in the parametrization $n = n(t, s)$ and $s_i = s_i(t, s)$ satisfying

$$\nu' \equiv \frac{\partial \nu}{\partial t} = 0 \quad \text{at } t = s = 0.$$

Then we obtain $s_i' = 0$ there, by

$$\nu \cdot s_i = 0, \quad |s_i| = 1.$$

Using notations similar to (7.20), we end up with (7.18), the Frenet Serret formula in §1.2 derived for a curve in $\mathbf{R}^3$, that is,

$$\frac{\partial \nu}{\partial s_i} = \kappa_i s_i, \quad 1 \leq i \leq N - 1 \tag{7.21}$$

and also (7.14) everywhere on Γ:

$$\frac{\partial \nu}{\partial \nu} = 0. \tag{7.22}$$

Henceforth, we put

$$a \otimes b = (a_i b_j) = ab^T$$

for vectors $a = (a_i)$ and $b = (b_i)$, and also

$$\nabla \nu = (\nabla \otimes \nu)^T = \left(\frac{\partial \nu^j}{\partial x_k} \right)_{jk} \qquad (7.23)$$

for $\nu = (\nu^j)$. Using the frame (7.16) as an orthonormal system in $\mathbf{R}^N$, we have

$$\begin{aligned}
(\nabla \nu)^T &= \sum_{i=1}^{N-1} \frac{\partial \nu}{\partial s_i} \otimes s_i + \frac{\partial \nu}{\partial \nu} \otimes \nu \\
&= \sum_{i=1}^{N-1} \kappa_i s_i \otimes s_i + \sum_{i=1}^{N-1} \sum_{j \neq i} (\frac{\partial \nu}{\partial s_i} \cdot s_j) s_j \otimes s_i \qquad (7.24)
\end{aligned}$$

and

$$\begin{aligned}
(\nabla s_j)^T &= \sum_{i=1}^{N-1} \frac{\partial s_j}{\partial s_i} \otimes s_i + \frac{\partial s_j}{\partial \nu} \otimes \nu \\
&= -\kappa_j \nu \otimes s_j - \sum_{i \neq j} (\frac{\partial \nu}{\partial s_i} \cdot s_j) \nu \otimes s_i \qquad (7.25)
\end{aligned}$$

by (7.21) and (7.22). It holds that

$$\nabla \cdot \nu = \operatorname{tr}(\nabla \nu) = \sum_{i=1}^{N-1} \kappa_i. \qquad (7.26)$$

This value, called *mean curvature*, is independent of the choice of frame.

Remark 7.1. By the notation $\nabla \otimes$ of (7.23) above, it holds that

$$\nabla \otimes (f a) = (\nabla f) \otimes a + f(\nabla \otimes a) = (\nabla f) \otimes a + f(\nabla a)^T,$$

where f and a are scalar and vector fields, respectively.

Lemma 7.1. *If $\Omega \subset \mathbf{R}^N$ is a $C^{1,1}$-domain and $f \in C^{1,1}(\overline{\Omega})$ then it holds that*

$$\begin{aligned}
\nabla^2 f &= \sum_{i=1}^{N-1} (\nabla s_i)^T \frac{\partial f}{\partial s_i} + (\nabla \nu)^T \frac{\partial f}{\partial \nu} + \sum_{i,j=1}^{N-1} s_i \otimes s_j \frac{\partial^2 f}{\partial s_i \partial s_j} \\
&\quad + \sum_{i=1}^{N-1} (s_i \otimes \nu + \nu \otimes s_i) \frac{\partial^2 f}{\partial s_i \partial \nu} + \nu \otimes \nu \frac{\partial^2 f}{\partial \nu^2} \quad \text{a.e. on } \Gamma, \quad (7.27)
\end{aligned}$$

where $\Gamma = \partial \Omega$ and

$$\nabla^2 f = \left(\frac{\partial^2 f}{\partial x_i \partial x_j} \right)$$

denotes the Hesse matrix of f.

Proof. Using the orthonormal coordinate (7.16), we have

$$\nabla^2 f = \nabla \otimes \nabla f = \nabla \otimes \left(\sum_{i=1}^{N-1} s_i \frac{\partial f}{\partial s_i} + \nu \frac{\partial f}{\partial \nu} \right)$$

$$= \sum_{i=1}^{N-1} (\nabla s_i)^T \frac{\partial f}{\partial s_i} + (\nabla \nu)^T \frac{\partial f}{\partial \nu} + \sum_{i=1}^{N-1} \nabla \frac{\partial f}{\partial s_i} \otimes s_i + \nabla \frac{\partial f}{\partial \nu} \otimes \nu.$$

Here the third and the fourth terms on the right-hand side are equal to

$$\sum_{i=1}^{N-1} \nabla \left(\frac{\partial f}{\partial s_i} \right) \otimes s_i = \sum_{i=1}^{N-1} \left(\sum_{j=1}^{N-1} s_j \frac{\partial^2 f}{\partial s_j \partial s_i} + \nu \frac{\partial^2 f}{\partial \nu \partial s_i} \right) \otimes s_i$$

$$= \sum_{i,j=1}^{N-1} [s_j \otimes s_i] \frac{\partial^2 f}{\partial s_i \partial s_j} + \sum_{i=1}^{N} [\nu \otimes s_i] \frac{\partial^2 f}{\partial s_i \partial \nu}$$

and

$$\nabla \left(\frac{\partial f}{\partial \nu} \right) \otimes \nu = \left(\sum_{i=1}^{N-1} s_i \frac{\partial^2 f}{\partial s_i \partial \nu} + \nu \frac{\partial^2 f}{\partial \nu^2} \right) \otimes \nu$$

$$= \sum_{i=1}^{N-1} (s_i \otimes s_j) \frac{\partial^2 f}{\partial s_i \partial \nu} + (\nu \otimes \nu) \frac{\partial^2 f}{\partial \nu^2},$$

respectively. Hence it follows that

$$\nabla^2 f = \sum_{i=1}^{N-1} (\nabla s_i)^T \frac{\partial f}{\partial s_i} + (\nabla \nu)^T \frac{\partial f}{\partial \nu} + \sum_{i,j=1}^{N-1} s_i \otimes s_j \frac{\partial^2 f}{\partial s_i \partial s_j}$$

$$+ \sum_{i=1}^{N-1} (s_i \otimes \nu + \nu \otimes s_i) \frac{\partial^2 f}{\partial s_i \partial \nu} + \nu \otimes \nu \frac{\partial^2 f}{\partial \nu^2},$$

the conclusion. $\qquad\square$

This lemma implies the following result.

Corollary 7.1. *If* $\Omega \subset \mathbf{R}^N$ *is a* $C^{1,1}$*-domain and* $f \in C^{1,1}(\overline{\Omega})$ *then it holds that*

$$\Delta f = (\nabla \cdot \nu) \frac{\partial f}{\partial \nu} + \frac{\partial^2 f}{\partial \nu^2} + \sum_{i=1}^{N-1} \frac{\partial^2 f}{\partial s_i^2} \qquad a.e. \ on \ \Gamma. \tag{7.28}$$

Proof. In (7.27) we have

$$s_k \otimes \nu + \nu \otimes s_k = (s_k^i \nu^j + \nu^i s_k^j)_{ij}$$

and hence

$$\text{tr}\,(s_k \otimes \nu + \nu \otimes s_k) = \sum_{i=1}^{N-1}(s_k^i \nu^i + \nu^i s_k^i) = 2\nu \cdot s_k = 0.$$

We have also

$$\text{tr}\,[(\nabla \nu)^T] = \nabla \cdot \nu, \;\; \text{tr}(\nu \otimes \nu) = |n|^2 = 1.$$

Finally, there arises that

$$\text{tr}\,[(\nabla s_i)^T] = 0$$

by

$$[(\nabla s_i)^T]s_k \cdot s_\ell = \frac{\partial s_i}{\partial s_k} \cdot s_\ell = -\kappa_i \delta_{ik}\nu \cdot s_\ell = 0$$

and

$$[(\nabla s_i)^T]\nu \cdot \nu = \frac{\partial s_i}{\partial \nu} \cdot \nu = 0,$$

which implies (7.28). $\square$

7.4   Convex Domains

The standard elliptic regularity is concerned on the bounded domain $\Omega \subset \mathbf{R}^N$ with C^2 boundary $\partial\Omega$. Given $f \in L^2(\Omega)$, thus, the solution $u = u(x) \in H_0^1(\Omega)$ to

$$-\Delta u = f \text{ in } \Omega, \;\; u|_{\partial\Omega} = 0$$

belongs to $H^2(\Omega)$. Convexity of the domain, however, compensates the lack of smoothness of the boundary for this H^2 regularity of the solution as is examined to the Laplace equation on the sector in the plane (§7.1).

Theorem 7.5 ([Kadlec (1964)]). *If $\Omega \subset \mathbf{R}^N$ is a convex bounded open set and $f \in L^2(\Omega)$, there is a unique $u \in H^2(\Omega)$ satisfying*

$$-\Delta u = f \text{ in } \Omega, \;\; u|_{\partial\Omega} = 0. \tag{7.29}$$

Theorem 7.6 ([Grisvard–Ioss (1975)]). *If $\Omega \subset \mathbf{R}^N$ is a convex bounded open set and $f \in L^2(\Omega)$, there is a unique $u \in H^2(\Omega)$ satisfying*

$$-\Delta u + u = f \text{ in } \Omega, \;\; \frac{\partial u}{\partial \nu}\bigg|_{\partial\Omega} = 0, \tag{7.30}$$

where ν denotes the outer normal unit vector.

Remark 7.2. Equations (7.29) and (7.30) mean

$$u \in H_0^1(\Omega), \quad (\nabla u, \nabla \varphi) = (f, \varphi), \ \forall \varphi \in H_0^1(\Omega)$$

and

$$u \in H^1(\Omega), \quad (\nabla u, \nabla \varphi) + (u, \varphi) = (f, \varphi), \ \forall \varphi \in H^1(\Omega),$$

respectively. The unique existence of these solutions is a consequence of the representation theorem of Riesz. Hence above theorems are concerned on the regularity of the solution.

These theorems are based on an identity valid for Lipschitz domains presented as equation (3.1.1.8) in [Grisvard (1985)], that is, Lemma 7.2 below. Noting that the outer unit normal vector ν belongs to $L^\infty(\partial\Omega)$ if Ω is Lipschitz continuous, we conclude this chapter to show this identity and its consequences.

Henceforth,

$$\gamma : H^1(\Omega) \ \to \ H^{1/2}(\partial\Omega)$$

denotes the trace operator assured by the Lipschitz continuity of $\partial\Omega$. To avoid confusion, for the moment we put

$$\{e_1, \cdots, e_{N-1}, \nu\}$$

for the frame of $\mathcal{M} = \partial\Omega$, and s_i, $1 \le i \le N-1$, for the arc-length parameter along $\mathcal{C}_i$, the curve made by the cross section of $\mathcal{M}$ and the plane generated by $\{e_i, \nu\}$.

If Ω is $C^{1,1}$, the second fundamental form on $\mathcal{M} = \partial\Omega$ is defined by

$$\mathcal{B}(\xi, \eta) = -\frac{\partial \nu}{\partial \xi} \cdot \eta \tag{7.31}$$

as in (1.9) in §1.3, where

$$\xi = \sum_{i=1}^{N-1} \xi_i e_i, \quad \eta = \sum_{i=1}^{N-1} \eta_i e_i \in \mathcal{M}.$$

Then it holds that

$$\mathcal{B}(\xi, \eta) = -\sum_{j,k=1}^{N-1} \frac{\partial \nu}{\partial s_j} \cdot e_k \xi_j \eta_k = -\sum_{i=1}^{N-1} \kappa_i \xi_i \eta_i, \tag{7.32}$$

where

$$\kappa_i, \quad 1 \le i \le N-1,$$

are the sectional curvatures satisfying

$$\frac{\partial \nu}{\partial s_j} \cdot e_k = \delta_{jk}\kappa_j$$

by (7.21) in §7.3. Trace of this bilinear form $\mathcal{B}$, therefore, is equal to the mean curvature,

$$\mathrm{tr}\, \mathcal{B} = -\sum_{i=1}^{N-1} \kappa_i = -\nabla \cdot \nu \qquad (7.33)$$

by (7.26).

Finally, the tangential part and the tangential derivative of the vector field ζ and the scalar field z on $\mathcal{M}$, denoted by ζ_T and $\nabla_T z$, respectively, are defined by

$$\zeta_T = \zeta - (\zeta \cdot \nu)\nu, \quad \nabla_T z = \nabla z - (\nu \cdot \nabla z)\nu. \qquad (7.34)$$

The following theorem is proven at the end of this section.

Theorem 7.7. *If $\Omega \subset \mathbf{R}^N$ is a bounded $C^{1,1}$-domain and $v \in H^1(\Omega)^N$, it holds that*

$$\int_\Omega (\nabla \cdot v)^2 - \sum_{i,j=1}^N \frac{\partial v_i}{\partial x_i}\frac{\partial v_j}{\partial x_i}\, dx = -2\,\langle (\gamma v)_T, \nabla_T(\gamma v \cdot \nu)\rangle$$

$$- \int_{\partial\Omega} \mathcal{B}((\gamma v)_T, (\gamma v)_T) + (\mathrm{tr}\,\mathcal{B})[(\gamma v \cdot \nu)^2]\, dS. \qquad (7.35)$$

Although we do not provide the proof of Theorem 7.5, the following result is its starting point.

Corollary 7.2. *If $\Omega \subset \mathbf{R}^N$ is a bounded convex $C^{1,1}$-domain and $u \in H^2 \cap H^1_0(\Omega)$, it holds that*

$$\sum_{i,j=1}^N \int_\Omega \left(\frac{\partial^2 u}{\partial x_i \partial x_j}\right)^2\, dx \le \int_\Omega (\Delta u)^2\, dx. \qquad (7.36)$$

Proof. We admit the previous theorem. Given $u \in H^2 \cap H^1_0(\Omega)$, we put $v = \nabla u \in H^1(\Omega)^N$. It holds that

$$(\gamma v)_T = 0.$$

From the assumption on Ω, we have

$$\kappa_i \ge 0, \quad 1 \le i \le N-1.$$

Then equality (7.35) implies

$$\sum_{i,j=1}^{N} \int_\Omega \left(\frac{\partial^2 u}{\partial x_i \partial x_j} \right)^2 dx = \sum_{i,j=1}^{N} \int_\Omega \frac{\partial v_i}{\partial x_j} \frac{\partial v_j}{\partial x_i} \, dx$$

$$\leq \int_\Omega (\nabla \cdot v)^2 \, dx = \int_\Omega (\Delta u)^2 \, dx,$$

and hence (7.36). □

Remark 7.3. Applying Theorem 7.7 to

$$v = \nabla u, \quad \left. \frac{\partial u}{\partial \nu} \right|_{\partial \Omega} = 0$$

similarly, we obtain

$$\int_\Omega (\Delta u)^2 - \sum_{i,j=1}^{N} \left(\frac{\partial^2 u}{\partial x_i \partial x_j} \right)^2 dx = - \int_{\partial \Omega} \mathcal{B}(\nabla_T u, \nabla_T u) \, dS \qquad (7.37)$$

by

$$\gamma v \cdot \nu = 0.$$

If Ω is convex, the bilinear form $\mathcal{B}(\cdot, \cdot)$ is non-positive by (7.32). Hence there arises (7.36), which is a base of the proof of Theorem 7.6, as in Corollary 7.2 for Theorem 7.5.

The key identity for the proof of Theorem 7.7 is stated as follows.

Lemma 7.2. *Under the assumption of Theorem 7.7 it holds that*

$$(v \cdot \nu)\nabla \cdot v - [(v \cdot \nabla)v] \cdot \nu$$
$$= \nabla_T \cdot ((v \cdot \nu)v_T) - (tr\, \mathcal{B})(v \cdot \nu)^2 - \mathcal{B}(v_T, v_T) - 2v_T \cdot \nabla_T(v \cdot \nu)$$

on $\mathcal{M} = \partial \Omega$.

Proof. We have (7.32),

$$\mathcal{B}(\xi, \eta) = - \sum_{j,k} \frac{\partial \nu}{\partial s_j} \cdot e_k \xi_j \eta_k, \quad \xi = \sum_j \xi_j e_j, \ \eta = \sum_j \eta_j e_j,$$

and

$$\nabla u = \nabla_T u + \frac{\partial u}{\partial \nu} \nu, \quad v = v_T + v_\nu \nu, \ v_\nu = v \cdot \nu \quad \text{in } \mathbf{R}^N,$$

where u and v are scalar and vector fields near $\mathcal{M} = \partial \Omega$, respectively. Using

$$v = \sum_k v_k e_k + v_\nu \nu,$$

we obtain

$$
\begin{aligned}
\nabla \cdot v &= \sum_j \frac{\partial v}{\partial s_j} \cdot e_j + \frac{\partial v}{\partial \nu} \cdot \nu \\
&= \sum_{j,k} \left(\frac{\partial v_k}{\partial s_j} e_k + v_k \frac{\partial e_k}{\partial s_j} \right) \cdot e_j + \sum_j \left(\frac{\partial v_\nu}{\partial s_j} \nu + v_\nu \frac{\partial \nu}{\partial s_j} \right) \cdot e_j \\
&\quad + \sum_k \left(\frac{\partial v_k}{\partial \nu} e_k + v_k \frac{\partial e_k}{\partial \nu} \right) \cdot \nu + \left(\frac{\partial v_\nu}{\partial \nu} \nu + v_\nu \frac{\partial \nu}{\partial \nu} \right) \cdot \nu
\end{aligned}
$$

with (7.22).

Here we recall (7.15):

$$
\frac{\partial e_k}{\partial \nu} \cdot \nu = 0, \ 1 \le k \le N - 1, \tag{7.38}
$$

to obtain

$$
\nabla \cdot v = \sum_j \frac{\partial v_j}{\partial s_j} + \sum_{j,k} v_k \frac{\partial e_k}{\partial s_j} \cdot e_j + v_\nu \sum_j \frac{\partial \nu}{\partial s_j} \cdot e_j + \frac{\partial v_\nu}{\partial \nu}. \tag{7.39}
$$

It holds also that

$$
\begin{aligned}
(v \cdot \nabla)v &= \sum_j v_j \frac{\partial v}{\partial s_j} + v_\nu \frac{\partial v}{\partial \nu} \\
&= \sum_{j,k} v_j \left(\frac{\partial v_k}{\partial s_j} e_k + v_k \frac{\partial e_k}{\partial s_j} \right) + \sum_j v_j \left(\frac{\partial v_\nu}{\partial s_j} \nu + v_\nu \frac{\partial \nu}{\partial s_j} \right) \\
&\quad + \sum_j v_n \left(\frac{\partial v_k}{\partial \nu} e_k + v_k \frac{\partial e_k}{\partial \nu} \right) + v_\nu \left(\frac{\partial v_\nu}{\partial \nu} \nu + v_k \frac{\partial \nu}{\partial \nu} \right),
\end{aligned}
$$

and therefore,

$$
((v \cdot \nabla)v) \cdot \nu = \sum_{j,k} v_j v_k \frac{\partial e_k}{\partial s_j} \cdot \nu + \sum_j v_j \frac{\partial v_\nu}{\partial s_j} + v_\nu \frac{\partial v_\nu}{\partial \nu} \tag{7.40}
$$

by (7.38), the equality

$$
\frac{\partial \nu}{\partial s_j} \cdot \nu = 0
$$

derived from (7.21), and (7.22). Then we obtain

$$v_\nu \nabla \cdot v - ((v \cdot \nabla)v) \cdot \nu$$

$$= v_\nu \sum_j \frac{\partial v_j}{\partial s_j} + v_\nu \sum_{j,k} v_k \frac{\partial e_k}{\partial s_j} \cdot e_j + v_\nu^2 \sum_j \frac{\partial \nu}{\partial s_j} \cdot e_j$$

$$- \sum_{j,k} v_j v_k \frac{\partial e_k}{\partial s_j} \cdot \nu - \sum_j v_j \frac{\partial v_\nu}{\partial s_j}$$

$$= v_\nu \sum_j \frac{\partial v_j}{\partial s_j} + v_\nu \sum_{j,k} v_k \frac{\partial e_k}{\partial s_j} \cdot e_j - (\text{tr } \mathcal{B}) v_\nu^2$$

$$- \mathcal{B}(v_T, v_T) - \sum_j v_j \frac{\partial v_\nu}{\partial s_j}. \tag{7.41}$$

Finally, there arises that

$$\nabla_T \cdot (v_\nu v_T) = \sum_j \frac{\partial (v_\nu v_T)}{\partial s_j} \cdot e_j = \sum_{j,k} \frac{\partial (v_\nu v_k e_k)}{\partial s_j} \cdot e_j$$

$$= \sum_{j,k} \left\{ \frac{\partial v_\nu}{\partial s_j} v_k e_k + v_\nu \frac{\partial v_k}{\partial s_j} e_k + v \nu v_k \frac{\partial e_k}{\partial s_j} \right\} \cdot e_j$$

$$= \sum_{j,k} \frac{\partial v_\nu}{\partial s_j} v_j + v_\nu \sum_j \frac{\partial v_j}{\partial s_j} + v_\nu \sum_{j,k} v_k \frac{\partial e_k}{\partial s_j} \cdot e_j. \tag{7.42}$$

Inequalities (7.40), (7.41), and (7.42) imply

$$v_\nu \nabla \cdot v - ((v \cdot \nabla)v)\nu$$

$$= \nabla_T \cdot (v_\nu v_T) - (\text{tr } \mathcal{B}) v_\nu^2 - \mathcal{B}(v_T, v_T) - 2 \sum_j v_j \frac{\partial v_\nu}{\partial s_j}$$

$$= \nabla_T \cdot (v_\nu v_T) - (\text{tr } \mathcal{B}) v_\nu^2 - \mathcal{B}(v_T, v_T) - 2 v_T \cdot \nabla_T v_\nu,$$

and hence the result. $\qquad\qquad\square$

We are ready to give the following proof.

Proof of Theorem 7.7. It suffices to show for $v \in H^2(\Omega)^N$. By Green's formula we have

$$\int_\Omega (\nabla \cdot v)^2 \, dx = \sum_{i,j=1}^N \int_\Omega \frac{\partial v_i}{\partial x_i} \frac{\partial v_j}{\partial x_j} \, dx$$

$$= - \sum_{i,j=1}^N \int_\Omega v_i \frac{\partial^2 v_i}{\partial x_i \partial x_j} \, dx + \sum_{i,j=1}^N \int_{\partial\Omega} v_i \frac{\partial v_j}{\partial x_j} \nu_j \, dS$$

$$= \sum_{i,j=1}^N \int_\Omega \frac{\partial v_i}{\partial x_j} \frac{\partial v_j}{\partial x_i} \, dx + \sum_{i,j=1}^N \int_{\partial\Omega} v_i \frac{\partial v_j}{\partial x_j} \nu_i - v_i \frac{\partial v_j}{\partial x_i} \nu_j \, dS,$$

which implies

$$I(v) \equiv \int_\Omega (\nabla \cdot v)^2 - \sum_{i,j=1}^{N} \frac{\partial v_i}{\partial x_j} \frac{\partial v_j}{\partial x_i} \, dx$$

$$= \int_{\partial\Omega} (v \cdot \nu)\nabla \cdot v - [(v \cdot \nabla)v] \cdot \nu \, dS.$$

Then Lemma 7.2 is applicable to the right-hand side of the above equality. If ζ is a vector field on $\mathcal{M} = \partial\Omega$, we obtain

$$\nabla_T \cdot \zeta = \sum_{j=1}^{N-1} \frac{\partial \zeta}{\partial s_j} \cdot s_j$$

$$= \sum_{j,k=1}^{N-1} \left(\frac{\partial}{\partial s_j} [(\zeta \cdot s_k)s_k] \right) \cdot s_j$$

$$= \sum_{j,k=1}^{N} \frac{\partial}{\partial s_j} [(\zeta \cdot s_k)s_k \cdot s_j]$$

by (7.20), and hence

$$\int_{\partial\Omega} \nabla_T \cdot ((v \cdot n)v_T) \, dS = 0.$$

Then the result follows.   □

Chapter 8

Liouville's Formulae

Theory of transformation formulates the variance of physical quantities under the material transport. Liouville's transformation theory is a set of variational formulae concerning volume and area integrals described in Lagrange coordinates. Its applications spread into wide fields of mathematical physics and engineering. In this chapter, first, we confirm the notion of deformation of domains (§8.1). Then we derive Liouville's formulae on Jacobian (§8.2) to reach the first and the second volume derivatives (§8.3). The former is applied in the proof of Lemma 3.9 in §3.13, while the latter is involved by Lemma 7.2 in §7.4 associated with the second fundamental form on the boundary. We next apply the first volume derivative to derive a numerical scheme to a problem in engineering, that is, the filtration (§8.4). Then we formulate flux of the material transport (§8.5) and Stefan condition associated with the latent heat (§8.6). Finally, we study area integrals, to give the first and the second area derivatives (§8.7).

8.1 Transformation of Variables

One-parameter family of transformations is sometimes used to derive mathematical models in physics. It is denoted by T_t, $|t| < \varepsilon$ for $0 < \varepsilon \ll 1$, on a bounded domain $\Omega \subset \mathbf{R}^N$, satisfying

$$T_t = I + tS + \frac{t^2}{2}R + o(t^2), \quad t \to 0 \tag{8.1}$$

uniformly on Ω, where $I : \Omega \to \Omega$ is the identity mapping.

Liouville's formulae are concerned with the asymptotics of several quantities under this family of transformations as $t \to 0$. Using the transformation of variables, they are reduced to the derivatives in t of the Jacobian of

$T_t : \Omega \to \Omega_t = T_t(\Omega)$, denoted by $J(T_t)$. Theorem 8.1 in §8.2 says that

$$J(T_t) = 1 + t\nabla \cdot S + \frac{t^2}{2}\left(\nabla \cdot R + (\nabla \cdot S)^2 - J(S) : J(S)^T\right) + o(t^2), \quad (8.2)$$

where

$$A : B = \sum_{i,j=1}^{N} a_{ij}b_{ij}, \quad A = (a_{ij}), \ B = (b_{ij}).$$

Then, equation (8.2) induces the first and the second variational formulae of the volume integral

$$\left.\frac{d}{dt}\int_{\Omega_t} c(x,t)\,dx\right|_{t=0}, \quad \left.\frac{d^2}{dt^2}\int_{\Omega_t} c(x,t)\,dx\right|_{t=0} \quad (8.3)$$

under the transformation (8.1) as in Theorems 8.2 and 8.3 in §8.3.

To be precise, we assume that $\Omega \subset \mathbf{R}^N$ is a Lipschitz domain and

$$T_t : \Omega \to \Omega_t = T_t(\Omega), \ |t| \ll 1 \quad (8.4)$$

is a family of bi-Lipschitz homeomorphisms satisfying (8.1). Hence the boundary $\partial\Omega$ of Ω is Lipschitz continuous, denoted by $C^{0,1}$.

Definition 8.1 (Differentiable Deformation). The family of deformations $\{T_t\}$ of Ω in (8.4) satisfying (8.1) is said to be differentiable if $T_t x$ is continuously differentiable in t for every $x \in \Omega$ and the mappings

$$\frac{\partial}{\partial t} DT_t, \ \frac{\partial}{\partial t}(DT_t)^{-1} : \Omega \to \mathbf{R}^N$$

are uniformly bounded, where DT_t denotes the Jacobi matrix of T_t:

$$DT_t = \left(\frac{\partial T_t^i}{\partial x_j}\right)_{1 \le i,j \le N}, \quad T_t = \left(T_t^1, \cdots, T_t^N\right)^T.$$

It is said to be twice differentiable if it is differentiable, $T_t x$ is continuously differentiable twice in t for every $x \in \Omega$ and the mappings

$$\frac{\partial^2}{\partial t^2} DT_t, \ \frac{\partial^2}{\partial t^2}(DT_t)^{-1} : \Omega \to \mathbf{R}^N$$

are uniformly bounded.

If $\{T_t\}$ is differentiable and twice differentiable, we put

$$\delta\rho = \left.\frac{\partial T_t}{\partial t}\right|_{t=0} \cdot \nu = S \cdot \nu \quad (8.5)$$

and

$$\delta^2\rho = \left.\frac{\partial^2 T_t}{\partial t^2}\right|_{t=0} \cdot \nu = R \cdot \nu, \quad (8.6)$$

respectively.

Example 8.1 (Dynamical Deformations). Let $v = v(x)$ be a Lipschitz vector field defined in a domain $\tilde{\Omega}$ containing $\overline{\Omega}$. Given $x \in \tilde{\Omega}$, we take the ordinary differential equation

$$\frac{dc(t)}{dt} = v(c(t)), \quad c(0) = x \tag{8.7}$$

to define the integral curve $c : (-\varepsilon, \varepsilon) \to \mathbf{R}^n$, $|t| < \varepsilon$. Writing $c(t) = T_t x$, we obtain a family of bi-Lipschitz homeomorphisms $T_t : \Omega \to \Omega_t$, $|t| \ll 1$. Then it holds that

$$Sx = \frac{\partial}{\partial t} T_t x \Big|_{t=0} = v(T_t x)|_{t=0} = v(x)$$

$$Rx = \frac{\partial^2}{\partial t^2} T_t x \Big|_{t=0} = \frac{\partial}{\partial t} v(T_t x) \Big|_{t=0} = [(v \cdot \nabla)v](x). \tag{8.8}$$

This $\{T_t\}$ is differentiable. It is twice differentiable if ∇v is furthermore Lipschitz continuous in $\tilde{\Omega}$. Then it holds that

$$T_t x = x + tv(x) + \frac{t^2}{2} [(v \cdot \nabla)v](x) + o(t^2)$$

uniformly in $x \in \overline{\Omega}$.

Example 8.2 (Normal Deformations). If $\partial\Omega$ is $C^{1,1}$, the unit normal vector $\nu = \nu_x$ is Lipschitz continuous on $\Gamma = \partial\Omega$. Then we can take a bi-Lipschitz deformation of Γ by

$$\Gamma_t : x + t \cdot \mu(x)\nu_x, \quad x \in \Gamma$$

for $|t| \ll 1$, where $\mu = \mu(x)$, $x \in \Gamma$, is a Lipschitz continuous function. Then there is a domain $\Omega_t \subset \mathbf{R}^N$ such that $\Gamma_t = \partial\Omega_t$. There is, also, a bi-Lipschitz mapping $T_t : \Omega \to \Omega_t$, $|t| \ll 1$, which satisfies

$$\delta\rho = \frac{\partial T_t}{\partial t} \Big|_{t=0} \cdot \nu = \mu, \quad \delta^2\rho = \frac{\partial^2 T_t}{\partial t^2} \Big|_{t=0} \cdot \nu = 0 \quad \text{on } \Gamma = \partial\Omega, \tag{8.9}$$

and hence

$$T_t x = x + t\mu(x)\nu_x + o(t^2), \quad x \in \Gamma = \partial\Omega.$$

This deformation used in the classical theory [Garabedian–Schiffer (1952–53)] does not work for the general Lipschitz domain, for example, if $\partial\Omega$ has a corner.

8.2   Variational Formulae of Jacobian

We continue to suppose that $\Omega \subset \mathbf{R}^N$ is a bounded Lipschitz domain and $\{T_t\}$, $|t| < \varepsilon$, is a family of bi-Lipschitz deformations of Ω satisfying (8.1), where S and R are vector fields on Ω. In what follows, E and O denote the $N \times N$ unit and zero matrices, respectively.

Lemma 8.1. *It holds that*

$$DT_t|_{t=0} = E, \quad \frac{\partial}{\partial t}(DT_t)\bigg|_{t=0} = DS, \quad \frac{\partial}{\partial t}(DT_t)^{-1}\bigg|_{t=0} = -DS \qquad (8.10)$$

uniformly on Ω if $\{T_t\}$ is differentiable, where DS and DR are the Jacobi matrices of S and R, respectively. It holds, furthermore, that

$$\frac{\partial^2}{\partial t^2}(DT_t)\bigg|_{t=0} = DR, \quad \frac{\partial^2}{\partial t^2}(DT_t)^{-1}\bigg|_{t=0} = 2(DS)^2 - DR \qquad (8.11)$$

uniformly on Ω if $\{T_t\}$ is twice differentiable.

Proof. The limits as $t \to 0$ below are uniform on Ω. The first and the second limits in (8.10), and the first limit in (8.11), are obvious by (8.1). Then we differentiate

$$E = (DT_t)(DT_t)^{-1} \qquad (8.12)$$

once and twice with respect to t, to obtain

$$O = \left(\frac{\partial}{\partial t}(DT_t)\right)(DT_t)^{-1} + (DT_t)\frac{\partial}{\partial t}(DT_t)^{-1}$$

$$O = \left(\frac{\partial^2}{\partial t^2}(DT_t)\right)(DT_t)^{-1} + 2\frac{\partial}{\partial t}(DT_t)\frac{\partial}{\partial t}(DT_t)^{-1} + (DT_t)\frac{\partial^2}{\partial t^2}(DT_t)^{-1}.$$

With $t \to 0$, there arises that

$$O = DS + \frac{\partial}{\partial t}(DT_t)^{-1}\bigg|_{t=0}$$

$$O = DR + 2(DS)(-DS) + \frac{\partial^2}{\partial t^2}(DT_t)^{-1}\bigg|_{t=0}$$

by

$$(DT_t)^{-1}\big|_{t=0} = E$$

derived from (8.12). Then the last equalities of (8.10) and (8.11) follow. □

Now we show (8.2) for the Jacobian

$$J(T_t) = \det DT_t$$

of the transformation $T_t : \Omega \to \Omega_t$.

Theorem 8.1. *If $\Omega \subset \mathbf{R}^N$ is a Lipschitz domain and $\{T_t\}$ is differentiable, it holds that*

$$\frac{\partial}{\partial t} \det DT_t \bigg|_{t=0} = \nabla \cdot S. \tag{8.13}$$

If $\{T_t\}$ is twice differentiable, furthermore, there arises that

$$\frac{\partial^2}{\partial t^2} \det DT_t \bigg|_{t=0} = \nabla \cdot R + (\nabla \cdot S)^2 - (DS)^T : DS. \tag{8.14}$$

Proof. We take the case that $\{T_t\}$ is twice differentiable to make the description simple. Then, both vector fields

$$S = (S^1, \ldots, S^N)^T, \quad R = (R^1, \cdots, R^N)^T \tag{8.15}$$

are Lipschitz continuous on Ω. Put

$$S_j^i = \frac{\partial S^i}{\partial x_j}, \quad R_j^i = \frac{\partial R^i}{\partial x_j}. \tag{8.16}$$

It holds that

$$DT_t = \begin{pmatrix} 1 + tS_1^1 + \frac{1}{2}t^2 R_1^1 & \cdots & tS_N^1 + \frac{1}{2}t^2 R_N^1 \\ \vdots & \vdots & \vdots \\ tS_1^N + \frac{1}{2}t^2 R_1^N & \cdots & 1 + tS_N^N + \frac{1}{2}t^2 R_N^N \end{pmatrix} + o(t^2)$$

and hence

$$\det DT_t = 1 + t\nabla \cdot S + t^2 \sum_{i<j}(S_i^i S_j^j - S_j^i S_i^j) + \frac{t^2}{2}\sum_i R_i^i + o(t^2).$$

Then we obtain (8.13) as

$$DT_t|_{t=0} = \nabla \cdot S,$$

and (8.14) by

$$\frac{\partial^2}{\partial t^2} \det DT_t \bigg|_{t=0} = 2\sum_{i<j}(S_i^i S_j^j - S_j^i S_i^j) + \nabla \cdot R$$

$$= \sum_{i \neq j}(S_i^i S_j^j - S_j^i S_i^j) + \nabla \cdot R = \sum_{i,j}(S_i^i S_j^j - S_j^i S_i^j) + \nabla \cdot R$$

$$= \nabla \cdot R + (\nabla \cdot S)^2 - (DS)^T : DS.$$

$\square$

8.3   Volume Derivatives

Here we formulate Liouville's volume derivatives up to the second order. Let $\Omega \subset \mathbf{R}^N$ and $\{T_t\}$, $|t| < \varepsilon$, be as in the previous section. Put

$$Q = \bigcup_{|t|<\varepsilon} \Omega_t \times \{t\} \tag{8.17}$$

and recall $\delta\rho = S \cdot \nu$ and $\delta^2\rho = R \cdot \nu$ in (8.5)–(8.6).

In the following theorem we say $c = c(x,t) \in C^{1,0}(Q)$ if it is continuously differentiable on a domain including $\overline{Q}$.

Theorem 8.2 (First Volume Derivative). *Assume that $\Omega \subset \mathbf{R}^N$ is a Lipschitz domain, $\{T_t\}$ is a differentiable deformation, and $c = c(x,t) \in C^{1,0}(Q)$. Then it holds that*

$$\frac{d}{dt} \int_{\Omega_t} c \, dx \bigg|_{t=0} = \int_{\Omega} c_t + \nabla \cdot (cS) \, dx \bigg|_{t=0}$$

$$= \int_{\Omega} \dot{c}_0 \, dx + \int_{\partial\Omega} c_0 \delta\rho \, dS, \tag{8.18}$$

where $\dot{c}_0 = c_t|_{t=0}$ and $c_0 = c|_{t=0}$.

Proof. We have

$$\int_{\Omega_t} c(y,t) \, dy = \int_{\Omega} c(T_t x, t) \, \det D(T_t x) \, dx$$

and hence

$$\frac{d}{dt} \int_{\Omega_t} c(y,t) \, dy \bigg|_{t=0} = \int_{\Omega} \left(c_t(T_t x, t) + \nabla c(T_t x, t) \cdot \frac{\partial T_t x}{\partial t} \right) \det DT_t(x)$$

$$+ c(T_t x, t) \frac{\partial}{\partial t} \det DT_t(x) \, dx \bigg|_{t=0} = \int_{\Omega} \dot{c}_0 + \nabla c_0 \cdot S + c_0 \nabla \cdot S \, dx$$

by (8.13). Then (8.18) follows from the divergence formula of Gauss.   □

Here we use the tangential derivative ∇_T and the tangential part S_T of the scalar field S defined by (7.34). Recall also the second fundamental form $\mathcal{B} = \mathcal{B}(\cdot, \cdot)$ defined by (7.31) on $\partial\Omega$.

In the following theorem we say $c = c(x,t) \in C^{2,0}(Q)$ if it is twice continuously differentiable on a domain including $\overline{Q}$.

Theorem 8.3 (Second Volume Derivative). *Let $\{T_t\}$ be twice differentiable, $\Omega \subset \mathbf{R}^N$ is $C^{1,1}$, and $c = c(x,t) \in C^{2,0}(Q)$ in the previous theorem. Then it holds that*

$$
\frac{d^2}{dt^2} \int_{\Omega_t} c\, dx \Big|_{t=0} = \int_\Omega \ddot{c}_0 dx + \int_{\partial\Omega} (2\dot{c}_0 + [(S \cdot \nabla)c_0])\delta\rho
$$
$$
+ c_0(\delta^2\rho - S_T \cdot \nabla_T \delta\rho + \delta\rho \nabla_T \cdot S_T)
$$
$$
- c_0((\mathrm{tr}\, \mathcal{B})(\delta\rho)^2 + \mathcal{B}(S_T, S_T))\, dS, \tag{8.19}
$$

where $\ddot{c}_0 = c_{tt}|_{t=0}$

Proof. We have

$$
\frac{\partial}{\partial t} c(T_t x, t) = \left(c_t + \nabla c \cdot \frac{\partial T_t}{\partial t} \right)(T_t x, t),
$$

and therefore,

$$
\frac{d^2}{dt^2} \int_{\Omega_t} c\, dx \Big|_{t=0} = \int_\Omega \frac{\partial^2}{\partial t^2} [c(T_t x, t)\, \det DT_t(x)]\, dx \Big|_{t=0}
$$
$$
= \int_\Omega \ddot{c}_0 + 2\nabla\dot{c}_0 \cdot S + (\nabla^2 c_0)[S, S] + \nabla c_0 \cdot R
$$
$$
+ c_0 \frac{\partial^2}{\partial t^2} \det DT_t + 2\left[\dot{c}_0 + \nabla c_0 \cdot S\right] \frac{\partial}{\partial t} \det DT_t \Big|_{t=0} dx.
$$

Then equalities (8.13) and (8.14) imply

$$
\frac{d^2}{dt^2} \int_{\Omega_t} c\, dx \Big|_{t=0} = \int_\Omega \ddot{c}_0 + 2\nabla\dot{c}_0 \cdot S + (\nabla^2 c_0)[S, S] + \nabla c_0 \cdot R
$$
$$
+ 2(\dot{c}_0 + \nabla c_0 \cdot S)\nabla \cdot S + c_0(\nabla \cdot R + (\nabla \cdot S)^2 - J(S)^T : J(S))\, dx
$$
$$
= \int_\Omega \ddot{c}_0 + 2\nabla \cdot (\dot{c}_0 S) + (\nabla^2 c_0)[S, S] + \nabla \cdot (c_0 R) + 2(\nabla c_0 \cdot S)(\nabla \cdot S)
$$
$$
+ c_0((\nabla \cdot S)^2 - J(S)^T : J(S))\, dx
$$
$$
= \int_\Omega \ddot{c}_0 + (\nabla^2 c_0)[S, S] + 2(\nabla c_0 \cdot S)\nabla \cdot S
$$
$$
+ c_0((\nabla \cdot S)^2 - J(S)^T : J(S))\, dx + \int_{\partial\Omega} 2\dot{c}_0 \delta\rho + c_0 \delta^2\rho\, dS
$$

We have also

$$
(\nabla \cdot S)^2 - J(S)^T : J(S) = \sum_{i,j}(S_i^i S_j^j - S_i^j S_j^i)
$$

by (8.16), and hence

$$\frac{d^2}{dt^2}\int_{\Omega_t} c\,dx\bigg|_{t=0} - \int_\Omega \ddot{c}_0\,dx - \int_{\partial\Omega} 2\dot{c}_0\delta\rho + c_0\delta^2\rho\,dS$$

$$= \sum_{i,j}\int_\Omega c_{0ij}S^iS^j + 2c_{0i}S^iS^j_j + c_0(S^i_iS^j_j - S^j_iS^i_j)\,dx,$$

where

$$c_{0i} = \frac{\partial c_0}{\partial x_i}, \quad c_{0ij} = \frac{\partial^2 c_0}{\partial x_i\partial x_j}.$$

Then we use

$$\sum_{i,j}\int_\Omega c_{0ij}S^iS^j + 2c_{0i}S^iS^j_j\,dx$$

$$= \int_{\partial\Omega} (\nabla c_0 \cdot S)(S \cdot \nu)\,dS + \sum_{i,j}\int_\Omega c_{0i}\left(-\frac{\partial}{\partial x_j}(S^iS^j) + 2S^iS^j_j\right)\,dx$$

$$= \int_{\partial\Omega} (\nabla c_0 \cdot S)\delta\rho\,dS + \sum_{i,j}\int_\Omega c_{0i}(S^iS^j_j - S^j_iS^j)\,dx$$

to obtain

$$\frac{d^2}{dt^2}\int_{\Omega_t} c\,dx\bigg|_{t=0} - \int_\Omega \ddot{c}_0\,dx - \int_{\partial\Omega} (2\dot{c}_0 + \nabla c_0 \cdot S)\delta\rho + c_0\delta^2\rho\,dS$$

$$= \sum_{i,j}\int_\Omega c_{0i}(S^iS^j_j - S^j_iS^j) + c_0(S^i_iS^j_j - S^j_iS^i_j)\,dx$$

$$= \sum_{i,j}\int_\Omega S^j_j\frac{\partial}{\partial x_i}(c_0S^i) - S^i_j\frac{\partial}{\partial x_i}(c_0S^j)\,dx \equiv I.$$

This integral is treated similarly as in §7.4, and thus we obtain

$$I = \sum_{i,j}\int_{\partial\Omega} S^j_jc_0S^i\nu_i - S^i_jc_0S^j\nu_i\,dS + \sum_{i,j}\int_\Omega -c_0S^iS^j_{ij} + c_0S^jS^i_{ij}\,dx$$

$$= \int_{\partial\Omega} c_0[(\nabla \cdot S)\delta\rho - (S \cdot \nabla)S \cdot \nu]\,dS.$$

Then we apply Lemma 7.2 to conclude

$$I = \int_{\partial\Omega} c_0((\nabla \cdot S)(S \cdot \nu) - [(S \cdot \nabla)S] \cdot \nu) \, dS$$

$$= \int_{\partial\Omega} c_0(\nabla_T \cdot [(S \cdot \nu)S_T] - (\text{tr } \mathcal{B})(S \cdot \nu)^2 - \mathcal{B}(S_T, S_T)$$
$$-2S_T \cdot \nabla_T(S \cdot \nu)) \, dS$$

$$= \int_{\partial\Omega} c_0(\nabla_T \cdot ((\delta\rho)S_T) - (\text{tr } \mathcal{B})(\delta\rho)^2 - \mathcal{B}(S_T, S_T) - 2S_T \cdot \nabla_T(\delta\rho)) \, dS$$

$$= -\int_{\partial\Omega} c_0\{(\text{tr } \mathcal{B})(\delta\rho)^2 + \mathcal{B}(S_T, S_T) + S_T \cdot \nabla_T(\delta\rho) - \delta\rho\nabla_T \cdot S_T\} \, dS,$$

and hence the conclusion. □

In the above proof we derive

$$I = \sum_{i,j} \int_\Omega c_{0i}(S^i S_j^i - S_j^i S^j) + c_0(S_i^i S_j^j - S_j^j S_i^i) \, dx$$

$$= -\int_{\partial\Omega} (\nabla_T c_0 \cdot S_T)\delta\rho + c_0[(\text{tr } \mathcal{B})(\delta\rho)^2 + \mathcal{B}(S_T, S_T) + 2S_T \cdot \nabla_T(\delta\rho)] \, dS$$

with $\delta\rho = S \cdot \nu$. If

$$S = \nabla u$$

this identity means

$$\sum_{i,j} \int_\Omega c_{0i} \left(\frac{\partial u}{\partial x_i} \frac{\partial^2 u}{\partial x_i \partial x_j} - \frac{\partial^2 u}{\partial x_i \partial x_j} \frac{\partial u}{\partial x_j} \right) + c_0 \left(\frac{\partial^2 u}{\partial x_i^2} \frac{\partial^2 u}{\partial x_j^2} - (\frac{\partial^2 u}{\partial x_i \partial x_j})^2 \right) \, dx$$

$$= -\int_{\partial\Omega} (\nabla_T c_0 \cdot S_T)\delta\rho + c_0[(\text{tr } \mathcal{B})(\delta\rho)^2 + \mathcal{B}(S_T, S_T)$$
$$+2S_T \cdot \nabla_T(\delta\rho)] \, dS. \tag{8.20}$$

If

$$u|_{\partial\Omega} = 0,$$

furthermore, it holds that $S_T = 0$ and

$$\delta\rho = S \cdot \nu = \nu \cdot \nabla u = \frac{\partial u}{\partial \nu}.$$

Then we obtain

$$II + \int_\Omega \left[(\Delta u)^2 - \sum_{i,j} \left(\frac{\partial^2 u}{\partial x_i \partial x_j} \right)^2 \right] c_0 \, dx = \int_{\partial\Omega} (\nabla \cdot \nu) \left(\frac{\partial u}{\partial \nu} \right)^2 c_0 \, dS$$

by

$$\nabla \cdot \nu = -\text{tr}\, \mathcal{B},$$

where

$$II \equiv \sum_{i,j} \int_\Omega c_{0i} \left(\frac{\partial u}{\partial x_i} \frac{\partial^2 u}{\partial x_i \partial x_j} - \frac{\partial^2 u}{\partial x_i \partial x_j} \frac{\partial u}{\partial x_j} \right) dx$$

$$= \sum_{i,j} \int_\Omega (c_{0i} - c_{0j}) \frac{\partial u}{\partial x_i} \frac{\partial^2 u}{\partial x_i \partial x_j} dx$$

$$= \frac{1}{2} \sum_{i,j} \int_\Omega (c_{0i} - c_{0j}) \left(\frac{\partial u}{\partial x_i} - \frac{\partial u}{\partial x_j} \right) \frac{\partial^2 u}{\partial x_i \partial x_j} dx.$$

It thus holds that

$$\frac{1}{2} \sum_{i,j} \int_\Omega (c_{0i} - c_{0j}) \left(\frac{\partial u}{\partial x_i} - \frac{\partial u}{\partial x_j} \right) \frac{\partial^2 u}{\partial x_i \partial x_j} dx$$

$$+ \int_\Omega \left[(\Delta u)^2 - \sum_{i,j} \left(\frac{\partial^2 u}{\partial x_i \partial x_j} \right)^2 \right] c_0 \, dx$$

$$= \int_{\partial \Omega} (\nabla \cdot \nu) \left(\frac{\partial u}{\partial \nu} \right)^2 c_0 \, dS. \tag{8.21}$$

Corollary 7.2 in §7.4 is a consequence of (8.21) for $c_0 \equiv 1$.
    If

$$\left. \frac{\partial u}{\partial \nu} \right|_{\partial \Omega} = 0$$

in (8.20), on the other hand, we have

$$S \cdot \nu = \delta \rho = \frac{\partial u}{\partial \nu} = 0$$

for $S = \nabla u$. Then it follows that

$$II + \int_\Omega \left[(\Delta u)^2 - \sum_{i,j} \left(\frac{\partial^2 u}{\partial x_i \partial x_j} \right)^2 \right] c \, dx = - \int_{\partial \Omega} c \mathcal{B}(S_T, S_T) \, dS, \tag{8.22}$$

where equality (7.37) in §7.4 is the case of $c \equiv 1$ in (8.22).

8.4 Filtration

In §8.4-§8.6 we apply Theorem 8.2 in §8.3 to formulate the problems in physics. This section is devoted to the *filtration*. It is a free boundary problem concerning

$$u(x) = p(x) + x_2$$

for $x = (x_1, x_2)$, where $p(x)$ denotes the pressure of water. The boundary of the domain Ω is composed of Γ_i, $i = 1, 2, 3, 4$, where Γ_1 is the base, Γ_3 is composed of disjoint sides according to up and down waves, respectively, Γ_4 is the main free boundary, and Γ_4 arises in the down wave with Γ_3. Furthermore, h_1 and h_2 are the height of water up and down waves, respectively.

Hence

$$v = -\kappa \nabla u$$

is the velocity, and the problem is formulated by the *quasi-variational problem*

$$\int_\Omega \nabla u \cdot \nabla \zeta \leq 0, \quad \forall \zeta \in H^1(\mathcal{D}), \quad \zeta \geq 0 \text{ on } S_2, \quad \zeta = 0 \text{ on } S_3,$$

where $\mathcal{D} \subset \mathbf{R}^2$ indicates the dam containing Ω with the boundary composed of S_i, $i = 1, 2, 3$. Here, $S_1 = \Gamma_1$ is a base curve with left and right end points denoted by ζ_1 and ζ_2, respectively. It is contained in $\{x_2 < h_1\}$. We thus obtain

$$S_2 = \partial \mathcal{D} \cap \{(x_1, x_2) \mid x_1 > a, \ x_2 > h_2\}, \quad S_3 = \partial \mathcal{D} \setminus (S_1 \cup S_2),$$

where a is the value of x_1 coordinate at the left point of $\partial \mathcal{D} \cap \{x_2 = h_1\}$, denoted by ζ_3. A constraint on $u \in H^1(\mathcal{D})$ is the condition

$$u = u^0 \quad \text{on } \Gamma_2 \cup \Gamma_3 \cup \Gamma_4,$$

where

$$u^0(x) = \begin{cases} h_1 & \text{in } \mathcal{D}_0 \\ x_2 & \text{in } \mathcal{D}_1 \\ h_2 & \text{in } \mathcal{D}_3. \end{cases}$$

Hence $\mathcal{D}_1$ is the upper portion of $\mathcal{D}$ higher than Γ_2 and

$$\mathcal{D}_0 = \mathcal{D} \cap \{x_1 < a\}, \quad \mathcal{D}_3 = (\mathcal{D} \cap \{x_2 < h_2\}) \setminus \mathcal{D}_0.$$

The Euler–Lagrange equation of this variational problem is given by

$$\Delta u = 0 \text{ in } \Omega, \quad \frac{\partial u}{\partial \nu} = 0 \text{ on } \Gamma_1, \quad u = u^0, \quad \frac{\partial u}{\partial \nu} = 0 \text{ on } \Gamma_2$$

$$u = u^0 \text{ on } \Gamma_3, \quad u = u^0, \quad \frac{\partial u}{\partial \nu} \leq 0 \text{ on } \Gamma_4$$

([Friedman (1982)]). To introduce a variational reformulation we define the set of admissible domains denoted by $\mathcal{A}(\mathcal{D})$. Hence we say $\Omega \in \mathcal{A}(\mathcal{D})$ if and only if $\Omega \subset \mathcal{D}$ is a Lipschitz simply-connected domain,

$$S_1 \cup S_3 \subset \partial\Omega,$$

and $\partial\Omega \setminus (S_1 \cup S_3)$ is a Lipschitz monotone-decreasing graph in the direction x_2. Then we define the closed cone

$$\mathcal{K} = \{v \in H^1(\Omega) \mid v = 0 \text{ on } \Gamma_1 \cup \Gamma_2 \cup \Gamma_3, \ v \geq 0 \text{ on } \Gamma_4\}$$

and its dual cone

$$\mathcal{K}^* = \{(\nabla v, \nabla \chi) \leq 0, \ \forall \chi \in \mathcal{K}(\Omega)\}.$$

Given $\chi \in H^2(\Omega)$, we have $\chi \in K^*(\Omega)$ if and only if

$$\frac{\partial \chi}{\partial \nu} \leq 0 \quad \text{on } \Gamma_4.$$

Let, first,

$$A(\Omega) = \{v \in K^*(\Omega) \mid v = u^0 \text{ on } \Gamma_2 \cup \Gamma_3 \cup \Gamma_4\}$$
$$a(\Omega) = \inf_{v \in A(\Omega)} D_\Omega(v)$$

for

$$D_\Omega(u) = \frac{1}{2} \int_\Omega |\nabla v|^2 dx.$$

This $a(\Omega)$ is attained by some $u = u_\Omega \in A(\Omega)$ satisfying

$$\Delta u = 0 \text{ in } \Omega, \quad u = u^0 \text{ on } \Gamma_2 \cup \Gamma_3 \cup \Gamma_4, \quad \left.\frac{\partial u}{\partial \nu}\right|_{\Gamma_4} \leq 0, \quad \left.\frac{\partial u}{\partial \nu}\right|_{\Gamma_1} = 0.$$

Let, second,

$$B(\Omega) = \{v \in H^1(\Omega) \mid v = u^0 \text{ on } \Gamma_3 \cup \Gamma_4\}$$
$$b(\Omega) = \inf_{v \in B(\Omega)} D_\Omega(v).$$

This $b(\Omega)$ is attained by some $w = w_\Omega \in B(\Omega)$ satisfying

$$\Delta w = 0 \text{ in } \Omega, \quad w = u^0 \text{ on } \Gamma_3 \cup \Gamma_4, \quad \left.\frac{\partial w}{\partial \nu}\right|_{\Gamma_1 \cup \Gamma_2} = 0.$$

The following fact is easy to see.

Theorem 8.4. *Let* $J(\Omega) = a(\Omega) - b(\Omega)$. *Then*

$$\inf_{\Omega \in \mathcal{A}(\mathcal{D})} J(\Omega) = 0$$

is attained by $\Omega \in \mathcal{A}(\mathcal{D})$ *if and only if it is a solution to the filtration problem.*

Each admissible domain $\Omega \in \mathcal{A}(\mathcal{D})$ is associated with the mapping

$$\varphi = \varphi_\Omega : B = B(0,1) \subset \mathbf{R}^2 \to \Omega,$$

which has a bi-Lipschitz continuous extension $\overline{B} \to \overline{\Omega}$ satisfying the thre-point condition, $\varphi(z_i) = \zeta_i$, $i = 1, 2, 3$, where

$$z_1 = 1, \ z_2 = e^{\imath 2\pi/3}, \ z_3 = e^{-\imath 2\pi/3}.$$

Then we obtain the following theorem.

Theorem 8.5 ([Suzuki–Tsuchiya (2005)]). *The set $\mathcal{A}(\mathcal{D}) \neq \emptyset$ is closed and sequentially compact with respect to the uniform convergence. If*

$$\Omega_k \to \Omega \text{ uniformly in } \mathcal{A}(\mathcal{D}), \quad \lim_{k \to \infty} J(\Omega_k) = 0,$$

it holds that $J(\Omega) = 0$.

To realize minimizing sequence for

$$\inf_{\mathcal{A}(\mathcal{D})} J,$$

the bi-Lipschitz deformation of the domain, $T_t : \Omega \to \Omega_t$ is useful. If

$$T_t x = x + tSx + \frac{t^2}{2}Rx + o(t^2) \quad \text{uniformly in } x \in \Omega$$

is valid as $t \to 0$, this deformation is said to satisfy the *NPO condition* if $v \in V_1(\Omega_t)$ is equivalent to $v \circ T_t \in V_1(\Omega)$ for $|t| \ll 1$, where

$$V_1(\Omega_t) = \{v \in H^1(\Omega_t) \mid v|_{\Gamma_3 \cup \Gamma_4^t} = 0\}, \ \Gamma_4^t = T_t(\Gamma_4),$$

and $\Omega_0 = \Omega$. In the following theorem, $\langle \ , \ \rangle_{\Gamma_2 \cup \Gamma_4}$ and $\langle \cdot, \cdot \rangle_{\Gamma_2}$ denote the pairings on $\Gamma_2 \cup \Gamma_4$ and Γ_4, respectively, and τ stands for the unit tangential vector.

Definition 8.2 (NPO Condition). We say that $\{T_t\}$ satisfies the *NPO condition* if $v \in V_1(\Omega_t)$ is equivalent to $v \circ T_t \in V_1(\Omega)$.

Theorem 8.6 ([Suzuki–Tsuchiya (2011)]). *The first variation of $a(\Omega_t)$ exists as*

$$\delta a(\Omega) \equiv \frac{d}{dt} a(\Omega_t)\Big|_{t=0} = \frac{1}{2}\left\langle 1 - (\frac{\partial p_\Omega}{\partial \nu})^2, \delta\rho \right\rangle_{\Gamma_2 \cup \Gamma_4},$$

where $\delta\rho = S \cdot \nu$ and

$$p_\Omega = u_\Omega - x_2$$

for $u = u_\Omega$ satisfying

$$\Delta u = 0 \ in \ \Omega, \quad u = u_0 \ on \ \Gamma_2 \cup \Gamma_3 \cup \Gamma_4, \quad \left.\frac{\partial u}{\partial \nu}\right|_{\Gamma_4} \le 0, \quad \left.\frac{\partial u}{\partial \nu}\right|_{\Gamma_1} = 0.$$

Hence $u = u_\Omega$ attains

$$a(\Omega) = \inf_{v \in \mathcal{A}(\Omega)} D_\Omega(v).$$

If $\{T_t\}$ satisfies the NPO condition, furthermore, the first variation of $b(\Omega_t)$ exists as

$$\delta b(\Omega) \equiv \left.\frac{d}{dt} b(\Omega_t)\right|_{t=0} = \frac{1}{2}\left\langle \left(\frac{\partial w_\Omega}{\partial \tau}\right)^2, \delta \rho \right\rangle_{\Gamma_2},$$

where $w = w_\Omega$ attains $b(\Omega)$. Hence it holds that

$$\Delta w = 0 \ in \ \Omega, \quad w = u^0 \ on \ \Gamma_3 \cup \Gamma_4, \quad \left.\frac{\partial w}{\partial \nu}\right|_{\Gamma_4 \cup \Gamma_2} = 0.$$

The first variation of the functional J is applied to design an iteration scheme for the filtration problem. Given $\mathcal{D} \subset \mathbf{R}^2$ and h_1, h_2, we can thus realize the flow region Ω and the potential function u by this iteration.

More precisely, let Ω^k and $\Gamma_2^k \subset \partial\Omega_k$ be the k-th guess of the flow region and the free boundary, respectively. Since the first variation of the functional $J : \mathcal{A}(\mathcal{D}) \to \mathbf{R}$ is

$$\delta J(\Omega^k) = \left\langle 1 - \left(\frac{\partial p_\Omega^k}{\partial \nu}\right)^2 - \left(\frac{\partial w_\Omega^k}{\partial \tau}\right)^2, \delta \rho \right\rangle_{\Gamma_2^k},$$

an intuitive iterative scheme is defined by

$$\Gamma_2^{k+1} = \{x + \varepsilon FV(x)\nu(x) \mid x \in \Gamma_2^k\}$$

for

$$FV(x) \equiv 1 - \left(\frac{\partial p_\Omega^k}{\partial \nu}\right)^2 - \left(\frac{\partial w_\Omega^k}{\partial \tau}\right)^2, \quad x \in \Gamma_2^k$$

with $0 < \varepsilon \ll 1$. This scheme of the *steepest descent method*, however, does not fit numerical experiments well. After several iterations, Γ^k becomes very jagged and computation cannot be carried out anymore.

So far, effectiveness of the H^1 *gradient method* [Azegami (2020)] is confirmed. It is given by

$$\Gamma_2^{k+1} = \{x - z^k(x)\nu(x) \mid x \in \Gamma_2^k\},$$

where $z^k \in H^1(\Omega^k)$ denotes the solution to the boundary value problem

$$\Delta z^k = 0 \ in \ \Omega^k, \quad z^k|_{\Gamma^3 \cup \Gamma_4^k} = 0, \quad \left.\frac{\partial z^k}{\partial \nu}\right|_{\Gamma^1} = 0, \quad \frac{\partial z^k}{\partial \nu} = FV \ on \ \Gamma^k.$$

8.5   Flux of the Flow

Partial differential equations (PDE) are used to describe averaged movement of particles. Then the outer force can be subject to gradient of the scalar field (Chapter 2 of [Suzuki (2022a)]). If Ω_t denotes the domain moving with the velocity $v = v(\cdot, t)$ and $\rho = \rho(x, t)$ is the smooth scalar field varying in t, then the first volume derivative guarantees

$$\frac{d}{dt} \int_{\Omega_t} \rho \, dx = \int_{\Omega_t} \rho_t + \nabla \cdot \rho v \, dx. \tag{8.23}$$

Mass conservation has two descriptions. First, the equation of conservation law,

$$\rho_t = -\nabla \cdot j, \tag{8.24}$$

is derived from the Euler coordinate, that is,

$$\frac{d}{dt} \int_\omega \rho \, dx = - \int_{\partial \omega} \nu \cdot j \, dS$$

and the divergence formula of Gauss,

$$\int_{\partial \omega} \nu \cdot j \, dS = \int_\omega \nabla \cdot j \, dx$$

for the fixed domain ω. Hence $j = j(x, t)$ stands for the *flux* of this flow. The integral form

$$\frac{d}{dt} \int_{\Omega_t} \rho \, dx = 0$$

described by the Lagrange coordinate, on other hand, implies

$$\rho_t + \nabla \cdot \rho v = 0 \tag{8.25}$$

by (8.23). Adjusting (8.24) and (8.25), we obtain

$$j = \rho v.$$

Hence flux is equal to mass times velocity, that is, momentum.

8.6   Stefan Condition as Heat Transfer

Here we treat *two phase Stefan problem* from the viewpoint of the above transformation theory.

Let θ be the relative temperature, and assume that the heat conductor is water or ice if $\theta > 0$ or $\theta < 0$, respectively. Then we obtain the *heat equation*

$$c\rho\theta_t = \nabla \cdot (\kappa \nabla \theta) \quad \text{in } \{\theta \neq 0\}. \tag{8.26}$$

The region $\{x \in \Omega \mid \theta(x,t) = 0\}$ is composed of an interface, denoted by Γ_t, where the *phase transition* occurs under the exchange of the latent heat. Usually, the density $\rho = \rho(\theta)$ depends continuously on $\theta \in \mathbf{R}$, but the *specific heat* $c = c(\theta)$ and the *conductivity* $\kappa = \kappa(\theta)$ have the discontinuity of the first kind at $\theta = 0$.

To describe the motion of Γ_t, we take the *level set* approach and introduce the C^1 function $\Phi = \Phi(x,t)$ satisfying

$$\Gamma_t : \Phi(\cdot,t) = 0. \tag{8.27}$$

Let ν be the outer unit normal vector of Γ_t from $\{\theta > 0\}$ at $x \in \Gamma_t$. If x moves $\nu\Delta h$ during the time interval Δt, then it holds that

$$\Phi(x + \nu\Delta h, t + \Delta t) = 0.$$

Taking the infinitesimal approximation of this relation, we obtain

$$(\nu \cdot \nabla\Phi)\Delta h + \Phi_t \Delta t = 0 \qquad \text{on } \Gamma_t. \tag{8.28}$$

Meanwhile, $\ell\Delta h$ is radiated from the unit area on Γ_t as the latent heat, where $\ell = \lambda\rho$ with λ standing for the latent heat per unit weight, and, therefore, Newton Fourier Fick's heat energy balance law guarantees the relation

$$\ell\Delta h = -\left[\kappa\frac{\partial\theta}{\partial\nu}\right]_-^+ \Delta t, \tag{8.29}$$

where

$$[A]_-^+ = A_+ - A_-, \quad A_\pm(x) = \lim_{y \in \{\pm\theta > 0\},\ y \to x} A(y).$$

Combining (8.28) with (8.29), thus, we obtain the *Stefan condition*

$$\ell\Phi_t = \frac{\partial\Phi}{\partial\nu} \cdot \left[\kappa\frac{\partial\theta}{\partial\nu}\right]_-^+ \qquad \text{on } \Gamma_t, \tag{8.30}$$

which comprises the *Stefan problem* with (8.26) and (8.27).

Then we use the *Kirchhoff transformation*

$$u = \int_0^\theta \kappa(\theta')d\theta'$$

and the enthalpy $H = H(u)$ defined by

$$H(u) = \begin{cases} \int_0^\theta \rho(\theta')c(\theta')d\theta' - \ell, & u < 0 \\ \int_0^\theta \rho(\theta')c(\theta')d\theta', & u > 0, \end{cases}$$

which satisfies

$$H'(u) = \frac{\rho(\theta)c(\theta)}{\kappa(\theta)}, \quad u \neq 0, \quad H(+0) - H(-0) = \ell.$$

Concerning $u = u(x, t)$, we obtain

$$\nabla u = \kappa \nabla \theta, \quad H(u)_t = \frac{\rho c}{\kappa} \cdot \theta_t \cdot \kappa$$

and, therefore,

$$H(u)_t = \Delta u \qquad \text{in } \bigcup_{0 < t < T} (\Omega \setminus \Gamma_t) \times \{t\} = Q \setminus \Gamma$$

$$\ell \Phi_t = [\nabla u \cdot \nabla \Phi]_-^+ \quad \text{on } \bigcup_{0 < t < T} \Gamma_t \times \{t\} \equiv \Gamma$$

$$u|_{t=0} = u_0(x) \tag{8.31}$$

by (8.26) and (8.29), where $Q = \Omega \times (0, T)$. Since

$$\Gamma = \{\Phi = 0\},$$

there arises

$$\Phi_t + v \cdot \nabla \Phi = 0,$$

where v stands for the velocity of the movement of Γ. Then we obtain

$$\Phi_t = -(v \cdot \nu)\frac{\partial \Phi}{\partial \nu}$$

by $v = (v \cdot \nu)\nu$. The Stefan condition (8.31), on the other hand, implies

$$\ell \Phi_t = \left[\frac{\partial u}{\partial \nu}\right]_-^+ \frac{\partial \Phi}{\partial \nu}.$$

We thus reach the system

$$H(u)_t = \Delta u \text{ in } Q \setminus \Gamma, \quad u|_{\partial\Omega} = 0, \quad u|_{t=0} = u_0(x)$$

$$[u]_-^+ = 0, \ [H(u)]_-^+ = \ell, \ v \cdot \nu = -\frac{1}{\ell}\left[\frac{\partial u}{\partial \nu}\right]_-^+ \quad \text{on } \Gamma. \tag{8.32}$$

Now we show the following theorem, which ensures the interface vanishing of

$$H(u)_t = \Delta u$$

in (8.32) under the Stefan condition.

Theorem 8.7. *Equalities (8.32) imply*

$$H(u)_t = \Delta u \quad \text{in } Q = \Omega \times (0, T) \tag{8.33}$$

in the sense of distributions.

Proof. We take $\varphi \in C_0^\infty(Q)$ and put

$$\Omega_t^\pm = \{\pm\Phi(\cdot, t) > 0\}.$$

Then, the first volume formula assures

$$\frac{d}{dt}\int_{\Omega_t^\pm} H(u)\varphi dx = \int_{\Omega_t^\pm} H(u)_t\varphi + H(u)\varphi_t \, dx \pm \int_{\partial\Omega_t^\pm} H(u)\varphi v \cdot \nu \, dS$$

$$= \pm\int_{\partial\Omega_t^\pm} \varphi\frac{\partial u}{\partial\nu} - \frac{\partial\varphi}{\partial\nu}u + H(u)v \cdot \nu\varphi \, dS + \int_{\Omega_t^\pm} u\Delta\varphi + H(u)\varphi_t \, dx.$$

It then follows that

$$\frac{d}{dt}\int_\Omega H(u)\varphi = \int_\Omega u\Delta\varphi + H(u)\varphi_t \, dx$$

from the second relations in (8.32), and hence

$$\iint_Q u\Delta\varphi + H(u)\varphi_t \, dxdt = 0$$

from $\varphi|_{t=0,T} = 0$, which means (8.33) in the sense of distributions in Q. $\square$

Here, $H = H(u)$ takes discontinuity of the first kind at $u = 0$. If ρ is constant and c and κ are piecewise constant, then it holds that

$$H'(u) = \begin{cases} c_+\kappa_+, \ u > 0 \\ c_-\kappa_-, \ u < 0 \end{cases}, \quad H(+0) - H(-0) = \ell.$$

In any case, we define the *maximum monotone graph*, still denoted by $H = H(u)$ in $\mathbf{R} \times \mathbf{R}$, putting

$$H(0) = [H(-0), H(+0)].$$

Then we obtain the unique existence of the weak solution

$$u = u(x, t) \in L^\infty(Q), \quad Q = \Omega \times (0, T))$$

to (8.33) with

$$u|_{\partial\Omega} = 0, \ u|_{t=0} = u_0(x)$$

global in time [Oleinik (1960); Kamenomostsukaja (1961)]. More precisely, there is $h = h(x, t) \in L^1(Q)$ such that $h \in H(u)$ a.e. and

$$\iint_Q u\Delta\varphi + h\varphi_t \, dxdt + \int_\Omega u_0\varphi(\cdot, 0)dx = 0$$

for any $\varphi = \varphi(x, t) \in C^{2,1}(\overline\Omega \times [0, T])$ with $\varphi|_{t=T} = 0$, where $\Gamma = \partial\Omega \times (0, T)$.

Using the single-valued maximal monotone graph $f(v) = H^{-1}(v)$, we can derive

$$v_t = \Delta f(v)$$

from (8.33), and then the theory of *nonlinear semigroup* is applicable. In more details, if $f : \mathbf{R} \to \mathbf{R}$ is a non-decreasing continuous function satisfying $f(0) = 0$, then the operator

$$Av = -\Delta f(v)$$

with the domain

$$D(A) = \left\{ v \in L^1(\Omega) \mid f(v) \in W_0^{1,1}(\Omega), \ \ \Delta f(v) \in L^1(\Omega) \right\}$$

is maximum monotone [Brezis–Strauss (1973)], and hence generates a contraction semigroup in $X = L^1(\Omega)$ ([Crandall–Ligget (1971)]), denoted by $\{T_t\}_{t \geq 0}$. Then

$$v(\cdot, t) = T_t v_0$$

is regarded as a solution to

$$v_t = \Delta f(v), \quad f(v)|_{\partial\Omega} = 0, \quad v|_{t=0} = v_0, \tag{8.34}$$

and in this sense the problem (8.34) is well-posed in $X = L^1(\Omega)$ global in time.

8.7   Area Derivatives

We conclude this chapter with the first and the second area derivatives. We continue to suppose that $\Omega \subset \mathbf{R}^N$ is a bounded Lipschitz domain, and $\{T_t\}$, $|t| < \varepsilon$ is its deformation. The outer unit normal vector $\nu = \nu(\cdot, t)$, $|t| < \varepsilon$, is defined almost everywhere on $\partial\Omega_t$ because Ω_t is a Lipschitz domain.

Put

$$\delta\rho = \frac{\partial T_t}{\partial t} \cdot \nu \bigg|_{t=0}, \quad \delta^2\rho = \frac{\partial^2 T_t}{\partial t^2} \cdot \nu \bigg|_{t=0}, \tag{8.35}$$

and

$$Q = \bigcup_{|t|<\varepsilon} \Omega_t \times \{t\}, \quad \Gamma = \bigcup_{|t|<\varepsilon} \partial\Omega_t \times \{t\}$$

for $\Omega_t = T_t\Omega$ as in (8.5)–(8.6) in §8.1 and (8.17) in §8.2. If dS_t is the area element of $\partial\Omega_t$, it holds that

$$\int_{\partial\Omega_t} \nu \cdot a \, dS_t = \sum_i \int_{\partial\Omega_t} a^i * dy^i \tag{8.36}$$

for

$$y^i = T_t^i(x), \quad T_t = (T_t^1, \cdots, T_t^N)^T$$

by equation (6.32) in §6.3.

Since this integral is equal to

$$\sum_i \int_{\partial\Omega} a^i(T_t x) * dT_t^i(x), \tag{8.37}$$

it is continuously differentiable in t if $\{T_t\}$ is differentiable and $a^i \in C^{1,0}(\Gamma)$ for $1 \le i \le N$. Here we say that $f \in C^{1,0}(\Gamma)$ if it has an extension as an element in $C^1(G)$, where G is an open set containing Γ.

Henceforth,

$$\langle\, \cdot\, ,\, \cdot\, \rangle_{\partial\Omega}$$

denotes the paring identified with the inner product on $\partial\Omega$. The first area derivative is formulated as follows, where $\nu = \nu_0$ for $\nu_0 = \nu|_{t=0}$ and

$$dS = dS_0.$$

Theorem 8.8 (First Area Derivative). *If Ω is a Lipschitz domain, $\{T_t\}$ is differentiable, and $c \in C^{1,0}(\Gamma)$, it holds that*

$$\frac{d}{dt} \int_{\partial\Omega_t} c \, dS_t \bigg|_{t=0} = \int_{\partial\Omega} \dot{c}_0 \, dS + \left\langle (\nabla \cdot \nu)c_0 + \frac{\partial c_0}{\partial \nu}, \delta\rho \right\rangle_{\partial\Omega},$$

where $\dot{c}_0 = c_t|_{t=0}$ and $c_0 = c|_{t=0}$.

We begin with the following lemma.

Lemma 8.2. *If $\Omega \subset \mathbf{R}^N$ is a bounded Lipschitz domain, $\{T_t\}$ is differentiable, and $a \in C^{1,0}(\Gamma; \mathbf{R}^N)$, it holds that*

$$\frac{d}{dt} \int_{\partial\Omega_t} \nu \cdot a \, dS_t \bigg|_{t=0} = \int_{\partial\Omega} (\nu \cdot \dot{a}_0) \, dS + \langle \nabla \cdot a_0, \delta\rho \rangle_{\partial\Omega}, \tag{8.38}$$

where $\dot{a}_0 = a_t|_{t=0}$ and $a_0 = a|_{t=0}$.

Proof. By the definition there is an extension of a, denoted by the same symbol, such that $a \in C^{0,1}(G; \mathbf{R}^N)$. Then, we can assume $a \in C^2(G)$ for (8.38) to ensure.

First, we show

$$\frac{d}{dt} \int_{\partial\Omega_t} \nu \cdot a \, dS_t = \int_{\Omega} \frac{\partial}{\partial t} \{\mathrm{tr}[(D_x b)(DT_t)^{-1}]\det(DT_t)\} \, dx \tag{8.39}$$

for

$$b(x,t) = a(y,t), \quad y = T_t x, \tag{8.40}$$

where $\operatorname{tr} X$ denotes the trace of the matrix X. In fact, equation (8.40) implies

$$D_x b(x,t) = D_y a(y,t) D T_t(x)$$

and hence

$$D_y a(y,t) = D_x b(x,t) (D T_t(x))^{-1},$$

where $D_x b$ and $D_y a$ denote the Jacobi matrices of b and a with respect to x and y, respectively. Then Green's formula implies

$$\int_{\partial \Omega_t} (\nu \cdot a)(y,t) \, dS_t = \int_{\Omega_t} \nabla_y \cdot a(y,t) \, dy = \int_{\Omega_t} \operatorname{tr}[D_y a(y,t)] \, dy$$

$$= \int_{\Omega} \operatorname{tr}[D_x b(x,t) (D_x T_t(x))^{-1}] \det(D T_t)(x) \, dx, \tag{8.41}$$

and hence (8.39).

Second, Lemma 8.1 and Theorem 8.1 in §8.2 imply

$$D_x b|_{t=0} = D a_0, \quad (\det(D T_t)_t|_{t=0} = \nabla \cdot S$$

and

$$\left[(D_x b_t)(D T_t)^{-1} + (D_x b) \frac{\partial}{\partial t}(D T_t)^{-1}\right]\Bigg|_{t=0} = [D_x b_t - (D_x a)(D S)]|_{t=0}$$

uniformly in Ω. Since

$$D_x b_t(x,t) = D_x \left(D_y a(T_t x, t) \frac{\partial T_t}{\partial t}(x) + a_t(T_t x, t) \right)$$

$$= (D_y^2 a(T_t x, t) D T_t(x)) \frac{\partial T_t}{\partial t}(x) + D_y a(T_t x, t) \frac{\partial}{\partial t}(D T_t(x))$$

$$+ D_y a_t(T_t x, t) D T_t(x)$$

it holds that

$$D_x b_t|_{t=0} = (D^2 a_0) S + (D a_0)(D S) + D \dot{a}_0, \tag{8.42}$$

where $D^2 a$ is the third-order tensor consisting of the second derivative of the components of a. Hence there arises that

$$(D^2 a_0) S = \left(\sum_k \frac{\partial^2 a_0^i}{\partial x_j \partial x_k} S^k \right)_{i,j=1,\cdots,N}$$

for $a = (a^i)$ and $S = (S^i)$.

Gathering these observations, we obtain

$$\frac{d}{dt}\int_{\partial\Omega_t}(\nu\cdot a)\,dS_t\bigg|_{t=0} = \frac{d}{dt}\int_{\Omega}\text{tr}[(D_x b)(DT_t)^{-1}]\det D_t\,dx\bigg|_{t=0}$$

$$= \int_{\Omega}\text{tr}[(D^2 a_0)S + (Da_0)(DS) + D\dot a_0] + \text{tr}[(Da_0)(-DS)]$$

$$+\text{tr}[Da_0](\nabla\cdot S)\,dx$$

$$= \int_{\Omega}\text{tr}[(D^2 a_0)S] + (\nabla\cdot a_0)(\nabla\cdot S) + (\nabla\cdot\dot a_0)\,dx$$

$$= \int_{\Omega}\nabla\cdot[(\nabla\cdot a_0)S] + (\nabla\cdot\dot a_0)\,dx = \int_{\partial\Omega}(\nu\cdot\dot a_0 + (\nabla\cdot a_0)(S\cdot\nu)\,dS,$$

and hence the conclusion. □

Now we show the following proof.

Proof of Theorem 8.8. We can assume $\nu = \nu(x,t) \in C^{0,1}(G;\mathbf{S}^{N-1})$ by the method of extension described at the begining of §7.3, where $\mathbf{S}^{N-1}$ denotes the unit surface in $\mathbf{R}^N$. Then we apply Lemma 8.2 to $a = \nu c \in C^{1,0}(\Gamma)$.

There arises that

$$\nu\cdot\nu_t = 0 \quad \text{on } \Gamma \tag{8.43}$$

by $|\nu|^2 = 1$ in $\tilde\Gamma$, and therefore, $\nu\cdot\dot a_0 = \dot c_0$. Now we obtain the result by

$$\nabla\cdot a = (\nabla\cdot\nu)c + \nu\cdot\nabla c \quad \text{on } \Gamma.$$

□

The second area derivative arises if $\Omega \subset \mathbf{R}^N$ is a bounded $C^{2,1}$ domain and $\{T_t\}$ is twice differentiable. In this case there arises that $\nu \in C^{1,1}(G : \mathbf{S}^{N-1})$. We have also $s_i \in C^{1,1}(\Gamma, S^{N-1})$, $1 \le i \le N-1$, such that

$$\{s_1(\cdot,t),\cdots,s_{N-1}(\cdot,t),\nu(\cdot,t)\}$$

forms a frame of $\partial\Omega_t$ for each t. We now show the following lemma.

Lemma 8.3. *If Ω is $C^{2,1}$, $\{T_t\}$ is twice differentiable in t, and $a \in C^{1,1}(\Gamma;\mathbf{R}^N)$, it holds that*

$$\frac{d^2}{dt^2}\int_{\partial\Omega_t}\nu\cdot a\,dS_t\bigg|_{t=0} = \int_{\partial\Omega}\nu\cdot\ddot a_0 dS + \langle 2\nabla\cdot\dot a_0 + \nabla\cdot[(\nabla\cdot a_0)S], \delta\rho\rangle_{\partial\Omega}$$

$$+\langle\nabla\cdot a_0, (R - (S\cdot\nabla)S)\cdot\nu\rangle_{\partial\Omega}, \tag{8.44}$$

where $\ddot a_0 = a_{tt}|_{t=0}$, $\dot a_0 = a_t|_{t=0}$, and $a_0 = a|_{t=0}$.

Proof. We may assume $a \in C^3(G; \mathbf{R}^N)$ to ensure (8.44). Use the transformation (8.40), $b(x,t) = a(y,t)$, $y = T_t x$, to reach (8.41):

$$\int_{\partial\Omega_t} \nu \cdot a \, dS_t = \int_{\Omega} \operatorname{tr}\left[(Db)(DT_t)^{-1}\right] \det(DT_t) \, dx.$$

Differentiating the right-hand side twice, here we obtain

$$\frac{\partial^2}{\partial t^2} \operatorname{tr}[(Db)(DT_t)^{-1}]\det(DT_t) = \operatorname{tr}[(Db_{tt})(DT_t)^{-1}]\det(DT_t)$$

$$+\operatorname{tr}[(Db)[(DT_t)^{-1}]_{tt}]\det(DT_t) + \operatorname{tr}[(Db)(DT_t)^{-1}][\det(DT_t)]_{tt}$$

$$+2\operatorname{tr}[(Db)[(DT_t)^{-1}]_t]\det(DT_t) + 2\operatorname{tr}[(Db)_t(DT_t)^{-1}][\det(DT_t)]_t$$

$$+2\operatorname{tr}[(Db)[(DT_t)^{-1}]_t][\det(DT_t)]_t.$$

Since

$$D_x b(x,t) = D_y a(T_t x, t)(DT_t(x)),$$

there arises that

$$D_x b_{tt}(x,t) = (D_y^3 a(T_t x, t) DT_t(x))(\frac{\partial T_t}{\partial t}(x))^2$$

$$+2(D_y^2 a(T_t x, t)(DT_t(x))_t)\frac{\partial T_t}{\partial t}(x) + 2(D_y^2 a_t(T_t x, t) DT_t(x))\frac{\partial T_t}{\partial t}(x)$$

$$+2D_y a_t(T_t x, t)(DT_t(x))_t + (D_y^2 a(T_t x, t) DT_t(x))\frac{\partial^2 T_t}{\partial t^2}(x)$$

$$+D_y a(T_t x, t)(DT_t(x))_{tt} + D_y a_{tt}(T_t x, t)(DT_t(x)),$$

and hence

$$Db_{tt}|_{t=0} = (D^3 a_0)S^2 + 2((D^2 a_0)DS)S + 2(D^2 \dot{a}_0)S$$

$$+2(D\dot{a}_0)DS + (D^2 a_0)R + (Da_0)DS + D\ddot{a}_0$$

uniformly on Ω, where $D^3 a$ denotes the fourth-order tensor which consists of the third derivatives of the elements of a. It thus follows that

$$\frac{\partial^2}{\partial t^2} \operatorname{tr}[(Db)(DT_t)^{-1}]\det(DT)\Big|_{t=0}$$

$$= \operatorname{tr}[(D^3 a_0)S^2 + 2((D^2 a_0)DS)S + 2D^2 \dot{a}_0 S]$$

$$+\operatorname{tr}[2(D\dot{a}_0)DS + (D^2 a_0)R + (Da_0)DR]$$

$$+\operatorname{tr}[D\ddot{a}_0 + (Da_0)(2(DS)^2 - DR)]$$

$$+\operatorname{tr}[Da_0](\nabla \cdot R + (\nabla \cdot S)^2 - DS^T : DS)$$

$$+2\operatorname{tr}[(D^2 a_0)S + (Da_0)DS + (D\dot{a}_0)(-DS)]$$

$$+2\operatorname{tr}[(D^2 a_0)S + (Da_0)DS + D\dot{a}_0](\nabla \cdot S) + 2\operatorname{tr}[(Da_0)(-DS)](\nabla \cdot S)$$

$$= \operatorname{tr}[(D^3 a_0)S^2 + 2(D^2 \dot{a}_0)S + (D^2 a_0)S + D\ddot{a}_0]$$

$$+\operatorname{tr}[Da_0](\nabla \cdot S + (\nabla \cdot S)^2 - DS^T : DS)$$

$$+2\operatorname{tr}[(D^2 a_0)S + D\dot{a}_0](\nabla \cdot S).$$

Hence we obtain

$$\frac{\partial^2}{\partial t^2} \int_\Omega \text{tr}[(Db)(DT_t)^{-1}]\det(DT_t)\, dx\bigg|_{t=0}$$

$$= \int_\Omega \nabla \cdot \ddot{a}_0 + \nabla \cdot [(\nabla \cdot a_0)R] + 2\nabla \cdot [(\nabla \cdot a_0)S] + \text{tr}[(D^3 a_0)S^2]$$

$$+ 2\text{tr}[D\dot{a}_0](\nabla \cdot S) + \text{tr}[Da_0]((\nabla \cdot S)^2 - DS^T : DS)\, dx, \qquad (8.45)$$

using

$$\text{tr}[(D^2 a_0)R] + \text{tr}[Da_0](\nabla \cdot R) = \nabla \cdot [((\nabla \cdot a_0))R]$$

and

$$\text{tr}[(D^2 \dot{a}_0)S] + \text{tr}[D\dot{a}_0](\nabla \cdot S) = \nabla \cdot [(\nabla \cdot \dot{a}_0)S].$$

We simplify the the last three terms of (8.45) further by the divergence formula. Let

$$a_0 = (a^1, \cdots, a^N)^T$$
$$S = (S^1, \cdots, S^N)^T$$
$$\nu = (\nu^1, \cdots, \nu^N)^T,$$

and put

$$f_i = \frac{\partial f}{\partial x_i}$$

$$f_{ij} = \frac{\partial^2 f}{\partial x_i \partial x_j}$$

$$f_{ijk} = \frac{\partial^3 f}{\partial x_i \partial x_j \partial x_k}.$$

Then it follows that

$$X \equiv \int_\Omega \text{tr}[(D^3 a_0)S^2]\, dx = \sum_{i,p,q} \int_\Omega a^i_{ipq} S^p S^q\, dx$$

$$= -\sum_{i,p,q} \int_\Omega a^i_{ip}(S^p_q S^q + S^p S^q_q)\, dx + \sum_{i,p,q} \langle a^i_{ip} S^p, S^q \nu^q \rangle_{\partial\Omega},$$

$$Y \equiv 2\int_\Omega \text{tr}[(D^2 a_0)S](\nabla \cdot S)\, dx = 2\sum_{i,p,q} \int_\Omega a^i_{ip} S^p S^q_q dx,$$

and

$$Z \equiv \int_\Omega \operatorname{tr}[(Da_0)]((\nabla \cdot S)^2 - DS^T : DS)\, dx$$

$$= \sum_{i,p,q} \int_\Omega a^i (S_p^p S_q^q - S_p^q S_q^p)\, dx$$

$$= \sum_{i,p,q} \int_\Omega -(a_{ip}^i S^p S_q^q + a^i S^p S_{pq}^q) + (a_{ip}^i S^q S_q^p + a_i^i S^q S_{pq}^p)\, dx$$

$$+ \sum_{i,p,q} \{ \langle a_i^i S_q^q, S^p \nu^p \rangle_{\partial\Omega} - \langle a_i^i S^q, S_q^p \nu^p \rangle_{\partial\Omega} \}.$$

We thus obtain

$$X + Y + Z = \sum_{i,p,q} \{ \langle a_{ip}^i S^p, S^q \nu^q \rangle_{\partial\Omega} + \langle a^i S_q^q, S^p \nu^p \rangle_{\partial\Omega} - \langle a_i^i S^q, S_q^p \nu^p \rangle_{\partial\Omega} \}$$

$$= \langle \nabla \cdot [(\nabla \cdot a_0)S], S \cdot \nu \rangle_{\partial\Omega} - \langle \nabla \cdot a_0, [(S \cdot \nabla)S] \cdot \nu \rangle_{\partial\Omega},$$

recalling $S \in C^{0,1}(\Omega)$.

Gathering all equations, we obtain the result. $\square$

The following lemma is a refinement of (8.43).

Lemma 8.4. *If Ω is $C^{1,1}$ and $\{T_t\}$ is differentiable, it holds that*

$$\nu_t = -\sum_{i=1}^{N-1} \left[\frac{\partial}{\partial s_i} \left(\frac{\partial T_t}{\partial t} \cdot \nu \right) \right] s_i, \quad a.e. \ on \ \Gamma.$$

Proof. We may fix $x_0 \in \Omega$ and assume that

$$\{s_1, \cdots, s_{N-1}, \nu\}$$

are differentiable at $(x,t) = (x_0, 0)$ to show the desired equality at this $(x_0, 0)$. Write

$$\nu(t) = \nu(x_0, t), \quad s_i = s_i(x_0, 0), 1 \le i \le N-1, \quad \nu = \nu(x_0, 0).$$

We take the *exponential mapping* aroud x_0:

$$\xi_1 s_1 + \cdots + \xi_{N-1} s_{N-1} \in T_{x_0}(\partial\Omega) \mapsto x(\xi) \in \partial\Omega \qquad (8.46)$$

for

$$\xi = (\xi_1, \cdots, \xi_{N-1}) \in \mathbf{R}^{N-1}.$$

This mapping is defined for $|\xi| \ll 1$, and satisfies $x(0) = x_0$. It is a local $C^{0,1}$ diffeomorphism, and there arises that

$$\frac{\partial x}{\partial \xi_i}\bigg|_{\xi=0} = s_i.$$

The perturbed boundary $\partial\Omega_t$ around $x_0(t) = T_t x_0$ is thus parametrized by ξ as $T_t(x(\xi))$, and furthermore, the tangent space

$$T_{x_0(t)}(\partial\Omega_t)$$

is spanned by

$$\{\tilde{s}_1(t), \cdots, \tilde{s}_{N-1}(t)\}, \quad \tilde{s}_i(t) = \frac{\partial}{\partial\xi_i} T_t(x(\xi))\Big|_{\xi=0}, \quad i = 1, \cdots, N-1,$$

although

$$\{\tilde{s}_1(t), \cdots, \tilde{s}_{N-1}(t), \nu(t)\}$$

does not necessarily form a frame at

$$x_0(t) = T_t x_0 \in \partial\Omega_t.$$

Since Ω is $C^{0,1}$ and $\{T_t\}$ is differentiable, these vectors are Lipschitz continuous in t, and it follows that

$$s_i(t) \cdot \nu(t) = 0, \quad \nu(t) \cdot \nu(t) = 1.$$

Then we obtain

$$\nu_t|_{t=0} \cdot s_i + \nu \cdot \frac{\partial s_i}{\partial t}\Big|_{t=0} = \nu_t|_{t=0} \cdot \nu = 0, \tag{8.47}$$

and furthermore,

$$\begin{aligned}
\frac{\partial s_i}{\partial t}\Big|_{t=0} &= \frac{\partial^2}{\partial t \partial\xi_i} T_t(x(\xi))\Big|_{t=0,\ \xi=0} = \frac{\partial x}{\partial\xi_i} S(x(\xi))\Big|_{\xi=0} = \nabla S \cdot s_i \\
&= \frac{\partial}{\partial s_i} \frac{\partial T_t}{\partial t}\Big|_{t=0}
\end{aligned}$$

by (8.35). Hence it follows that

$$\nu_t = -\sum_{i=1}^{N-1}\left(\nu \cdot \frac{\partial s_i}{\partial t}\right)s_i = -\sum_{i=1}^{N-1}\left[\frac{\partial}{\partial s_i}\left(\frac{\partial T_t}{\partial t} \cdot \nu\right)\right]s_i$$

at $(x,t) = (x_0, 0)$. $\qquad\square$

We are ready to show the following theorem.

Theorem 8.9 (Second Area Derivative). *If Ω is $C^{2,1}$, $\{T_t\}$ is twice differentiable, and $c \in C^{1,1}(\Gamma)$, it holds that*

$$\begin{aligned}
\frac{d^2}{dt^2}\int_{\partial\Omega_t} c\, dS_t\Big|_{t=0} &= \int_{\partial\Omega} \ddot{c}_0\, dS - \langle c_0, |\nabla_\tau \delta\rho|^2\rangle_{\partial\Omega} \\
&+ \langle -2(\Delta_\tau\delta\rho)c_0 + 2(\nabla\cdot\nu)\dot{c}_0 + \nabla\cdot[((\nabla\cdot\nu)c_0 + \frac{\partial c_0}{\partial\nu})S], \delta\rho\rangle_{\partial\Omega} \\
&- \langle\nabla_\tau^2 c_0, (\delta\rho)^2\rangle_{\partial\Omega} + \langle(\nabla\cdot\nu)c_0 + \frac{\partial c_0}{\partial\nu}, \delta^2\rho - ((S\cdot\nabla)S)\cdot\nu\rangle_{\partial\Omega},
\end{aligned}$$

where $c_0 = c|_{t=0}$, $\dot{c}_0 = c_t|_{t=0}$, $\ddot{c}_0 = c_{tt}|_{t=0}$, and

$$\Delta_\tau = \nabla_\tau \cdot \nabla_\tau, \quad \nabla_\tau = \left(\frac{\partial}{\partial s_1}, \cdots, \frac{\partial}{\partial s_{N-1}} \right)^T. \tag{8.48}$$

Proof. By the assumption it holds that

$$\nu \in C^{1,1}(\Gamma; S^{N-1}).$$

Then we apply Lemma 8.3 to $a = \nu c$.

First, it follows that

$$a_{tt} = \nu_{tt}c + 2\nu_t c_t + \nu_{tt}, \quad a_t = \nu_t c + \nu c_t.$$

Second, $|\nu|^2 = 1$ in $\tilde{\Gamma}$ implies

$$\nu \cdot \nu_{tt} = -|\nu_t|^2 \quad \text{on } \Gamma$$

as well as (8.43). Then we obtain

$$\nu \cdot a_{tt} = -|\alpha|^2 c + c_{tt}$$

with the continuity of its right-hand side on Γ, where

$$\alpha = \left(\frac{\partial T_t}{\partial s_1} \cdot \nu, \cdots, \frac{\partial T_t}{\partial s_{N-1}} \cdot \nu \right)^T.$$

It holds also that

$$\begin{aligned}
\nabla \cdot a_t &= [(\nabla \cdot \nu)c + \nu \cdot \nabla c]_t = [(\nabla \cdot \nu)c]_t + \nu_t \cdot \nabla c + \nu \cdot \nabla c_t \\
&= [(\nabla \cdot \nu)c]_t + \sum_{i=1}^{N-1} \left(\frac{\partial S}{\partial s_i} \cdot \nu \right) \frac{\partial c}{\partial s_i} + \frac{\partial c_t}{\partial \nu} \\
&= (\nabla \cdot \nu)c]_t + \alpha \cdot \nabla_\tau c + \frac{\partial c_t}{\partial \nu}
\end{aligned}$$

with the continuity of its right-hand side on Γ.

Since Ω is $C^{2,1}$ and $\{T_t\}$ is twice differentiable, ν_t is continuously differentiable on Γ and it holds that

$$\nabla \cdot \nu_t = -\sum_{i=1}^{N-1} \frac{\partial^2}{\partial s_i^2} \left(\frac{\partial T_t}{\partial t} \cdot \nu \right).$$

We thus obtain

$$[(\nabla \cdot \nu)c]_t = (\nabla \cdot \nu_t)c + (\nabla \cdot \nu)c_t = -\left(\Delta_\tau \left(\frac{\partial T_t}{\partial t} \cdot \nu \right) \right) c + (\nabla \cdot \nu)c_t.$$

Then the result follows from

$$\nabla \cdot a = (\nabla \cdot \nu)c + \frac{\partial c}{\partial \nu}$$

$$|\alpha|^2\Big|_{t=0} = \sum_{i=1}^{N-1}\left(\frac{\partial \delta \rho}{\partial s_i}\right)^2 = |\nabla_\tau \delta \rho|^2,$$

and

$$2\left\langle \alpha \cdot \nabla_\tau c, \frac{\partial T_t}{\partial t} \cdot \nu\right\rangle_{\partial \Omega}\Big|_{t=0} = \langle \nabla_\tau c_0, \nabla_\tau (\delta \rho)^2\rangle = -\langle \nabla_\tau^2 c_0, (\delta \rho)^2\rangle$$

obtained by

$$\alpha \cdot \nabla_\tau c = \sum_{i=1}^{N-1}\left(\frac{\partial T_t}{\partial s_i} \cdot \nu\right)\frac{\partial c}{\partial s_i} = \nabla_\tau\left(\frac{\partial T_t}{\partial t} \cdot \nu\right) \cdot \nabla_\tau c.$$

$\square$

Remark 8.1. Lemma 8.2 and Lemma 8.3 can be derived directly from (8.36)–(8.37), that is,

$$\int_{\partial \Omega_t} \nu \cdot a \, dS_t = \sum_i \int_{\partial \Omega_t} a^i * dy^i = \sum_i \int_{\partial \Omega} a^i(T_t x) * dT_t^i(x)$$

([Suzuki–Tsuchiya (2023a)]).

Chapter 9

Hadamard's Variational Formula

This chapter is devoted to Hadamard's variational formula of Green's function for $-\Delta$ concerning the deformation of domains. First, we intoduce an abstract theory which guarantees the existence of Hadamard's variation (§9.1). Then we define the Green function using the fundamental solution to $-\Delta$ and the Laplace equation (§9.2). Second, existence of the first and the second variational formulae are confirmed for Dirichlet boundary condtion (§9.3–§9.4). Then, the first variational formula of this case is given using the frame on the boundary (§9.5). The second formula is provided similarly, in accordance with the second fundamental form (§9.6–§9.7). Finally, we study the Neumann boundary condition (§9.8). The first variational formula is associated with the tangential derivative of the Green function (§9.9). The second variational formula is given using the first variation as in [Garabedian–Schiffer (1952–53)] for three dimension, where the second fundamental form on the boundary is involved (§9.10–§9.11).

9.1   An Abstract Theorem

Derivation of Hadamard's variational formulae is hard even formally. Fundamental obstruction is the change of domain where Green's function is defined. These variations are taken only inside the domain, while Green's function is defined by prescribing its boundary behavior. This discrepancy is essential if one uses the Euler coordinate. Then we use the Lagrange coordinate to ensure the differentiability, where abstract theory is applicable.

Let V be a Hilbert space over $\mathbf{R}$, and

$$a_t : V \times V \to \mathbf{R}, \quad t \in (-\varepsilon, \varepsilon) \equiv I,$$

be a family of symmetric bilinear forms. We assume their uniform boundedness and coercivity. Hence there are $\delta > 0$ and $M > 0$ such that

$$a_t(u, u) \geq \delta \|u\|_V^2, \quad |a_t(u, v)| \leq M\|u\|_V \|v\|_V \tag{9.1}$$

for any $u, v \in V$ and $t \in I$. Given $f : I \to V'$, we have a unique $u = u(t) \in V$, $t \in I$, such that

$$a_t(u(t), v) = \langle v, f(t) \rangle, \quad \forall v \in V, \tag{9.2}$$

by the representation theorem of Riesz. Then we obtain the following theorem [Suzuki (1982)].

Theorem 9.1. *Let the above $a_t(\cdot, \cdot)$ and $f(t)$ be strongly differentiable with respect to t. Namely, first, there is a bounded symmetric bilinear form denoted by*

$$\dot{a}_t : V \times V \to \mathbf{R}$$

such that

$$\lim_{h \to 0} \sup_{\|u\| \leq 1, \ \|v\| \leq 1} \left| \frac{a_{t+h}(u, v) - a_t(u, v)}{h} - \dot{a}_t(u, v) \right| = 0, \quad \forall t \in I. \tag{9.3}$$

Second, there is

$$\dot{f}(t) \in V'$$

such that

$$\lim_{h \to 0} \sup_{\|v\| \leq 1} \left| \frac{\langle v, f(t+h) \rangle - \langle v, f(t) \rangle}{h} - \langle v, \dot{f}(t) \rangle \right| = 0, \quad \forall t \in I. \tag{9.4}$$

Then $u = u(t) \in V$ defined by (9.2) is strongly differentiable in t, that is, there is

$$\dot{u}(t) \in V$$

such that

$$\lim_{h \to 0} \left\| \frac{u(t+h) - u(t)}{h} - \dot{u}(t) \right\|_V = 0, \quad \forall t \in I, \tag{9.5}$$

and it holds that

$$a_t(\dot{u}(t), v) + \dot{a}_t(u(t), v)) = \langle v, \dot{f}(t) \rangle, \quad \forall v \in V, \ t \in I. \tag{9.6}$$

The second strong differentiability of $u = u(t)$ arises similarly if

$$a_t : V \times V \to \mathbf{R}$$

and

$$f = f(t) : I \to V'$$

are twice differentiable. Hence if there are bounded symmetric bilinear form

$$\ddot{a}_t : V \times V \to \mathbf{R}$$

and $\ddot{f}(t) \in V'$ such that

$$\lim_{h \to 0} \sup_{\|u\| \leq 1, \ \|v\| \leq 1} \left| \frac{\dot{a}_{t+h}(u,v) - \dot{a}_t(u,v)}{h} - \ddot{a}_t(u,v) \right| = 0 \qquad (9.7)$$

and

$$\lim_{h \to 0} \sup_{\|v\| \leq 1} \left| \frac{\langle v, \dot{f}(t+h) \rangle - \langle v, \dot{f}(t) \rangle}{h} - \langle v, \ddot{f}(t) \rangle \right| = 0, \qquad (9.8)$$

then there is

$$\ddot{u}(t) \in V$$

such that

$$\lim_{h \to 0} \left\| \frac{\dot{u}(t+h) - \dot{u}(t)}{h} - \ddot{u}(t) \right\|_V = 0. \qquad (9.9)$$

It holds also that

$$a_t(\ddot{u}(t), v) + 2\dot{a}_t(\dot{u}(t), v) + \ddot{a}_t(u(t), v) = \langle v, \ddot{f}(t) \rangle, \quad \forall v \in V, \ t \in I. \quad (9.10)$$

Remark 9.1. To prove strong differentiability, for example, first, we define $\dot{u}$ by (9.6) and then derive (9.5) using (9.1). The second differentiability is shown similarly.

Remark 9.2. If equalities (9.3)–(9.4) hold for fixed $t \in I$, equality (9.5) arises at this t for $\dot{u}$ defined by (9.6). If equalities (9.3)–(9.4) hold for any $t \in I$ and equalities (9.7)–(9.8) hold for fixed $t \in I$, there arises that (9.9) at this t for $\ddot{u}$ defined by (9.10). We develop a refined argument on these differentiabilities in Chapter 10 (Theorem 10.12 and Theorem 10.15 in §10.8).

9.2   Green's Function

Let $\Omega \subset \mathbf{R}^N$ be a Lipschitz domain. The Green's function for $-\Delta$ with $\cdot|_{\partial\Omega} = 0$, denoted by $G = G(x,y)$, $(x,y) \in \overline{\Omega} \times \Omega$, is defined by

$$-\Delta G(\cdot, y) = \delta(\cdot - y) \quad \text{in } \Omega, \quad G(\cdot, y)|_{\partial\Omega} = 0,$$

where $\delta = \delta(x)$ is the delta function. The fundamental solution of $-\Delta$,

$$\Gamma(x) = \gamma(|x|), \quad \gamma(r) = \begin{cases} \frac{1}{2\pi} \log \frac{1}{r}, & N = 2 \\ \frac{1}{(N-2)\omega_N} r^{2-N}, & N \geq 3 \end{cases} \tag{9.11}$$

takes the property

$$-\Delta\Gamma(\cdot - y) = \delta(\cdot - y),$$

where ω_N denotes the $(N-1)$-dimensional volume of the N-dimensional unit ball. Hence the function $u = u(x)$ defined by

$$G(\cdot, y) = \Gamma(\cdot - y) + u \tag{9.12}$$

satisfies

$$\Delta u = 0 \quad \text{in } \Omega, \qquad u = -\Gamma(\cdot - y) \quad \text{on } \partial\Omega. \tag{9.13}$$

Remark 9.3. We use the notation

$$\Gamma(x) = \frac{1}{2\pi} \log \frac{1}{|x|}$$

slightly different from (9.11) for $N = 2$ in Chapter 2 and Chapter 3.

We reduce this boundary value problem of the Laplace equation to that of the Poisson equation, taking a pair of open sets $\omega \subset\subset \hat{\omega}$ satisfying $\partial\Omega \subset \omega$, $y \notin \hat{\omega}$ and $\varphi = \varphi(x) \in C_0^\infty(\mathbf{R}^N)$ such that $0 \leq \varphi \leq 1$ and

$$\varphi(x) = \begin{cases} 1, & x \in \omega \\ 0, & x \notin \hat{\omega}. \end{cases}$$

In fact, then (9.13) is reduced to

$$-\Delta w = g \quad \text{in } \Omega, \qquad w = 0 \quad \text{on } \partial\Omega \tag{9.14}$$

for

$$w = u - \varphi\Gamma(\cdot - y), \quad g = \Delta(\varphi\Gamma(\cdot - y)).$$

This $g = g(x) \in C_0^\infty(\mathbf{R}^N)$ is regarded as an element of $H^{-1}(\Omega)$ by

$$\langle v, g \rangle = -(\nabla v, \nabla(\varphi\Gamma(\cdot - y))), \quad v \in H_0^1(\Omega).$$

The boundary value problem (9.14) admits a unique solution $w \in H_0^1(\Omega)$. Then the solution to (9.13) is obtained by

$$u = w + \varphi\Gamma(\cdot - y) \in H^1(\Omega)$$

because $\Gamma(\cdot - y)$ is smooth in $\hat{\omega}$. Hence (9.12)–(9.13) implies

$$0 = G(\cdot, y)|_{\partial\Omega} \in H^{1/2}(\partial\Omega), \quad \left.\frac{\partial G(\cdot, y)}{\partial\nu}\right|_{\partial\Omega} \in H^{-1/2}(\partial\Omega) \qquad (9.15)$$

by Theorem 7.3 in §7.1. In the following theorem, the pairing

$$\langle \, , \, \rangle$$

is taken between $H^{1/2}(\partial\Omega)$ and $H^{-1/2}(\partial\Omega)$.

Theorem 9.2. *If $\Omega \subset \mathbf{R}^N$ is a bounded Lipschitz domain and $f \in H^1(\Omega)$ is harmonic in Ω, then it holds that*

$$f(y) = -\left\langle f, \frac{\partial G(\cdot, y)}{\partial\nu} \right\rangle, \quad y \in \Omega. \qquad (9.16)$$

Proof. Since $\Delta f = 0 \in H^1(\Omega)'$ we have

$$(\nabla f, \nabla \varphi) = \left\langle \varphi, \frac{\partial f}{\partial\nu} \right\rangle, \quad \varphi \in H^1(\Omega) \qquad (9.17)$$

by Theorem 7.3. Given $y \in \Omega$, we take

$$\varphi_\varepsilon(x) = \begin{cases} \gamma(|x - y|), & |x - y| \geq \varepsilon \\ \gamma(\varepsilon), & |x - y| < \varepsilon \end{cases}$$

for $0 < \varepsilon \ll 1$, which belongs to $H^1(\Omega)$. Then we obtain

$$(\nabla f, \nabla \varphi_\varepsilon) = \int_{\Omega \setminus B(y,\varepsilon)} \nabla f(x) \cdot \nabla \Gamma(x - y) \, dx$$

$$= \int_{\partial\Omega} f(x) \frac{\partial \Gamma(x - y)}{\partial\nu_x} \, dS_x - \int_{\partial B(y,\varepsilon)} f(x) \frac{\partial \Gamma(x - y)}{\partial\nu_x} \, dS_x,$$

using the traces

$$f|_{\partial\Omega} \in H^{1/2}(\partial\Omega), \quad f|_{\partial B(y,\varepsilon)} \in H^{1/2}(\partial B(y, \varepsilon)).$$

Since $f = f(x)$ is smooth in Ω by Weyl's lemma, the left-hand side of (9.17) for $\varphi = \varphi_\varepsilon$, that is, $(\nabla f, \nabla \varphi_\varepsilon)$, converges to

$$\int_{\partial\Omega} f(x) \frac{\partial \Gamma(x - y)}{\partial\nu_x} \, dS_x + f(y)$$

as $\varepsilon \downarrow 0$ by a classical argument. Since its right-hand side,

$$\left\langle \varphi_\varepsilon, \frac{\partial f}{\partial\nu} \right\rangle,$$

is independent of $0 < \varepsilon \ll 1$, we end up with

$$\left\langle \Gamma(\cdot - y), \frac{\partial f}{\partial \nu} \right\rangle = \left\langle f, \frac{\partial \Gamma(\cdot - y)}{\partial \nu} \right\rangle + f(y). \tag{9.18}$$

We define $u = u(x)$ by (9.12), and note

$$\Delta f = \Delta u = 0 \in H^1(\Omega)'.$$

Then it follows that

$$(\nabla u, \nabla f) = \left\langle u, \frac{\partial f}{\partial \nu} \right\rangle = \left\langle f, \frac{\partial u}{\partial \nu} \right\rangle \tag{9.19}$$

from Theorem 7.3. Since (9.15) implies

$$\left\langle u + \Gamma(\cdot - y), \frac{\partial f}{\partial \nu} \right\rangle = \left\langle G(\cdot, y), \frac{\partial f}{\partial \nu} \right\rangle = 0$$

there arises that

$$\left\langle \Gamma(\cdot - y), \frac{\partial f}{\partial \nu} \right\rangle = -\left\langle u, \frac{\partial f}{\partial \nu} \right\rangle = -\left\langle f, \frac{\partial u}{\partial \nu} \right\rangle$$

by (9.19). Then we obtain

$$f(y) = -\left\langle f, \frac{\partial \Gamma(\cdot - y)}{\partial \nu} + \frac{\partial u}{\partial \nu} \right\rangle = -\left\langle f, \frac{\partial G(\cdot, y)}{\partial \nu} \right\rangle$$

by (9.18). $\qquad\qquad\qquad\qquad\qquad\qquad\qquad\qquad\qquad\qquad\qquad\square$

9.3 Lagrange Derivatives

We show two formulations of the variational formulae, using Euler and Lagrange coordinates in §9.3 and §9.4, respectively. In the Lagrange coordinate, we do not have any derivative losses in space variables. The variational formulae, however, are not clear compared with the ones in Euler coordinates.

We recall that $\Omega \subset \mathbf{R}^N$ is a bounded Lipschitz domain and $\{T_t\}$ is either once or twice differentiable deformation of it. Green's function for $-\Delta$ in Ω_t with $\cdot|_{\partial\Omega_t} = 0$ is denoted by $G_t = G_t(x, y)$. We fix $y \in \Omega$ to define $u = u(\cdot, t)$ by (9.12):

$$G_t(x, y) = \Gamma(x - y) + u(x, t), \tag{9.20}$$

which satisfies

$$\Delta u(\cdot, t) = 0 \text{ in } \Omega_t, \quad u(\cdot, t) = -\Gamma(\cdot - y) \text{ on } \partial\Omega_t \tag{9.21}$$

for $|t| \ll 1$.

As in the previous section this problem is reduced to the boundary value problem of the Poisson equation

$$-\Delta w(\cdot, t) = g \text{ in } \Omega_t, \quad w(\cdot, t) = 0 \text{ on } \partial\Omega_t \tag{9.22}$$

where

$$g = \Delta(\varphi\Gamma(\cdot - y))$$

for some $\varphi \in C_0^\infty(\mathbf{R}^N)$. Given $y \in \Omega$, this $\varphi = \varphi(x)$ is uniform in $|t| \ll 1$.

Lemma 9.1. *Let $\Omega \subset \mathbf{R}^N$ be a bounded Lipschitz domain and $\{T_t\}$, $|t| < \varepsilon$ be once or twice differentiable deformation. Then, according to its differentiabilities we have the existence of*

$$\frac{\partial}{\partial t}u(T_tx, t), \quad \frac{\partial^2}{\partial t^2}u(T_tx, t)$$

strongly in $H^1(\Omega)$.

Proof. The weak form of (9.22) is

$$\int_{\Omega_t} \nabla w(\cdot, t) \cdot \nabla\varphi \, dy = \int_{\Omega_t} \varphi g \, dy, \quad \forall\varphi \in H_0^1(\Omega_t). \tag{9.23}$$

Given $\psi \in H_0^1(\Omega)$, we define

$$v = v(\cdot, t)$$

by

$$\varphi(y, t) = \psi(T_t^{-1}y), \quad y \in \Omega_t, \quad |t| \ll 1.$$

It holds that

$$v(\cdot, t) \in H_0^1(\Omega_t)$$

because $T_t : \Omega \to \Omega_t$ is bi-Lipschitz. Then (9.23) implies

$$\int_\Omega [J_t(x)^{-1}\nabla v(x, t) \cdot J_t(x)^{-1}\nabla\psi(x)] \det J_t(x) \, dx$$

$$= \int_\Omega \psi(x)g(T_tx) \det J_t(x) \, dx$$

for

$$J_t = J(T_t), \quad v(x, t) = w(T_tx, t),$$

where $J(T_t) = DT_t$ denotes the Jacobi matrix of $T_t : \Omega \to \Omega_t$.

Now we apply Theorem 9.1 for $V = H_0^1(\Omega)$,

$$f(t) = h(\cdot, t) \det J_t(\cdot), \quad h(x, t) = g(T_t x),$$

and

$$a_t(v, \psi) = \int_\Omega [J_t^{-1} \nabla v \cdot J_t^{-1} \nabla \psi] \det J_t \, dx.$$

From the dominated convergence theorem the strong differentiability of $v(\cdot, t) \in H_0^1(\Omega)$ once or twice in t follows, which implies those of $u(T_t x, t)$ and $G_t(T_t x, y)$. $\qquad\square$

To conclude this section, we show concrete forms of the Lagrange derivative of $G(\cdot, y)$. Let $\dot{v} = \dot{v}(0) \in H_0^1(\Omega)$ and $\ddot{v} = \ddot{v}(0) \in H_0^1(\Omega)$. Put also

$$\dot{f} = \dot{f}(0), \; \ddot{f} = \ddot{f}(0) \in V' = H^{-1}(\Omega)$$

for $f = f(t)$ used in the proof of Lemma 9.1.

First,

$$f(x, t) = g(T_t x) \det J_t(x)$$

satisfies

$$\frac{\partial f}{\partial t}(x, t) = \frac{\partial T_t x}{\partial t} \cdot \nabla g(T_t x) + g(T_t x) \frac{\partial}{\partial t} \det J_t(x)$$

and hence

$$\dot{f} = S \nabla g + g \nabla \cdot S = \nabla \cdot (Sg).$$

We have also

$$\ddot{f} = R \cdot \nabla g + [\nabla^2 g] S \cdot S + (S \cdot \nabla g) \nabla \cdot S$$
$$+ g \left[\nabla \cdot R + (\nabla \cdot S)^2 - J(S) : {}^t J(S) \right]$$
$$= \nabla \cdot (gR) + [\nabla^2 g] S \cdot S + (S \cdot \nabla g) \nabla \cdot S + g \left[(\nabla \cdot S)^2 - J(S)^T : J(S) \right].$$

Second, the strong derivatives

$$\dot{a}_t = \dot{a}_t(v, \psi), \; \ddot{a}_t = \ddot{a}_t(v, \psi), \quad v, \psi \in H_0^1(\Omega), \; |t| \ll 1$$

of $a_t = a_t(v, \psi)$ are given by

$$\dot{a}_t(v, \psi) = \int_\Omega \left[\frac{\partial J_t^{-1}}{\partial t} \nabla v \cdot J_t^{-1} \nabla \psi + J_t^{-1} \nabla v \cdot \frac{\partial J_t^{-1}}{\partial t} \nabla \psi \right] \det J_t$$

$$+ [J_t^{-1} \nabla v \cdot J_t^{-1} \nabla \psi] \frac{\partial}{\partial t} \det J_t \, dx$$

and
$$\ddot{a}_t(v,\psi) = \int_\Omega \Big[\frac{\partial^2 J_t^{-1}}{\partial t^2} \nabla v \cdot J_t^{-1}\nabla\psi + 2\frac{\partial J_t^{-1}}{\partial t}\nabla v \cdot \frac{\partial J_t^{-1}}{\partial t}\nabla\psi$$
$$+ J_t^{-1}\nabla v \cdot \frac{\partial^2 J_t^{-1}}{\partial t^2}\nabla\psi\Big] \det J_t$$
$$+ 2\Big[\frac{\partial J_t^{-1}}{\partial t}\nabla v \cdot J_t^{-1}\nabla\psi + J_t^{-1}\nabla v \cdot \frac{\partial J_t^{-1}}{\partial t}\nabla\psi\Big]\frac{\partial}{\partial t}\det J_t$$
$$+ \big[J_t^{-1}\nabla v \cdot J_t^{-1}\nabla\psi\big]\frac{\partial^2}{\partial t^2}\det J_t\, dx,$$
respectively.

Here we have
$$\frac{\partial J_t^{-1}}{\partial t}\Big|_{t=0} = -J(S), \qquad \frac{\partial^2 J_t^{-1}}{\partial t^2}\Big|_{t=0} = -J(R) + 2J(S)^2$$
and
$$J_0 = I, \qquad \frac{\partial J_t}{\partial t}\Big|_{t=0} = J(S), \qquad \frac{\partial^2 J_t}{\partial t^2}\Big|_{t=0} = J(R)$$
by Lemma 8.1 in §8.2, where I denotes the unit matrix. Then it follows that
$$\dot{a}_0(v,\psi) = -\int_\Omega \big(J(S) + J(S)^T - (\nabla \cdot S)I\big)\nabla v \cdot \nabla\psi\, dx$$
and
$$\ddot{a}_0(v,\psi) = \int_\Omega \big[-J(R) + 2J(S)^2\big]\nabla v \cdot \nabla\psi + 2J(S)\nabla v \cdot J(S)\nabla\psi$$
$$+ \nabla v \cdot \big[-J(R) + 2J(S)^2\big]\nabla\psi$$
$$-2\big[J(S)\nabla v \cdot \nabla\psi + \nabla v \cdot J(S)\nabla\psi\big]\nabla \cdot S$$
$$+ \nabla v \cdot \nabla\psi\big[\nabla \cdot R + (\nabla \cdot S)^2 - J(S)^T : J(S)\big]\, dx$$
$$= \int_\Omega \big(\big[-J(R) + 2J(S)^2\big] + \big[-J(R) + 2J(S)^2\big]^T\big)\nabla v \cdot \nabla\psi$$
$$+ 2\big(J(S)^T \cdot J(S) - \big[J(S) + J(S)^T\big]\nabla \cdot S\big)\nabla v \cdot \nabla\psi$$
$$+ \big(\nabla \cdot R + (\nabla \cdot S)^2 - J(S)^T : J(S)\big)\nabla v \cdot \nabla\psi\, dx.$$

The strong derivatives $\dot{v}$ and $\ddot{v}$ in H_0^1 are, therefore, defined by
$$a_0(\dot{v},\psi) + \dot{a}_0(v,\psi) = \langle\psi,\dot{f}\rangle, \quad \forall\psi \in H_0^1(\Omega)$$
$$a_0(\ddot{v},\psi) + 2\dot{a}_0(\dot{v},\psi) + \ddot{a}_0(v,\psi) = \langle\psi,\ddot{f}\rangle, \quad \forall\psi \in H_0^1(\Omega),$$
where $\langle\ ,\ \rangle$ denotes the paring between $V = H_0^1(\Omega)$ and $V' = H^{-1}(\Omega)$.

Remark 9.4. The complicated formulae above is reduced to simpler ones in the Euler coordinate under the cost of the regularity of u in §9.4.

9.4   Euler Derivatives

Lemma 9.1 in §9.3 guarantees the differentiability of $u = u(\cdot, t)$ in $\mathcal{D}'(\Omega)$ for any Lipschitz domain Ω. To prescribe the boundary value of $\dot{u}$, however, we require additional regularity of $u_0 = u|_{t=0}$.

First, we show the following theorem. Let $\Omega \subset \mathbf{R}^N$ be a Lipschitz domain and $G = G(\cdot, y)$, $y \in \Omega$, be Green's function of $-\Delta$ under the Dirichlet boundary condition. Define $u = u(\cdot, t)$ by (9.20) in §9.3.

Lemma 9.2. *If $\{T_t\}$ is differentiable, the first variation*

$$\dot{u}_0 = \delta G(\cdot, y) = \frac{\partial G_t}{\partial t}(\cdot, y)\Big|_{t=0}$$

of

$$G_t = G_t(x, y)$$

exists in the sense of distributions in Ω. It belongs to $H^1(\Omega)$ if and only if

$$S \cdot \nabla u_0 \in H^1(\Omega).$$

If this condition is satisfied then it holds that

$$S \cdot \nabla G(\cdot, y) \in H^{1/2}(\partial\Omega)$$

and

$$\Delta \dot{u}_0 = 0 \; in \; \Omega, \quad \dot{u}_0 = -S \cdot \nabla G(\cdot, y) \; on \; \partial\Omega. \tag{9.24}$$

Proof. We apply the method used in §8.3. Let

$$J_t(x) = J(T_t x), \quad z = T_t x.$$

Given $\varphi \in C_0^\infty(\Omega)$, we have

$$\int_\Omega u(z,t)\varphi(z)\,dz = \int_{\Omega_t} u(z,t)\varphi(z)\,dz$$
$$= \int_\Omega u(T_t x, t)\varphi(T_t x)\det J_t(x)\,dx$$

for $|t| \ll 1$ with the well-definedness of its left-hand side. Lemma 9.1 guarantees the existence of

$$\frac{\partial u}{\partial t}(T_t x, t)$$

in $H^1(\Omega)$, while the existence of

$$\frac{\partial}{\partial t}\left[\varphi(T_t x)\ \det\ J_t(x)\right]$$

in $L^2(\Omega)$ is immediate. Then Theorem 8.1 in §8.2 implies

$$\frac{d}{dt}\int_\Omega u(z,t)\varphi(z)\ dz\bigg|_{t=0}$$

$$=\int_\Omega\left[\frac{\partial u}{\partial t}(T_t x,t)\cdot\varphi(T_t x)+u(T_t x,t)\frac{\partial T_t x}{\partial t}\cdot\nabla\varphi(T_t x)\right]\det\ J_t(x)$$

$$+\ u(T_t x,t)\varphi(T_t x)\frac{\partial}{\partial t}\det\ J_t(x)\ dx\bigg|_{t=0}$$

$$=\int_\Omega\frac{\partial u}{\partial t}(T_t x,t)\bigg|_{t=0}\varphi+u(S\cdot\nabla\varphi+\varphi\nabla\cdot S)\ dx$$

$$=\int_\Omega\frac{\partial u}{\partial t}(T_t x,t)\bigg|_{t=0}\varphi+u\nabla\cdot(S\varphi)\ dx$$

because S is a Lipschitz continuous vector field on Ω.

We obtain

$$\frac{d}{dt}\int_\Omega u(z,t)\varphi(z)\ dz\bigg|_{t=0}=\int_\Omega\left[\frac{\partial u}{\partial t}(T_t x,t)\bigg|_{t=0}-S\cdot\nabla u\right]\varphi\ dx$$

by $u\in H^1(\Omega)$, which means the existence of

$$\dot u_0=u_t|_{t=0}$$

in $\mathcal{D}'(\Omega)$ with the equality

$$\dot u_0=\frac{\partial u}{\partial t}(T_t\cdot,t)\bigg|_{t=0}-S\cdot\nabla u.$$

The formal derivative

$$\frac{\partial u}{\partial t}(T_t\cdot,t)\bigg|_{t=0}=\dot u_0+S\cdot\nabla u_0 \tag{9.25}$$

is thus justified in $\mathcal{D}'(\Omega)$ and its right-hand side belongs to $L^2(\Omega)$. Here, we obtain

$$\frac{\partial u}{\partial t}(T_t\cdot,t)\bigg|_{t=0}\in H^1(\Omega)$$

by Lemma 9.1, and therefore, $\dot u_0=u_t|_{t=0}$ is in $H^1(\Omega)$ if and only if

$$S\cdot\nabla u_0\in H^1(\Omega).$$

We have also

$$\frac{\partial u}{\partial t}(T_t\cdot, t)\Big|_{t=0} = -\frac{\partial \Gamma}{\partial t}(T_t\cdot -y)\Big|_{t=0} = -S\cdot\nabla\Gamma(\cdot - y) \quad \text{in } H^{1/2}(\partial\Omega)$$

by

$$G_t(T_t\cdot, y)|_{\partial\Omega} = 0 \quad \text{in } H^{1/2}(\partial\Omega).$$

It holds, therefore, that

$$\dot{u}_0 = -S\cdot\nabla(\Gamma(\cdot - y) + u_0) = -S\cdot\nabla G(\cdot, y) \quad \text{in } H^{1/2}(\partial\Omega).,$$

and there arise the result because

$$\Delta\dot{u}_0 = 0 \quad \text{in } \mathcal{D}'(\Omega)$$

is obvious.   □

The result on the second formula is as follows.

Lemma 9.3. *Under the assumption of the previous theorem, if $\{T_t\}$ is twice differentiable the second variation*

$$\ddot{u}_0 = \delta^2 G(\cdot, y) = \frac{\partial^2}{\partial t^2}G_t(\cdot, y)\Big|_{t=0}$$

of $G_t = G_t(x, y)$ exists in the sense of distributions in Ω. It belongs to $H^1(\Omega)$ if and only if

$$2S\cdot\nabla\dot{u}_0 + R\cdot\nabla u_0 + (\nabla^2 u_0)[S, S] \in H^1(\Omega).$$

If this condition is satisfied it holds that $H \in H^{1/2}(\partial\Omega)$ for

$$H = 2S\cdot\nabla\dot{u}_0 + R\cdot\nabla G(\cdot, y) + (\nabla^2 G(\cdot, y))[S, S], \quad (9.26)$$

and furthermore,

$$\Delta\ddot{u}_0 = 0 \text{ in } \Omega, \quad \ddot{u}_0 = -H \text{ on } \partial\Omega.$$

Proof. The existence of $\ddot{u}_0 = u_{tt}|_{t=0}$ in $\mathcal{D}'(\Omega)$ is proven similarly to the previous theorem. First, the formal equality

$$\frac{\partial^2 u}{\partial t^2}(T_t\cdot, t)\Big|_{t=0} = \ddot{u}_0 + 2S\cdot\nabla\dot{u}_0 + R\cdot\nabla u_0 + [\nabla^2 u_0]S\cdot S \quad (9.27)$$

arises with the $(2,2)$-symmetric tensor

$$\nabla^2 u_0 = \left(\frac{\partial^2 u_0}{\partial x_i \partial x_j} \right)$$

indicating the Hesse matrix of u_0. In fact, we have, formally,

$$\frac{\partial^2 u}{\partial t^2}(T_t x, t) = \frac{\partial}{\partial t} \left\{ u_t(T_t x, t) + \frac{\partial T_t x}{\partial t} \cdot \nabla u(T_t x, t) \right\}$$

with

$$\frac{\partial}{\partial t} u_t(T_t x, t) = u_{tt}(T_t x, t) + \frac{\partial T_t x}{\partial t} \cdot \nabla u_t(T_t x, t)$$

and

$$\begin{aligned}
\frac{\partial}{\partial t} \left[\frac{\partial T_t x}{\partial t} \cdot \nabla u(T_t x, t) \right] &= \frac{\partial^2 T_t x}{\partial t^2} \cdot \nabla u(T_t x, t) + \frac{\partial T_t x}{\partial t} \cdot \frac{\partial}{\partial t} \nabla u(T_t x, t) \\
&= \frac{\partial^2 T_t x}{\partial t^2} \cdot \nabla u(T_t x, t) + \frac{\partial T_t x}{\partial t} \cdot \left\{ \nabla u_t(T_t x, t) + \frac{\partial T_t x}{\partial t} \nabla^2 u(T_t x, t) \right\} \\
&= \frac{\partial^2 T_t x}{\partial t^2} \cdot \nabla u(T_t x, t) + \frac{\partial T_t}{\partial t} \cdot \nabla u_t(T_t x, t) + [\nabla^2 u(T_t x, t)] \frac{\partial T_t x}{\partial t} \cdot \frac{\partial T_t x}{\partial t}.
\end{aligned}$$

Putting $t = 0$, we obtain (9.27).

Next, we confirm that the last three terms on the right-hand side of (9.27) are regarded as distributions in Ω. In fact, first, S and R are Lipschitz continuous. Then $u_0 \in H^1(\Omega)$ implies

$$R \cdot \nabla u_0 \in L^2(\Omega).$$

Second, the distributions

$$S \cdot \nabla \dot{u}_0, \ (\nabla^2 u_0)[S, S] \in \mathcal{D}'(\Omega)$$

are defined by

$$\langle \varphi, S \cdot \nabla \dot{u}_0 \rangle = -(\dot{u}_0, \nabla \cdot (S\varphi))$$

$$\left\langle \varphi, \frac{\partial^2 u_0}{\partial x_i \partial x_j} S^i S^j \right\rangle = -\left(\frac{\partial u_0}{\partial x_i}, \frac{\partial}{\partial x_j}(S^i S^j \varphi) \right), \quad \varphi \in C_0^\infty(\Omega),$$

using (8.15), $\dot{u}_0 \in L^2(\Omega)$, and $u_0 \in H^1(\Omega)$. Hence the last three terms on the right-hand side of (9.27) are actually distributions.

Then we justify equality (9.27) in the sense of distributions. This process is the same as in the case of the first variation. Taking $\varphi \in C_0^\infty(\Omega)$, thus, we show the existence of

$$\frac{d^2}{dt^2} \langle \varphi, u(\cdot, t) \rangle \bigg|_{t=0}$$

with the equality

$$\frac{d^2}{dt^2} \int_\Omega u(z,t)\varphi(z) \, dz \Big|_{t=0} = \int_\Omega \ddot{u}_0 \varphi$$
$$- 2\dot{u}_0 \nabla \cdot (S\varphi) - \sum_{i,j} \frac{\partial u_0}{\partial x_i} \frac{\partial}{\partial x_j} (S^i S^j \varphi) \, dx, \quad z = T_t x,$$

similarly.

To complete the proof, we regard

$$u(T_t x, t) = -\Gamma(T_t x - y)$$

as a function of $x \in \partial\Omega$. This function is twice differentiable in t strongly in $H^{1/2}(\partial\Omega)$, and it holds that

$$\frac{\partial \Gamma}{\partial t}(T_t \cdot -y) \Big|_{t=0} = S \cdot \nabla\Gamma(\cdot - y)$$
$$\frac{\partial^2 \Gamma}{\partial t^2}(T_t \cdot -y) \Big|_{t=0} = R \cdot \nabla\Gamma(\cdot - y) + (\nabla^2 \Gamma(\cdot - y))[S, S].$$

It is obvious that $\ddot{u}_0$ is harmonic in Ω, and hence we obtain the result. $\square$

9.5 $C^{1,1}$ Domains

If $\Omega \subset \mathbf{R}^N$ is a bounded $C^{1,1}$ domain there is a Lipschitz continuous frame (7.16) on $\partial\Omega$. The vector field ν, furthermore, has a $C^{1,1}$ extension in a neighbourhood of $\partial\Omega$ denoted by ω.

Given $v \in H^2(\Omega)$, the function $\nu \cdot \nabla v$ belongs to $H^{1/2}(\partial\Omega)$. By the identification given in Theorem 7.4 in §7.2 it thus holds that $\partial v/\partial\nu \in H^{1/2}(\partial\Omega)$. We have also

$$\frac{\partial v}{\partial s_i} = s_i \cdot \nabla v \in H^{1/2}(\partial\Omega)$$

because s_i, $1 \le i \le N-1$, are Lipschitz continuous.

If $v \in H_0^1(\Omega)$, furthermore, we obtain

$$0 = \int_{\partial\Omega} (s_i \cdot \nabla)(v\varphi) \, ds = \int_{\partial\Omega} [(s_i \cdot \nabla)v]\varphi \, ds$$

for any $\varphi \in C_0^\infty(\omega)$. Hence it follows that

$$\frac{\partial v}{\partial s_i} = (s_i \cdot \nabla)v = 0 \quad \text{in } H^{1/2}(\partial\Omega), \quad 1 \le i \le N-1,$$

which implies

$$S \cdot \nabla v = \left[(S \cdot \nu)\nu + \sum_{i=1}^{N-1} (S \cdot s_i)s_i) \right] \cdot \nabla v = (S \cdot \nu)(\nu \cdot \nabla v)$$

$$= \delta\rho \frac{\partial v}{\partial \nu} \in H^{1/2}(\partial\Omega), \quad \forall v \in H^2(\Omega) \cap H_0^1(\Omega)$$

because $S \cdot \nabla v \in H^1(\Omega)$ holds by the Lipschitz continuity of S. If Ω is $C^{1,1}$, therefore, the first variational formula of the Green's funtions is represented in Euler coordinates in $H^1(\Omega)$ by Lemma 9.2 in §9.4.

Here we use H^2-regularity of $\dot{u}_0$ in (9.24) in §9.4, which arises because $\Omega \subset \mathbf{R}^N$ is $C^{1,1}$. In the following theorem, the pairing $\langle \ , \ \rangle$ is identified with the L^2 inner product on $\partial\Omega$. Recall

$$\delta\rho = S \cdot \nu, \quad S = \left. \frac{\partial T_t}{\partial t} \right|_{t=0}.$$

Theorem 9.3 (Hadamard). *If $\Omega \subset \mathbf{R}^N$ is a bounded $C^{1,1}$-domain and the bi-Lipschitz deformation $\{T_t\}$ is differentiable, it holds that*

$$\dot{u}_0 = \delta G(\cdot, y) \in H^1(\Omega)$$

in Lemma 9.2 in §9.4. This function is harmonic in Ω and it holds that

$$\delta G(x, y) = \left\langle \delta\rho \frac{\partial G(\cdot, x)}{\partial \nu}, \frac{\partial G(\cdot, y)}{\partial \nu} \right\rangle, \quad x, y \in \Omega. \tag{9.28}$$

Proof. Let $y \in \Omega$ be given. Since Ω is C^2 we have

$$u = G(\cdot, y) - \Gamma(\cdot - y) \in H^2(\Omega)$$

by the elliptic regularity. Then it follows that

$$\nu \cdot \nabla G(\cdot, y) = \frac{\partial G(\cdot, y)}{\partial \nu} \in H^{1/2}(\partial\Omega) \tag{9.29}$$

and hence

$$S \cdot \nabla G(\cdot, y) = \delta\rho \frac{\partial G(\cdot, y)}{\partial \nu} \in H^{1/2}(\partial\Omega). \tag{9.30}$$

Then we obtain (9.28) by Theorem 9.2 and Lemma 9.2. $\qquad\square$

The second variational formula in Lemma 9.3 contains $\dot{u}$ in the first term of H defined by (9.26). The following lemma is used to reduce this term to an integration on Ω. The pairing $\langle \cdot, \cdot \rangle$ on the right-hand side of (9.31) is taken between $H^{1/2}(\partial\Omega)$ and $H^{-1/2}(\partial\Omega)$.

Lemma 9.4. *If $\Omega \subset \mathbf{R}^N$ is a $C^{1,1}$-domain, it holds that*

$$(\nabla \delta G(\cdot, x), \nabla \delta G(\cdot, y)) = -\left\langle \delta\rho \frac{\partial G(\cdot, x)}{\partial \nu}, \frac{\partial \delta G(\cdot, y)}{\partial \nu} \right\rangle, \quad x, y \in \Omega. \quad (9.31)$$

Proof. From the assumption it follows that $u \in H^2(\Omega)$, (9.29), (9.30), and

$$\dot{u}_0 = \delta G(\cdot, y) \in H^1(\Omega).$$

Given $\varphi \in H^1(\Omega)$ with $\Delta\varphi = 0$, therefore, we have

$$(\nabla \delta G(\cdot, y), \nabla\varphi) = (\nabla \dot{u}_0, \nabla\varphi) = \left\langle \dot{u}_0, \frac{\partial\varphi}{\partial\nu} \right\rangle$$
$$= -\left\langle \delta\rho \frac{\partial G(\cdot, y)}{\partial\nu}, \frac{\partial\varphi}{\partial\nu} \right\rangle.$$

Putting $\varphi = \delta G(\cdot, x) \in H^1(\Omega)$ for $x \in \Omega$ and changing x and y, we obtain (9.31). $\qquad\square$

9.6 $C^{2,\theta}$ Domains

If Ω is $C^{2,\theta}$, $0 < \theta < 1$, there arises $C^{2,\theta}$ regularity of the solution u to (7.2) because $\Gamma(\cdot - y)$ is C^∞ in ω, a neighborhood of $\partial\Omega$, provided that $y \notin \omega$. It follows that $u \in C^{2,\theta}(\overline{\Omega})$, which implies the same regularity of $G(\cdot, y)$ near $\partial\Omega$. In particular, we obtain

$$\frac{\partial^2 G(\cdot, y)}{\partial\nu^2} = -(\nabla \cdot \nu)\frac{\partial G(\cdot, y)}{\partial\nu} \quad \text{on } \partial\Omega \quad (9.32)$$

by Corollary 7.1 in §7.3.

Lemma 9.5. *If $\Omega \subset \mathbf{R}^N$ is a bounded $C^{2,\theta}$-domain and $\{T_t\}$ is a twice differentiable, it holds that*

$$(\nabla^2 G(\cdot, y))[S, S] = 2\delta\rho \sum_{i=1}^{N-1} \mu_i \frac{\partial^2 G(\cdot, y)}{\partial s_i \partial\nu}$$
$$+ \sum_{i=1}^{N-1} \kappa_i(\mu_i^2 - (\delta\rho)^2)\frac{\partial G(\cdot, y)}{\partial\nu} \quad \text{on } \partial\Omega, \quad (9.33)$$

where $\mu_i = S \cdot s_i$ and κ_i is the sectional curvature defined by (7.17).

Proof. It holds that (9.32) and

$$G(\cdot,y) = \frac{\partial G(\cdot,y)}{\partial s_i} = \frac{\partial^2 G(\cdot,y)}{\partial s_i \partial s_j} = 0 \quad \text{on } \partial\Omega, \quad 1 \le i,j \le N-1,$$

which implies

$$\nabla^2 G(\cdot,y) = \frac{\partial G(\cdot,y)}{\partial \nu}(\nabla \nu)^T + \sum_{i=1}^{N-1}(s_i \otimes \nu + \nu \otimes s_i)\frac{\partial^2 G(\cdot,y)}{\partial s_i \partial \nu}$$

$$+\nu \otimes \nu \frac{\partial^2 G(\cdot,y)}{\partial \nu^2} = \frac{\partial G(\cdot,y)}{\partial \nu}((\nabla \nu)^T - (\nabla \cdot \nu)\nu \otimes \nu)$$

$$+ \sum_{i=1}^{N-1}\frac{\partial^2 G(\cdot,y)}{\partial s_i \partial \nu}(s_i \otimes \nu + \nu \otimes s_i) \quad \text{on } \partial\Omega. \tag{9.34}$$

Here we have

$$S = \sum_{i=1}^{N-1}\mu_i s_i + (\delta\rho)\nu. \tag{9.35}$$

First, the relations

$$[\nu \otimes \nu]\nu \cdot \nu = \sum_{i,j=1}^{N} \nu^i \nu^j \nu^j \nu^i = |\nu|^2 \cdot |\nu|^2 = 1$$

$$[\nu \otimes \nu]s_k \cdot \nu = \sum_{i,j=1}^{N} \nu^i \nu^j s_k^j \nu^i = (\nu \cdot s_k)|\nu|^2 = 0$$

$$[\nu \otimes \nu]s_k \cdot s_\ell = \sum_{i,j=1}^{N} \nu^i \nu^j s_k^i s_\ell^j = (\nu \cdot s_k)(\nu \cdot s_\ell) = 0$$

imply

$$[\nu \otimes \nu]S \cdot S = (\delta\rho)^2.$$

Second, it holds that

$$[(\nabla\nu)^T]\nu \cdot \nu = \sum_{i,j=1}^{N} \frac{\partial \nu^i}{\partial x_j}\nu^j \nu^i = \frac{1}{2}\sum_{i,j=1}^{N}\frac{\partial (\nu^i)^2}{\partial x_j}\nu^j = \frac{1}{2}(\nu \cdot \nabla)|\nu|^2 = 0$$

$$[(\nabla\nu)^T]s_k \cdot \nu = \sum_{i,j=1}^{N} \frac{\partial \nu^i}{\partial x_j}s_k^j \nu^i = \frac{1}{2}\sum_{i,j=1}^{N}\frac{\partial (\nu^i)^2}{\partial x_k}s_k^j = \frac{1}{2}(s_k \cdot \nabla)|\nu|^2 = 0$$

by (7.24). Then, since

$$[s_m \otimes s_m](s_k \cdot s_\ell) = \sum_{i,j=1}^{N} s_m^i s_m^j s_k^j s_\ell^i = (s_m \cdot s_k)(s_m \cdot s_\ell)$$

$$= \delta_{mk}\delta_{m\ell}$$

we obtain

$$[(\nabla \nu)^T]s_k \cdot s_\ell = \sum_{m=1}^{N-1} \kappa_m \delta_{mk} \delta_{m\ell} = \kappa_k \delta_{k\ell},$$

which implies

$$[(\nabla \nu)^T]S \cdot S = \sum_{k,\ell=1}^{N-1} \mu_k \mu_\ell [(\nabla \nu)^T]s_k \cdot s_\ell = \sum_{k,\ell=1}^{N-1} \mu_k \mu_\ell \kappa_k \delta_{k\ell}$$

$$= \sum_{\ell=1}^{N-1} \mu_\ell^2 \kappa_\ell$$

by (7.24).

For the second term of the right-hand side of (9.34), we use

$$[s_k \otimes \nu]\nu \cdot \nu = \sum_{i,j=1}^{N} s_k^i \nu^j \nu^j \nu^i = (s_k \cdot \nu)|\nu|^2 = 0$$

$$[s_k \otimes \nu]s_\ell \cdot \nu = \sum_{i,j=1}^{N} s_k^i \nu^j s_\ell^j \nu^i = (\nu \cdot s_\ell)(s_k \cdot \nu) = 0$$

$$[s_k \otimes \nu]s_\ell \cdot s_m = \sum_{i,j=1}^{N} s_k^i \nu^j s_\ell^j s_m^i = (\nu \cdot s_\ell)(s_k \cdot s_m) = 0$$

$$[s_k \otimes \nu]\nu \cdot s_\ell = \sum_{i,j=1}^{N} s_k^i \nu^j \nu^j s_\ell^i = (s_k \cdot s_\ell)|\nu|^2 = \delta_{k\ell},$$

and hence

$$[\nu \otimes s_k]\nu \cdot \nu = [\nu \otimes s_k]s_\ell \cdot s_m = [\nu \otimes s_k]\nu \cdot s_\ell = 0$$

$$[\nu \otimes s_k]s_\ell \cdot \nu = \delta_{k\ell}.$$

Then it follows that

$$[s_i \otimes \nu + \nu \otimes s_i]S \cdot S = 2(\delta\rho)\mu_i.$$

We thus end up with (9.33). $\qquad\qquad\qquad\qquad\qquad\qquad\qquad\qquad\qquad\square$

Remark 9.5. The right-hand side of (9.33) is independent of the choice of the frame $\{s_1, \cdots, s_{N-1}\}$. First, we have

$$\sum_{i=1}^{N-1} \mu_i \frac{\partial^2 G(\cdot, y)}{\partial s_i \partial \nu} = (S \cdot \nabla) \frac{\partial G(\cdot, y)}{\partial \nu}.$$

Second, equality (7.21) implies

$$S \cdot \frac{\partial \nu}{\partial s_i} = \kappa_i \mu_i.$$

Therefore, there arises that

$$\sum_{i=1}^{N-1} \kappa_i \mu_i^2 = \sum_{i=1}^{N-1} (S \cdot \frac{\partial \nu}{\partial s_i}) s_i \cdot S = S \cdot \left[\sum_{i=1}^{N-1} s_i \frac{\partial \nu}{\partial s_i} \right] \cdot S$$

$$= S \cdot \left[(\nabla - \frac{\partial}{\partial \nu}) \nu \right] \cdot S = (S \cdot (\nabla \nu)) \cdot S = (\nabla \nu)^T [S, S].$$

We thus obtain

$$(\nabla^2 G(\cdot, y))[S, S] = 2(\delta \rho)(S \cdot \nabla) \frac{\partial G(\cdot, y)}{\partial \nu}$$

$$+ (\nabla \nu)^T [S, S] - (\nabla \cdot \nu)(\delta \rho)^2 \frac{\partial G(\cdot, y)}{\partial \nu} \quad \text{on } \partial \Omega.$$

9.7 $C^{2,1}$ Domains

In the following theorem, $\langle \ , \ \rangle$ is identified with the L^2-inner product on $\partial \Omega$. Recall also $\mu_i = S \cdot s_i$.

Theorem 9.4 ([Suzuki–Tsuchiya (2016)]). *Let $\Omega \subset \mathbf{R}^N$ be a bounded $C^{2,1}$-domain and $\{T_t\}$ be twice differentiable such that*

$$S = \frac{\partial T_t}{\partial t}\bigg|_{t=0} \in C^{1,1}(\overline{\Omega}, \mathbf{R}^N). \tag{9.36}$$

Then it holds that $\ddot{u}_0 = \delta^2 G(\cdot, y) \in H^1(\Omega)$ in Lemma 9.3 in §9.4. This $\ddot{u}$ is harmonic in Ω and it holds that

$$\delta^2 G(x, y) = -2(\nabla \delta G(\cdot, x), \nabla \delta G(\cdot, y)) + \left\langle \chi \frac{\partial G(\cdot, x)}{\partial \nu}, \frac{\partial G(\cdot, y)}{\partial \nu} \right\rangle \tag{9.37}$$

for $x, y \in \Omega$, where

$$\chi = \delta^2 \rho + \sum_{i=1}^{N-1} \{ \kappa_i (\mu_i^2 - (\delta \rho)^2) - 2\mu_i \frac{\partial \delta \rho}{\partial s_i} \} \tag{9.38}$$

is a Lipschitz continuous function on $\partial \Omega$.

Proof. By the elliptic regularity to (9.14), we have

$$u_0 = u|_{t=0} \in H^3(\Omega)$$

and hence $S \cdot \nabla u_0 \in H^2(\Omega)$ by (9.36). There arises that

$$S \cdot \nabla G(\cdot, y) = \delta\rho \frac{\partial G(\cdot, y)}{\partial \nu} \in H^{3/2}(\partial\Omega),$$

which means the existence of $g \in H^2(\Omega)$ such that

$$g = S \cdot \nabla G(\cdot, y) \quad \text{on } \partial\Omega.$$

Then the elliptic regularity applied to (9.24) implies

$$\dot{u}_0 = \delta G(\cdot, y) \in H^2(\Omega).$$

Thus we obtain $S \cdot \nabla \dot{u}_0 \in H^1(\Omega)$, and therefore,

$$S \cdot \nabla \delta G(\cdot, y) = \sum_{i=1}^{N-1} \mu_i \frac{\partial \delta G(\cdot, y)}{\partial s_i} + \delta\rho \frac{\partial \delta G(\cdot, y)}{\partial \nu} \quad \text{on } \partial\Omega$$

belongs to $H^{1/2}(\partial\Omega)$.

We have also

$$R \cdot \nabla \dot{u}_0 \in H^1(\Omega)$$

and hence

$$R \cdot \nabla G(\cdot, y) = \delta^2\rho \frac{\partial G(\cdot, y)}{\partial \nu} \quad \text{on } \partial\Omega$$

is in $H^{1/2}(\partial\Omega)$ similarly. Then it follows that

$$\ddot{u}_0 = -H = -2\sum_{i=1}^{N-1} \mu_i \left(\frac{\partial \delta G(\cdot, y)}{\partial s_i} + \delta\rho \frac{\partial^2 G(\cdot, y)}{\partial s_i \partial \nu} \right) - 2\delta\rho \frac{\partial \delta G(\cdot, y)}{\partial \nu}$$

$$- \left[\delta^2\rho + \sum_{i=1}^{N-1} \kappa_i(\mu_i^2 - (\delta\rho)^2) \right] \frac{\partial G(\cdot, y)}{\partial \nu} \quad \text{on } \partial\Omega$$

from Theorem 9.2. Since

$$\delta G(\cdot, y) = \dot{u}_0 = -\delta\rho \frac{\partial G(\cdot, y)}{\partial \nu} \quad \text{on } \partial\Omega$$

we have

$$\frac{\partial \delta G(\cdot, y)}{\partial s_i} + \delta\rho \frac{\partial^2 G(\cdot, y)}{\partial s_i \partial \nu} = -\frac{\partial}{\partial s_i} \left(\delta\rho \frac{\partial G(\cdot, y)}{\partial \nu} \right) + \delta\rho \frac{\partial^2 G(\cdot, y)}{\partial s_i \partial \nu}$$

$$= -\frac{\partial \delta\rho}{\partial s_i} \frac{\partial G(\cdot, y)}{\partial \nu} \quad \text{on } \partial\Omega.$$

We thus reach

$$\ddot{u} = -2\delta\rho \frac{\partial \dot{u}}{\partial \nu} - \chi \frac{\partial G(\cdot, y)}{\partial \nu} \quad \text{on } \partial\Omega$$

with (9.38) and the proof is complete by Lemma 9.4. $\qquad \square$

An immediate consequence is the following formula concerning normal deformation, Example 8.2 in §8.1.

Corollary 9.1 ([Garabedian–Schiffer (1952–53)]). *Assume that Ω is a bounded $C^{2,1}$ domain, and let $\{T_t\}$ be a family of normal deformations associated with $\delta\rho \in C^{1,1}(\partial\Omega)$. Then it holds that*

$$\delta^2 G(x,y) = -2\left(\nabla\delta G(\cdot,x), \nabla\delta G(\cdot,y)\right)$$
$$-\left\langle (\nabla\cdot\nu)(\delta\rho)^2 \frac{\partial G(\cdot,x)}{\partial\nu}, \frac{\partial G(\cdot,y)}{\partial\nu} \right\rangle, \quad x,y \in \Omega. \qquad (9.39)$$

Proof. The formula follows from the previous theorem because in this case it holds that

$$S\cdot\nu = \delta\rho, \quad \delta^2\rho = 0, \quad \mu_i = S\cdot s_i = 0, \ 1 \le i \le N-1.$$

$\square$

Here we represent χ in (9.38) independent of the choice of the frame.

Lemma 9.6. *It holds that*

$$\chi = \delta^2\rho - (\nabla\cdot\nu)(\delta\rho)^2 + \frac{\partial(\delta\rho)^2}{\partial\nu} - (S\cdot\nabla S)\cdot\nu - S\cdot\nabla\delta\rho \qquad (9.40)$$

in Theorem 9.4.

Proof. By (7.26) in §7.2 it holds that

$$\chi = \delta^2\rho - (\nabla\cdot\nu)(\delta\rho)^2 + \sum_{i=1}^{N-1}\left(\kappa_i\mu_i^2 - 2\mu_i\frac{\partial\delta\rho}{\partial s_i}\right).$$

Then equality (7.21) implies

$$\kappa_i\mu_i^2 = \kappa_i(S\cdot s_i)^2 = (S\cdot s_i)\left(S\cdot\frac{\partial\nu}{\partial s_i}\right),$$

where

$$S\cdot\frac{\partial\nu}{\partial s_i} = \frac{\partial}{\partial s_i}(S\cdot\nu) - \frac{\partial S}{\partial s_i}\cdot\nu = \frac{\partial\delta\rho}{\partial s_i} - \frac{\partial S}{\partial s_i}\cdot\nu.$$

Then we obtain

$$\chi = \delta^2\rho - (\nabla\cdot\nu)(\delta\rho)^2 - \sum_{i=1}^{N-1}(S\cdot s_i)\left\{\frac{\partial\delta\rho}{\partial s_i} + \frac{\partial S}{\partial s_i}\cdot\nu\right\} \qquad (9.41)$$

by $\mu_i = S\cdot s_i$.

The third term on the right-hand side of (9.41) is treated as follows. First, we have

$$\sum_{i=1}^{N-1}(S\cdot s_i)(\frac{\partial S}{\partial s_i}\cdot\nu)=\left[\sum_{i=1}^{N-1}(S\cdot s_i)\frac{\partial S}{\partial s_i}\right]\cdot\nu=\left[\sum_{i=1}^{N-1}(S\cdot s_i)s_i\cdot\nabla S\right]\cdot\nu$$

$$=[(S-(\delta\rho)\nu)\cdot\nabla S]\cdot\nu=(S\cdot\nabla S)\cdot\nu-\delta\rho\frac{\partial S}{\partial\nu}\cdot\nu.$$

Here, we have (7.24) so that

$$\frac{\partial S}{\partial\nu}\cdot\nu=\frac{\partial}{\partial\nu}(S\cdot\nu)=\frac{\partial\delta\rho}{\partial\nu}.$$

Hence it follows that

$$\sum_{i=1}^{N-1}(S\cdot s_i)\left(\frac{\partial S}{\partial s_i}\cdot\nu\right)=(S\cdot\nabla S)\cdot\nu-\frac{1}{2}\frac{\partial(\delta\rho)^2}{\partial\nu}.$$

We have, similarly,

$$\sum_{i=1}^{N-1}(S\cdot s_i)\frac{\partial\delta\rho}{\partial s_i}=\sum_{i=1}^{N-1}(S\cdot s_i)s_i\cdot\nabla\delta\rho=(S-(\delta\rho)\nu)\cdot\nabla\delta\rho$$

$$=S\cdot\nabla\delta\rho-\frac{1}{2}\frac{\partial}{\partial\nu}(\delta\rho)^2$$

and then (9.40) follows. □

Here we show the result on the dynamical deformation in Example 8.1.

Corollary 9.2. *Assume that Ω is a bounded $C^{2,1}$ domain, and let $\{T_t\}$ be a family of dynamical perturbations associated with the $C^{1,1}$ vector field v. Then it holds that*

$$\delta^2 G(x,y)=-2(\nabla\delta G(\cdot,x),\nabla\delta G(\cdot,y))-\left\langle\sigma\frac{\partial G(\cdot,x)}{\partial\nu},\frac{\partial G(\cdot,y)}{\partial\nu}\right\rangle\quad(9.42)$$

for

$$\sigma=(\nabla\cdot\nu)(v\cdot\nu)^2-\frac{\partial(v\cdot\nu)^2}{\partial\nu}+(v\cdot\nabla)(v\cdot\nu),\quad x,y\in\Omega.\quad(9.43)$$

Proof. In this case there arises that $S\cdot\nabla S=R$ and hence

$$\chi=-(\nabla\cdot\nu)(\delta\rho)^2+\frac{\partial(\delta\rho)^2}{\partial\nu}-S\cdot\nabla\delta\rho.$$

Then (9.42)–(9.43) follows from $S=v$ and $\delta\rho=v\cdot\nu$. □

9.8 Neumann Problems

We turn to Hadamard's variational formulae for the Neuman problem.

Let $\Omega \subset \mathbf{R}^N$ be a bounded Lipschitz domain and suppose that its boundary $\partial\Omega$ is divided into two non-overlapped closed sets γ_0 and γ_1 satisfying

$$\gamma_0 \cup \gamma_1 = \partial\Omega, \quad \gamma_0 \cap \gamma_1 = \emptyset. \tag{9.44}$$

This assumption yields that γ_i, $i = 0, 1$, do not have their boundaries in $\partial\Omega$. We study the Poisson problem with the mixed boundary condition:

$$-\Delta z = f \text{ in } \Omega, \quad z = \varphi \text{ on } \gamma_0, \quad \frac{\partial z}{\partial \nu} = \psi \text{ on } \gamma_1. \tag{9.45}$$

The standard elliptic theory tells us that if $\partial\Omega$ is $C^{1,1}$, $f \in L^2(\Omega)$, $\varphi \in H^2(\Omega)$, and $\psi \in H^1(\Omega)$, equation (9.45) admits a unique solution $z \in H^2(\Omega)$. This solution is represented by

$$z(y) = \int_\Omega N(x,y)f(x) \, dx - \int_{\gamma^0} \varphi(x)\frac{\partial N}{\partial \nu_x}(x,y) \, dS_x$$

$$+ \int_{\gamma^1} N(x,y)\psi(x) \, dS_x, \quad y \in \Omega, \tag{9.46}$$

where dS_x is the surface element and $N(x,y)$ is the Green's function. Thus, given $y \in \Omega$, if $\Gamma(x)$ denotes the fundamental solution of Δ defined by (9.11), and if $u = u(x)$ is the solution to

$$\Delta u = 0 \text{ in } \Omega, \quad u = -\Gamma(\cdot - y) \text{ on } \gamma_0, \quad \frac{\partial u}{\partial \nu} = -\frac{\partial}{\partial \nu}\Gamma(\cdot - y) \text{ on } \gamma_1, \tag{9.47}$$

this $N(x,y)$ is given by

$$N(x,y) = \Gamma(x - y) + u(x). \tag{9.48}$$

We develop H^1-theory as in §7.2, using the following results derived from Theorem 7.1 and Theorem 7.3 in §7.2. We write

$$\langle \, \cdot \, , \, \cdot \, \rangle_{\gamma_1}$$

for the paring between $H^{1/2}(\gamma_1)$ and $H^{-1/2}(\gamma_1)$.

Theorem 9.5. *If $\Omega \subset \mathbf{R}^N$ is a bounded Lipschitz domain satisfying (9.44), the trace operator*

$$v \in C^\infty(\overline{\Omega}) \mapsto v|_{\partial\Omega} \in C^{0,1}(\partial\Omega)$$

is extented as a bounded linear operator

$$v \in H^1(\Omega) \mapsto v|_{\partial\Omega} \in H^{1/2}(\partial\Omega).$$

There arises, furthermore, the isomorphism

$$v \in V/H_0^1(\Omega) \mapsto v|_{\gamma_1} \in H^{1/2}(\gamma_1)$$

for

$$V = \{v \in H^1(\Omega) \mid v|_{\gamma_0} = 0\}.$$

Theorem 9.6. *If*

$$\Delta v \in V'$$

holds in the previous theorem, the normal derivative of $v \in V$ on γ_1 is defined in the sense of

$$\left. \frac{\partial v}{\partial \nu} \right|_{\gamma_1} \in H^{-1/2}(\gamma_1),$$

and it holds that

$$\left\langle \varphi, \frac{\partial v}{\partial \nu} \right\rangle_{\gamma_1} = (\nabla v, \nabla \varphi)_{L^2(\Omega)} + \langle \varphi, \Delta v \rangle_{V,V'}, \quad \forall \varphi \in V.$$

Then we obtain the unique existence of H^1 solution to (9.47) on the Lipschitz domain.

Theorem 9.7. *Under the above situation of Ω, given*

$$f \in V', \ \varphi \in H^{1/2}(\gamma_0), \ \psi \in H^{-1/2}(\gamma_1),$$

there is a unique solution $z \in H^1(\Omega)$ to (9.45). Namely, it holds that

$$z|_{\gamma_0} = \varphi, \quad (\nabla z, \nabla \zeta) = (f, \zeta) + \langle \zeta, \psi \rangle_{\gamma_1}, \ \forall \zeta \in V.$$

Now we take a deformation of Ω, $\{T_t\}$, $|t| < \varepsilon$, which maps the boundaries γ^i, $i = 0, 1$, onto γ_{it}, $i = 0, 1$:

$$T_t(\gamma_i) = \gamma_{it}, \quad i = 0, 1.$$

Then the Green's function on Ω_t is defined by

$$N(x, y, t) = \Gamma(x - y) + u(x, t), \tag{9.49}$$

where $u = u(x, t)$ is the solution to

$$\Delta u(\cdot, t) = 0 \text{ in } \Omega_t \tag{9.50}$$

with

$$u(\cdot, t) = -\Gamma(\cdot - y) \text{ on } \gamma_{0t}, \quad \frac{\partial u}{\partial \nu}(\cdot, t) = -\frac{\partial}{\partial \nu}\Gamma(\cdot - y) \text{ on } \gamma_{1t}. \tag{9.51}$$

By this definition, it holds that

$$\Omega_0 = \Omega, \ N|_{t=0} = N(x, y), \ u|_{t=0} = u(x).$$

Given $x, y \in \Omega$, we have $x, y \in \Omega_t$ for $|t| \ll 1$. Then the Hadamard variation of the Green's function $N = N(\cdot, y)$ is defined by

$$\delta N(\cdot, y) = \frac{\partial N}{\partial t}(\cdot, y, t)\Big|_{t=0} = \dot{u}_0, \quad y \in \Omega \tag{9.52}$$

for $\dot{u}_0 = u_t|_{t=0}$. The second variation is defined similarly:

$$\delta^2 N(\cdot, y) = \frac{\partial^2 N}{\partial t^2}(\cdot, y, t)\Big|_{t=0} = \ddot{u}_0, \quad y \in \Omega \tag{9.53}$$

for $\ddot{u}_0 = u_{tt}|_{t=0}$. We show their existence and representation formulae in accordance with the smoothness of Ω and $\{T_t\}$, using Liouville's volume and area derivatives.

9.9 First Variational Formula

Existence of Hadamard variation (9.52) on the Green's function to (9.45) in §9.4 is assured by the Lagrange coordinate similarly. Recall that $N(x, y, t)$ and $N(x, y)$ denote the Green's function on Ω_t and Ω defined by (9.49) with (9.50) and (9.48) with (9.47), respectively.

Lemma 9.7. *Let $\Omega \subset \mathbf{R}^N$ be a bounded Lipschitz domain and fix $y \in \Omega$. Then, if $\{T_t\}$, $|t| < \varepsilon$, is differentiable, it holds that*

$$v \in C^1(-\varepsilon, \varepsilon; H^1(\Omega)) \tag{9.54}$$

for

$$v(x, t) = u(T_t x, t).$$

In particular, the first Hadamard variation in (9.52),

$$\delta N(\cdot, y) = \dot{u}_0 \equiv \frac{\partial u}{\partial t}\Big|_{t=0},$$

exists in the sense of distributions in Ω. There arises that

$$\Delta \delta N(\cdot, y) = \Delta \dot{u}_0 = 0 \quad in \ \Omega, \tag{9.55}$$

and furthermore,

$$\delta N(\cdot, y) = \frac{\partial v}{\partial t}\Big|_{t=0} - (S \cdot \nabla) u_0 \in L^2(\Omega) \tag{9.56}$$

for $u_0 = u|_{t=0}$ and

$$S = \frac{\partial T_t}{\partial t}\Big|_{t=0} \in C^{0,1}(\overline{\Omega}, \mathbf{R}^N).$$

Proof. To show (9.54), we take

$$\varphi \in C_0^\infty(\mathbf{R}^N), \quad 0 \le \varphi = \varphi(x) \le 1,$$

such that

$$\varphi(x) = \begin{cases} 0, \, x \in B(y,r) \\ 1, \, x \in \mathbf{R}^N \setminus B(y,2r) \end{cases}$$

for $0 < r \ll 1$. Then $\tilde{\Gamma} = \Gamma(\cdot - y)\varphi$ is independent of t, and

$$w = u + \tilde{\Gamma}$$

satisfies

$$-\Delta w = h \text{ in } \Omega_t, \quad w = 0 \text{ on } \gamma_{0t}, \quad \frac{\partial w}{\partial \nu} = 0 \text{ on } \gamma_{1t} \qquad (9.57)$$

for $h = -\Delta\tilde{\Gamma}$. The Poisson problem (9.57) takes the weak form

$$w \in V_t, \quad \int_{\Omega_t} \nabla w \cdot \nabla \varphi \, dx = \int_{\Omega_t} h\varphi \, dx, \, \forall \varphi \in V_t,$$

where

$$V_t = \{v \in H^1(\Omega_t) \mid v|_{\gamma_{1t}} = 0\}.$$

Then we obtain the result as in the proof of Theorem 9.1. $\qquad\square$

If Ω is $C^{1,1}$, we have $u_0 \in H^2(\Omega)$ by the elliptic regularity, recalling (9.44). Then it holds that

$$\delta N(\cdot, y) \in H^1(\Omega)$$

by (9.55)–(9.56), and then H^1 theory is applicable to (9.47) for $u = \delta N(\cdot, y)$. We have also

$$N(\cdot, y) \in H^2(\Omega),$$

which implies

$$\nabla N(\cdot, y)|_{\partial\Omega} \in H^{1/2}(\partial\Omega). \qquad (9.58)$$

Thus we have the well-definedness of the right-hand side of (9.59) in the following theorem. Recall $\delta\rho = S \cdot \nu$ and the tangential gradient ∇_τ on $\partial\Omega$ defined by (8.48). In the following theorem,

$$\langle \, , \, \rangle_{\gamma_i}, \quad i = 0, 1$$

denotes the paring between $H^{1/2}(\gamma_i)$ and $H^{-1/2}(\gamma_i)$. Here we show the *first variational formula.*

Theorem 9.8 ([Suzuki–Tsuchiya (2023a)]). *If Ω is $C^{1,1}$ and $\{T_t\}$ is differentiable, it holds that*

$$\delta N(x,y) = \left\langle \delta\rho\frac{\partial N}{\partial\nu}(\cdot,x), \frac{\partial N}{\partial\nu}(\cdot,y)\right\rangle_{\gamma_0} - \langle\delta\rho\nabla_\tau N(\cdot,x), \nabla_\tau N(\cdot,y)\rangle_{\gamma_1} \quad (9.59)$$

for $x,y \in \Omega$,

Proof. Continue to fix $y \in \Omega$. We have readily confirmed

$$u_0 = N(\cdot,y) \in H^2(\Omega), \quad \dot{u}_0 = \delta N(\cdot,y) \in H^1(\Omega).$$

Now we show that $\dot{u} = \delta N(\cdot,y) \in H^1(\Omega)$ solves

$$\Delta\dot{u}_0 = 0 \text{ in } \Omega \quad (9.60)$$

with

$$\dot{u}_0 = -\delta\rho\frac{\partial N}{\partial\nu}(\cdot,y) \text{ on } \gamma_0, \quad \frac{\partial\dot{u}_0}{\partial\nu} = \nabla_\tau \cdot (\delta\rho\nabla_\tau N(\cdot,y)) \text{ on } \gamma_1, \quad (9.61)$$

where

$$\nabla_\tau \cdot (\delta\rho\nabla_\tau N(\cdot,y)) = \sum_{i=1}^{N-1} \frac{\partial}{\partial s_i}(\delta\rho\frac{\partial N}{\partial s_i}N(\cdot,y)).$$

Once (9.61) is shown, Theorem 9.7 in §9.8 is applicable to this Poisson equation, because (9.58) implies

$$\left.\delta\rho\frac{\partial N}{\partial\nu}(\cdot,y)\right|_{\gamma_0} \in H^{1/2}(\gamma_0), \quad \nabla_\tau \cdot (\delta\rho\nabla_\tau N(\cdot,y))|_{\gamma_1} \in H^{-1/2}(\gamma_1) \quad (9.62)$$

by $\delta\rho \in C^{0,1}(\partial\Omega)$. Then the desired equality follows from the representation formula (9.46) in §9.8 of the solution $\dot{u} = \dot{u}(x)$ to (9.61), regarding

$$N(x,y) = N(y,x).$$

Since the boundary condition of $v = \delta N(\cdot,y)$ on γ_0 in (9.61) is assured by Theorem 9.3 in §9.5 on the Dirichlet boundary condition, we have only to confirm the boundary condition on γ_1 in (9.62). To this end, we take an open neighbourhood of γ^1, denoted by $\tilde{\Omega}$, satisfying

$$\tilde{\Omega} \cap \gamma^0 = \emptyset, \quad y \notin \tilde{\Omega}.$$

Let $\varphi \in C_0^\infty(\tilde{\Omega})$. Then, for $|t| \ll 1$ it holds that

$$\int_{\Omega_t} \nabla N(\cdot,y,t) \cdot \nabla\varphi \, dx = 0 \quad (9.63)$$

by

$$\Delta N(\cdot,y,t) = 0 \text{ in } \Omega_t \setminus \{y\}, \quad \frac{\partial N}{\partial \nu}(\cdot,y,t)\Big|_{\gamma_{1t}} = 0,$$

where the normal derivative of $N(\cdot,y,t)$ on γ_{1t} belongs to $H^{-1/2}(\gamma_{1t})$.

We apply Liouville's first volume formula, Theorem 8.2 in §8.2, to (9.63) for

$$c(\cdot,t) = \nabla N(\cdot,y,t) \cdot \nabla\varphi.$$

Then it follows that

$$0 = \frac{d}{dt}\int_{\Omega_t} \nabla N(\cdot,y,t)\cdot\nabla\varphi\,dx\Big|_{t=0} = \int_\Omega \dot{c}_0 + \nabla\cdot(c_0 S)\,dx$$

$$= \int_\Omega \nabla\dot{u}_0 \cdot \nabla\varphi + \nabla\cdot([\nabla N(\cdot,y)\cdot\nabla\varphi]S)\,dx$$

$$= \left\langle \varphi, \frac{\partial\dot{u}_0}{\partial\nu}\right\rangle_{\gamma_1} + \int_{\gamma_1} \delta\rho\nabla N(\cdot,y)\cdot\nabla\varphi\,dS$$

$$= \left\langle \varphi, \frac{\partial\dot{u}_0}{\partial\nu}\right\rangle_{\gamma_1} + \int_{\gamma_1} \delta\rho\nabla_\tau N(\cdot,y)\cdot\nabla_\tau\varphi\,dS$$

$$= \left\langle \varphi, \frac{\partial\dot{u}_0}{\partial\nu}\right\rangle_{\gamma_1} - \langle\varphi, \nabla_\tau(\delta\rho\nabla_\tau N(\cdot,y))\rangle_{\gamma_1}$$

by $\varphi \in C_0^\infty(\tilde\Omega)$ and (9.55). Then we obtain

$$\frac{\partial\dot{u}_0}{\partial\nu} = \nabla_\tau\cdot(\delta\rho\nabla_\tau N(\cdot,y)) \quad \text{on } \gamma_1$$

as an element in $H^{-1/2}(\gamma_1)$, because $\varphi \in C_0^\infty(\tilde\Omega)$ is arbitrary. $\qquad\square$

9.10 Second Variational Formula

We begin with the existence of the second variation $\delta^2 N(x,y)$ in (9.53) as in Lemma 9.7 in §9.8. Recall $v(x,t) = u(T_t x, t)$.

Lemma 9.8. *Let $\Omega \subset \mathbf{R}^N$ be a bounded Lipschitz domain, $\{T_t\}$ be twice differentiable, and $u \in H^1(\Omega_t)$ be the solution to (9.50) with (9.51) for $y \in \Omega$. Then it holds that*

$$v \in C^2(-\varepsilon,\varepsilon; H^1(\Omega)). \tag{9.64}$$

In particular, $\ddot{u}_0 = \delta^2 N(\cdot,y)$ in (9.53) exists in the sense of distributions in Ω, and there arises that

$$\Delta\ddot{u}_0 = 0 \quad \text{in } \Omega.$$

If Ω is $C^{1,1}$ and $C^{2,1}$, furthermore, this $\ddot{u}$ belongs to $L^2(\Omega)$ and $H^1(\Omega)$, respectively.

Proof. All the results except for the regularity of $\ddot{u}_0$ follow from the weak form (9.57). There arises also that

$$\left.\frac{\partial^2 v}{\partial t^2}\right|_{t=0} = \ddot{u}_0 + 2S \cdot \nabla \dot{u}_0 + R \cdot \nabla u_0 + (\nabla^2 u_0)[S, S] \in H^1(\Omega). \qquad (9.65)$$

If Ω is $C^{1,1}$ we have $u \in H^2(\Omega)$ and hence $\dot{u} \in H^1(\Omega)$ by (9.56). Then $\ddot{u} \in L^2(\Omega)$ follows from (9.65) and $S, R \in C^{0,1}(\overline{\Omega})$. If Ω is $C^{2,1}$ there arises that $u_0 \in H^3(\Omega)$ and hence $\dot{u}_0 \in H^2(\Omega)$ by (9.61) and (9.48). Then (9.65) implies $\ddot{u}_0 \in H^1(\Omega)$, similarly. $\qquad \square$

Lemma 9.9. *If Ω is $C^{2,1}$ and $\{T_t\}$ is twice differentiable,*

$$\ddot{u}_0 = \delta^2 N(\cdot, y) \in H^1(\Omega)$$

satisfies

$$\Delta \ddot{u}_0 = 0 \text{ in } \Omega, \quad \ddot{u}_0 = g \text{ on } \gamma_0, \quad \frac{\partial \ddot{u}_0}{\partial \nu} = h \text{ on } \gamma_1 \qquad (9.66)$$

for

$$g = -\chi \frac{\partial N}{\partial \nu}(\cdot, y) + 2\delta\rho \frac{\partial \dot{u}}{\partial \nu}$$
$$h = \nabla_\tau \cdot (\sigma N_\tau(\cdot, y)) + 2\nabla_\tau \cdot (\delta\rho \nabla_\tau \dot{u}), \qquad (9.67)$$

where $\delta\rho = S \cdot \nu$ and

$$\chi = \delta^2 \rho + \delta\rho \frac{\partial \delta\rho}{\partial \nu} + (\nabla\nu)[S, S] - (\delta\rho)^2 \nabla \cdot \nu - (S \cdot \nabla)\delta\rho$$
$$\sigma = \delta^2 \rho - 2(S_\tau \cdot \nabla_\tau)\delta\rho + (\nabla\nu)[S, S], \qquad (9.68)$$

for $S_\tau = S - (\delta\rho)\nu$.

Proof. We have readily obtained

$$u_0 = N(\cdot, y) - \Gamma(\cdot - y) \in H^3(\Omega) \qquad (9.69)$$

with

$$\dot{u}_0 = \delta N(\cdot, y) \in H^2(\Omega), \quad \ddot{u}_0 = \delta^2 N(\cdot, y) \in H^1(\Omega) \qquad (9.70)$$

because Ω is $C^{2,1}$ and $\{T_t\}$ is twice differentiable. By the same reason there arises that

$$\chi \in C^{1,1}(\gamma_0), \quad \sigma \in C^{0,1}(\gamma_1)$$

for χ and σ in (9.68). Hence it follows that

$$g \in H^{1/2}(\gamma_0), \quad h \in H^{-1/2}(\gamma_1)$$

for g and h defined by (9.67).

We have confirmed

$$\Delta \ddot{u}_0 = 0 \quad \text{in } \Omega \tag{9.71}$$

in the previous theorem. It is also shown that

$$\ddot{u}_0 = g = -\chi \frac{\partial N}{\partial \nu}(\cdot, y) + 2\delta\rho \frac{\partial \dot{u}_0}{\partial \nu} \quad \text{on } \gamma_0$$

for

$$\chi = (R - (S \cdot \nabla)S) \cdot \nu - (\delta\rho)^2 (\nabla \cdot \nu) - (S \cdot \nabla)\delta\rho + \frac{\partial(\delta\rho)^2}{\partial \nu} \tag{9.72}$$

by Theorem 9.4 in §9.6. The first equality of (9.68) arises by

$$R \cdot \nu = \delta^2 \rho,$$

$$[(S \cdot \nabla)S] \cdot \nu = (S \cdot \nabla)(S \cdot \nu) - [(S \cdot \nabla)\nu] \cdot S$$
$$= (S_\tau \cdot \nabla_\tau)\delta\rho + \delta\rho \frac{\partial \delta\rho}{\partial \nu} - S \cdot [(S \cdot \nabla)\nu], \tag{9.73}$$

and

$$S \cdot [(S \cdot \nabla)\nu] = (\nabla\nu)[S, S]. \tag{9.74}$$

It thus suffices to ensure

$$\frac{\partial \ddot{u}_0}{\partial \nu} = h \quad \text{on } \gamma_1. \tag{9.75}$$

First, Theorem 8.3 in §8.3 implies

$$0 = \frac{d^2}{dt^2} \int_{\Omega_t} \nabla N(\cdot, y, t) \cdot \nabla\varphi \, dx \Big|_{t=0}$$
$$= \int_\Omega \nabla\delta^2 N(\cdot, y) \cdot \nabla\varphi \, dx + 2 \langle \nabla\varphi, \delta\rho\nabla\delta N(\cdot, y) \rangle_{\partial\Omega}$$
$$+ \langle \nabla \cdot [(\nabla_x N(\cdot, y) \cdot \nabla\varphi)S], \delta\rho \rangle_{\partial\Omega}$$
$$+ \langle \nabla\varphi, [(R - (S \cdot \nabla)S) \cdot \nu]\nabla N(\cdot, y) \rangle_{\partial\Omega}$$

We examine each term on the right-hand side, recalling

$$\varphi \in C_0^\infty(\tilde{\Omega}).$$

First, it follows that

$$\int_\Omega \nabla\delta^2 N(\cdot, y) \cdot \nabla\varphi \, dx = \left\langle \varphi, \frac{\partial}{\partial \nu}\delta^2 N(\cdot, y) \right\rangle_{\gamma_1}$$

from (9.71). Second, we have

$$\langle \nabla\varphi, \nabla N(\cdot, y)(R - (S \cdot \nabla)S) \cdot \nu \rangle_{\partial\Omega}$$
$$= \langle \nabla_\tau\varphi, [(R - (S \cdot \nabla)S) \cdot \nu]\nabla_\tau N(\cdot, y) \rangle_{\gamma_1}$$
$$= -\langle \varphi, \nabla_\tau \cdot ([(R - (S \cdot \nabla)S) \cdot \nu]\nabla_\tau N(\cdot, y)) \rangle_{\gamma_1}$$

for

$$F = \varphi, \quad g = (R - (S \cdot \nabla)S) \cdot \nu, \quad H = N(\cdot, y).$$

Third, there arises that

$$\langle \nabla \cdot [(\nabla N(\cdot, y) \cdot \nabla \varphi)S], \delta \rho \rangle_{\partial \Omega}$$

$$= \left\langle \sum_{i=1}^{N-1} \frac{\partial}{\partial s_i}([S \cdot s_i] \nabla N(\cdot, y) \cdot \nabla \varphi) + \frac{\partial}{\partial \nu}([S \cdot \nu] \nabla N(\cdot, y) \cdot \nabla \varphi), S \cdot \nu \right\rangle_{\gamma_1}$$

$$= \left\langle \nabla N(\cdot, y) \cdot \nabla \varphi, (S \cdot \nu) \frac{\partial (S \cdot \nu)}{\partial \nu} - \sum_{i=1}^{N-1}(S \cdot s_i) \frac{\partial (S \cdot \nu)}{\partial s_i} \right\rangle_{\gamma_1}$$

$$+ \left\langle \frac{\partial}{\partial \nu}(\nabla N(\cdot, y) \cdot \nabla \varphi), (S \cdot \nu)^2 \right\rangle_{\gamma_1}$$

$$= \left\langle \nabla_\tau N(\cdot, y) \cdot \nabla_\tau \varphi, \delta \rho \frac{\partial \delta \rho}{\partial \nu} - (S \cdot \nabla_\tau) \delta \rho \right\rangle_{\gamma_1}$$

$$+ \sum_{i=1}^{N-1} \left\langle \frac{\partial^2 N}{\partial \nu^2} \frac{\partial \varphi}{\partial \nu} + \frac{\partial N}{\partial s_i} \frac{\partial^2 \varphi}{\partial s_i \partial \nu}, (\delta \rho)^2 \right\rangle_{\gamma_1}.$$

Here we have

$$\left\langle \nabla_\tau N(\cdot, y) \cdot \nabla_\tau \varphi, \delta \rho \frac{\partial \delta \rho}{\partial \nu} - (S \cdot \nabla_\tau) \delta \rho \right\rangle_{\gamma_1}$$

$$= - \left\langle \varphi, \nabla_\tau \cdot [(\delta \rho \frac{\partial \delta \rho}{\partial \nu} - (S \cdot \nabla_\tau) \delta \rho) \nabla_\tau N(\cdot, y)] \right\rangle_{\gamma_1}$$

and also

$$\left\langle \sum_{i=1}^{N-1} (\frac{\partial^2 N}{\partial \nu^2} \frac{\partial \varphi}{\partial \nu} + \frac{\partial N}{\partial s_i} \frac{\partial^2 \varphi}{\partial s_i \partial \nu}), (\delta \rho)^2 \right\rangle_{\gamma_1}$$

$$= - \left\langle \frac{\partial \varphi}{\partial \nu}, (\delta \rho)^2 (\nabla_\tau)^2 N(\cdot, y) \rangle_{\gamma_1} - \langle \frac{\partial \varphi}{\partial \nu}, \nabla_\tau (\delta \rho)^2 \cdot \nabla_\tau N(\cdot, y) \right\rangle$$

$$= -2 \left\langle \frac{\partial \varphi}{\partial \nu}, (\delta \rho)^2 (\nabla_\tau)^2 N(\cdot, y) \rangle_{\gamma_1} - \langle \frac{\partial \varphi}{\partial \nu}, \nabla_\tau (\delta \rho)^2 \cdot \nabla_\tau N(\cdot, y) \right\rangle_{\gamma_1}$$

by

$$\frac{\partial N}{\partial \nu}(\cdot, y) = \frac{\partial^2 N}{\partial s_i \partial \nu}(\cdot, y) = 0, \quad \sum_{i=1}^{N-1} \frac{\partial^2 N}{\partial s_i^2}(\cdot, y) + \frac{\partial^2 N}{\partial \nu^2}(\cdot, y) = 0 \quad \text{on } \gamma_1.$$

Finally, we notice (9.61) in §9.9, to deduce

$$2 \langle \nabla\varphi, \delta\rho\nabla\delta N(\cdot, y)\rangle_{\gamma_1}$$

$$= 2 \left\langle \frac{\partial\varphi}{\partial\nu}, \delta\rho\frac{\partial\delta N}{\partial\nu}(\cdot, y)\right\rangle_{\gamma_1} + 2 \langle \nabla_\tau\varphi, \delta\rho\nabla_\tau\delta N(\cdot, y)\rangle_{\gamma_1}$$

$$= 2 \left\langle \frac{\partial\varphi}{\partial\nu}, \delta\rho\nabla_\tau(\delta\rho)\cdot\nabla_\tau N(\cdot, y)\right\rangle_{\gamma_1} + 2 \left\langle \frac{\partial\varphi}{\partial\nu}, (\delta\rho)^2\nabla_\tau^2 N(\cdot, y)\right\rangle_{\gamma_1}$$

$$-2 \langle \varphi, \nabla_\tau\cdot(\delta\rho\nabla_\tau\delta N(\cdot, y)))\rangle_{\gamma_1}.$$

Gathering theese equalities, we obtain

$$0 = \left\langle \varphi, \frac{\partial}{\partial\nu}\delta^2 N(\cdot, y)\right\rangle_{\gamma_1} - \left\langle \varphi, \nabla_\tau\cdot[(\delta\rho\frac{\partial\delta\rho}{\partial\nu} - (S\cdot\nabla_\tau)\delta\rho)\nabla_\tau N(\cdot, y)]\right\rangle_{\gamma_1}$$

$$- \langle\varphi, \nabla_\tau\cdot([(R - (S\cdot\nabla)S)\cdot\nu]\nabla_\tau N(\cdot, y)))\rangle_{\gamma_1}$$

$$-2 \langle\varphi, \nabla_\tau\cdot(\delta\rho\nabla_\tau\delta N(\cdot, y)))\rangle_{\gamma_1}.$$

Since $\varphi \in C_0^\infty(\tilde\Omega)$ is arbitrary we obtain (9.67) for

$$\sigma = \delta\rho\frac{\partial\delta\rho}{\partial\nu} - (S_\tau\cdot\nabla_\tau)\delta\rho + (R - (S\cdot\nabla)S)\cdot\nu \qquad (9.76)$$

and then, the second equality of (9.68) follows from

$$R\cdot\nu = \delta^2\rho,$$

(9.73), and (9.74).   □

Now we give the second variational formula in the following form.

Theorem 9.9 ([Suzuki–Tsuchiya (2023a)]). *If Ω is $C^{2,1}$ and $\{T_t\}$ is twice differentiable, it holds that*

$$\delta^2 N(x, y) = -2(\nabla\delta N(\cdot, x), \nabla\delta N(\cdot, y)) + \left\langle \chi\frac{\partial N}{\partial\nu}(\cdot, x), \frac{\partial N}{\partial\nu}(\cdot, y)\right\rangle_{\gamma_0}$$

$$- \langle\sigma\nabla_\tau N(\cdot, x), \nabla_\tau N(\cdot, y)\rangle_{\gamma_1}$$

for $x, y \in \Omega$ and χ, σ defined by (9.68), where (,) denotes the inner product in $L^2(\Omega)$.

Proof. From Lemma 9.9 and the representation formula (9.46) in §9.8, it follows that

$$\delta^2 N(x, y) = \ddot{u}_0(x) = -\left\langle g, \frac{\partial N}{\partial\nu}(\cdot, x)\right\rangle_{\gamma_0} + \langle N(\cdot, x), h\rangle_{\gamma_1} \qquad (9.77)$$

for g, h defined by (9.67)–(9.68), where $x, y \in \Omega$.

By (9.61), we obtain

$$
0 = \left\langle \frac{\partial \delta N}{\partial \nu}(\cdot, x), \, \delta N(\cdot, y) + \delta\rho \frac{\partial N}{\partial \nu}(\cdot, y) \right\rangle_{\gamma_0}
$$
$$
+ \left\langle \delta N(\cdot, y), \, \frac{\partial \delta N}{\partial \nu}(\cdot, x) - \nabla_\tau \cdot (\delta\rho \nabla_\tau N(\cdot, x)) \right\rangle_{\gamma_1}
$$
$$
= \left\langle \delta N(\cdot, y), \, \frac{\partial \delta N}{\partial \nu}(\cdot, x) \right\rangle_{\partial\Omega} + \left\langle \frac{\partial \delta N}{\partial \nu}(\cdot, x), \, \delta\rho \frac{\partial N}{\partial \nu}(\cdot, y) \right\rangle_{\gamma_0}
$$
$$
+ \left\langle \nabla_\tau \delta N(\cdot, y), \, \delta\rho \nabla_\tau N(\cdot, x) \right\rangle_{\gamma_1}
$$
$$
= (\nabla \delta N(\cdot, y), \nabla \delta N(\cdot, x)) + \left\langle \frac{\partial \delta N}{\partial \nu}(\cdot, x), \, \delta\rho \frac{\partial N}{\partial \nu}(\cdot, y) \right\rangle_{\gamma_0}
$$
$$
- \left\langle N(\cdot, x), \nabla_\tau \cdot (\delta\rho \nabla_\tau \delta N(\cdot, y)) \right\rangle_{\gamma_1}
$$

for $x \in \Omega$, and hence

$$
(\nabla \delta N(\cdot, x), \nabla \delta N(\cdot, y)) = - \left\langle \delta\rho \frac{\partial N}{\partial \nu}(\cdot, y), \, \frac{\partial \delta N}{\partial \nu}(\cdot, x) \right\rangle_{\gamma_0}
$$
$$
+ \left\langle N(\cdot, x), \nabla_\tau \cdot (\delta\rho \nabla_\tau \delta N(\cdot, y)) \right\rangle_{\gamma_1} . \tag{9.78}
$$

Then the result follows from (9.77)–(9.78) as

$$
\delta^2 N(x, y) - \left\langle \chi \frac{\partial N}{\partial \nu}(\cdot, x), \, \frac{\partial N}{\partial \nu}(\cdot, y) \right\rangle_{\gamma^0} + \left\langle \sigma \nabla_\tau N(\cdot, x), \nabla_\tau N(\cdot, y) \right\rangle_{\gamma^1}
$$
$$
= -2 \left\langle \delta\rho \frac{\partial N}{\partial \nu}(\cdot, x), \, \frac{\partial N}{\partial \nu}(\cdot, y) \right\rangle_{\gamma^0} + 2 \left\langle N(\cdot, x), \nabla_\tau (\delta\rho \nabla_\tau \delta N(\cdot, y)) \right\rangle_{\gamma_1}
$$
$$
= -2(\nabla \delta N(\cdot, x), \nabla \delta N(\cdot, y)).
$$

□

9.11 Second Fundamental Form on $\partial\Omega$

In (9.68) if $\{T_t\}$ is the normal perturbation in Example 8.2, we have

$$
(\nabla \nu)[S, S] = \delta\rho \frac{\partial \nu}{\partial \nu} = 0
$$

by (8.9) in §8.1 and hence

$$
\chi = -(\delta\rho)^2 \nabla \cdot \nu, \quad \sigma = 0.
$$

If $\{T_t\}$ is the dynamical perturbation using (8.7), it follows that

$$
\chi = -(\delta\rho)^2 \nabla \cdot \nu + (v \cdot \nu \frac{\partial}{\partial \nu} - v \cdot \nabla_\tau) \delta\rho
$$
$$
\sigma = (v \cdot \nu \frac{\partial}{\partial \nu} - v \cdot \nabla_\tau) \delta\rho
$$

from (9.72), (9.76), and (8.8). So far, the second Hadamard variation for $N = 3$ is described in accordance with the second fundamental form of $\partial\Omega$ concerning the normal perturbation with (8.9) ([Garabedian–Schiffer (1952–53)]).

These values χ, σ used in Theorem 9.9 are actually associated with the second fundamental form on $\partial\Omega$, defined by (7.31). Recall

$$\delta\rho = S \cdot \nu, \quad \operatorname{tr} \mathcal{B} = -\nabla \cdot \nu, \quad S_\tau = (S \cdot \tau)\tau.$$

Lemma 9.10. *It holds that*

$$[(S \cdot \nabla S)] \cdot \nu + (-\delta\rho\frac{\partial}{\partial\nu} + S_\tau \cdot \nabla_\tau)\delta\rho = \mathcal{B}(S_\tau, S_\tau) + 2(S_\tau \cdot \nabla_\tau)\delta\rho.$$

Proof. The result follows from a direct computation. First, Lemma 7.2 in §7.4 applied to $v = S$ implies

$$\delta\rho\nabla \cdot S - [(S \cdot \nabla)S] \cdot \nu = \nabla_\tau \cdot (\delta\rho S_\tau) + (\nabla \cdot \nu)(\delta\rho)^2$$
$$-\mathcal{B}(S_\tau, S_\tau) - 2S_\tau \cdot \nabla_\tau(\delta\rho).$$

Then it follows that

$$(S_\tau \cdot \nabla_\tau)\delta\rho + \mathcal{B}(S_\tau, S_\tau) + (\nabla \cdot S)\delta\rho$$
$$= [(S \cdot \nabla)S] \cdot \nu + \delta\rho\nabla_\tau \cdot S_\tau + (\nabla \cdot \nu)(\delta\rho)^2$$

and hence

$$[(S \cdot \nabla)S] \cdot \nu + (\nabla \cdot \nu)(\delta\rho)^2 + (S \cdot \nabla)\delta\rho$$
$$= -\delta\rho\nabla_\tau \cdot S_\tau + (S_\tau \cdot \nabla_\tau)\delta\rho + \mathcal{B}(S_\tau, S_\tau) + (\nabla \cdot S)\delta\rho + (S \cdot \nabla)\delta\rho$$
$$= (S_\tau \cdot \nabla_\tau)\delta\rho + \mathcal{B}(S_\tau, S_\tau) - \delta\rho\nabla_\tau \cdot S_\tau + \nabla \cdot (S\delta\rho). \qquad (9.79)$$

Here, we have

$$(S \cdot \nabla)\delta\rho = (\delta\rho\frac{\partial}{\partial\nu} + S_\tau \cdot \nabla_\tau)\delta\rho$$

and

$$\nabla \cdot (S\delta\rho) = \nabla \cdot (S_\tau\delta\rho) + \nabla \cdot ((\delta\rho)^2\nu)$$
$$= \nabla_\tau \cdot (S_\tau\delta\rho) + \frac{\partial}{\partial\nu}((S_\tau \cdot \nu)\delta\rho) + \frac{\partial}{\partial\nu}(\delta\rho)^2 + (\delta\rho)^2\nabla \cdot \nu$$
$$= \delta\rho\nabla_\tau \cdot S_\tau + (S_\tau \cdot \nabla_\tau)\delta\rho + 2\delta\rho\frac{\partial\delta\rho}{\partial\nu} + (\delta\rho)^2\nabla \cdot \nu$$

by $S_\tau \cdot \nu = 0$ derived from

$$S_\tau = \sum_{j=1}^{N-1} \mu_j s_j, \quad \mu_j = (S, s_j),$$

and therefore,

$$[(S \cdot \nabla)S] \cdot \nu = \mathcal{B}(S_\tau, S_\tau) + (S_\tau \cdot \nabla_\tau + \delta\rho\frac{\partial}{\partial\nu})\delta\rho$$

by (9.79). Then the result follows. □

Theorem 9.10 ([Suzuki–Tsuchiya (2023a)]). *It holds that*

$$\chi = \delta^2\rho - 2(S_\tau \cdot \nabla_\tau)\delta\rho - \mathcal{B}(S_\tau, S_\tau) - (\delta\rho)^2(\nabla \cdot \nu)$$
$$\sigma = \delta^2\rho - 2(S_\tau \cdot \nabla_\tau)\delta\rho - \mathcal{B}(S_\tau, S_\tau)$$

in Theorem 9.9.

Proof. Equalities (9.72) and (9.76) imply

$$\chi = [R - (S \cdot \nabla)S] \cdot \nu - (\delta\rho)^2\nabla \cdot \nu + (\delta\rho\frac{\partial}{\partial\nu} - (S_\tau \cdot \nabla_\tau))\delta\rho$$

and

$$\sigma = [R - (S \cdot \nabla)S] \cdot \nu + (\delta\rho\frac{\partial}{\partial\nu} - (S_\tau \cdot \nabla_\tau))\delta\rho,$$

respectively. Then we obtain the result by Lemma 9.10. □

Chapter 10

Perturbation of Eigenvalues

We deal with the Hadamard variation of eigenvalues of the Poisson problem with respect to general perturbation of domains studied by [Suzuki–Tsuchiya (2023b)]. Rearrangement of multiple eigenvalues is necessary for them to be smooth with respect to the deformation parameter. So far, C^1 and analytic categories have been discussed in details. Here we execute the process by the transversal rearrangement. First, we introduce an abstract theory in accordance with the domain deformation (§10.1). Then existence of unilateral derivaties is formulated, where C^1 rearrangement ensures C^2 dependence with respect to the parameter (§10.2). Next, we characterize these unilateral derivatives by finite dimensional eigenvalue problems up to the second order (§10.3) and reduce these results to an abstract setting (§10.4). Having confirmed the continuity of eigenvalues and eigenspaces (§10.5), existence of the first unilateral derivatives and their characterization are proven (§10.6). Then, rearrangement of multiple eigenvalues to make them smooth are introduced (§10.7) and shown to be C^2 (§10.8). These results are applicable to general perturbation of symmetric bilinear forms. They are derived from the principle of cancellation of singularities via the symmtery of the associated bilinear forms. Without this symmetry, the differentiability of eigenvalues does not arise.

10.1 Eigenvalue Problems

We study Hadamard variation of eigenvalues of the Laplace operator with mixed boundary conditions. We characterize the first and the second derivatives, using associated finite dimensional eigenvalue problems, particularly, for multiple eigenvalues. So far, C^1 and analytic rearrangements of multiple eigenvalues of self-adjoint operator have been discussed [Rellich (1953);

Kato (1976); Chow–Hale (1982)]. Here we examine this process of re-arrangement in details, using the above described characterization of the derivatives, to ensure their efficiencies in C^2 category.

We continue to take the bounded Lipschitz domain $\Omega \subset \mathbf{R}^N$ and suppose that its boundary $\partial\Omega$ is divided into two non-overlapped closed sets γ_0 and γ_1 satisfying (9.44). We study the eigenvalue problem with mixed boundary condition,

$$-\Delta u = \lambda u \text{ in } \Omega, \quad u = 0 \text{ on } \gamma_0, \quad \frac{\partial u}{\partial \nu} = 0 \text{ on } \gamma_1. \qquad (10.1)$$

This problem takes the weak form as in the Poisson equation described in §9.8, that is, finding u satisfying

$$u \in V, \ B(u,u) = 1, \quad A(u,v) = \lambda B(u,v), \ \forall v \in V \qquad (10.2)$$

defined for

$$A(u,v) = \int_\Omega \nabla u \cdot \nabla v \, dx, \quad B(u,v) = \int_\Omega uv \, dx$$

and

$$V = \{v \in H^1(\Omega) \mid v|_{\gamma_0} = 0\}. \qquad (10.3)$$

This V is a closed subspace of $H^1(\Omega)$ under the norm

$$\|v\|_V = \sqrt{\|\nabla v\|_2^2 + \|v\|_2^2}.$$

To study (10.2), we note that if $\gamma_0 \neq \emptyset$, there is coercivity of $A : V \times V \to \mathbf{R}$, which means the existence of $\delta > 0$ such that

$$A(v,v) \geq \delta \|v\|_V^2, \quad \forall v \in V. \qquad (10.4)$$

If $\gamma_0 = \emptyset$ we replace A by $A + B$, denoted by $\tilde{A}$. Then this $\tilde{A} : V \times V \to \mathbf{R}$ is coercive, and the eigenvalue problem

$$u \in V, \ B(u,u) = 1, \quad \tilde{A}(u,v) = \tilde{\lambda} B(u,v), \ \forall v \in V,$$

is equivalent to (6.19) by $\tilde{\lambda} = \lambda + 1$. Henceforth, we assume (10.4), using this reduction if it is necessary.

These $A : V \times V \to \mathbf{R}$ and $B : X \times X \to \mathbf{R}$ are bounded, coercive, and symmetric bilinear forms for $X = L^2(\Omega)$. Since $V \hookrightarrow X$ is compact, there is a sequence of eigenvalues to (6.19), denoted by

$$0 < \lambda_1 \leq \lambda_2 \leq \cdots \to +\infty.$$

The associated eigenfunctions, $u_1, u_2, \cdots$, furthermore, form a complete ortho-normal system in X, provided with the inner product induced by $B = B(\cdot,\cdot)$:

$$B(u_i, u_j) = \delta_{ij}, \quad A(u_j, v) = \lambda_j B(u_j, v), \ \forall v \in V, \quad i,j = 1,2,\cdots.$$

The j-th eigenvalue of (10.2) is given by the mini-max principle

$$\lambda_j = \min_{L_j} \max_{v \in L_j \setminus \{0\}} R[v] = \max_{W_j} \min_{v \in W_j \setminus \{0\}} R[v], \qquad (10.5)$$

where

$$R[v] = \frac{A(v,v)}{B(v,v)}$$

is the Rayleigh quotient, and $\{L_j\}$ and $\{W_j\}$ denote the families of all subspaces of V with dimension and codimension j and $j-1$, respectively (Chapter 7 of [Suzuki (2022a)]).

Let

$$T_t : \Omega \to \Omega_t = T_t(\Omega), \quad |t| < \varepsilon_0 \qquad (10.6)$$

be a family of bi-Lipschitz homeomorphisms for $\varepsilon_0 > 0$, satisfying $T_0 = I$, the identity mapping. We assume that $T_t x$ is continuous in t uniformly in $x \in \Omega$, and recall the following definition used in §8.1.

Definition 10.1. The family $\{T_t\}$ of bi-Lipschitz homeomorphisms is said to be p-differentiable in t for $p \geq 1$, if $T_t x$ is p-times differentiable in t for any $x \in \Omega$ and the mappings

$$\frac{\partial^\ell}{\partial t^\ell} DT_t, \quad \frac{\partial^\ell}{\partial t^\ell}(DT_t)^{-1} : \Omega \to M_N(\mathbf{R}), \quad 0 \leq \ell \leq p$$

are uniformly bounded in $(x,t) \in \Omega \times (-\varepsilon_0, \varepsilon_0)$, where DT_t denotes the Jacobi matrix of $T_t : \Omega \to \Omega_t$ and $M_N(\mathbf{R})$ stands for the set of real $N \times N$ matrices. This $\{T_t\}$ is said to be continuously p-differentiable in t if it is p-differentiable and the mappings

$$t \in (-\varepsilon_0, \varepsilon_0) \mapsto \frac{\partial^\ell}{\partial t^\ell} DT_t, \quad \frac{\partial^\ell}{\partial t^\ell}(DT_t)^{-1} \in L^\infty(\Omega \to M_N(\mathbf{R})), \quad 0 \leq \ell \leq p$$

are continuous. For $p = 1$, we say differentiable or continuosly differentiable in short.

Putting

$$T_t(\gamma_i) = \gamma_{it}, \quad i = 0, 1, \qquad (10.7)$$

we introduce the other eigenvalue problem

$$-\Delta u = \lambda u \text{ in } \Omega_t, \quad u = 0 \text{ on } \gamma_{0t}, \quad \frac{\partial u}{\partial \nu} = 0 \text{ on } \gamma_{1t}, \qquad (10.8)$$

which is reduced to finding

$$u \in V_t, \int_{\Omega_t} u^2 \, dx = 1, \quad \int_{\Omega_t} \nabla u \cdot \nabla v \, dx = \lambda \int_{\Omega_t} uv \, dx, \, \forall v \in V_t \qquad (10.9)$$

for

$$V_t = \{v \in H^1(\Omega_t) \mid v|_{\gamma_{0t}} = 0\}. \tag{10.10}$$

Let $\lambda_j(t)$ be the j-th eigenvalue of the eigenvalue problem (10.8). In §10.4 we confirm that the eigenvalue problem (10.9)–(10.10) is reduced to

$$u \in V, \ B_t(u, u) = 1, \quad A_t(u, v) = \lambda B_t(u, v), \ \forall v \in V \tag{10.11}$$

by the transformation of variables $y = T_t x$ for $V \subset H^1(\Omega)$ defined by (10.3), where

$$B_t(u, v) = \int_\Omega uva_t \ dx, \quad A_t(u, v) = \int_\Omega Q_t[\nabla u, \nabla v]a_t \ dx, \tag{10.12}$$

and

$$a_t = \det DT_t, \quad Q_t = (DT_t)^{-1}(DT_t)^{-1T}. \tag{10.13}$$

Several representation formulae for

$$\lambda_j'(t) = \lim_{h \to 0} \frac{1}{h}(\lambda_j(t + h) - \lambda_j(t))$$

and

$$\lambda_j''(t) = \lim_{h \to 0} \frac{1}{h}(\lambda_j'(t + h) - \lambda_j'(t))$$

have been derived so far [Garabedian–Schiffer (1952–53)]. We characterize these derivatives, using associated finite dimensional eigenvalue problems (Theorem 10.1 and Theorem 10.4 in §). We prove also the existence of derivatives, particularly, the second derivatives of multiple eigenvalues. These results follow from the differentiability of A_t and B_t in t, defined by

$$\dot{A}_t(u, v) = \frac{d}{dt}A_t(u, v), \ \dot{B}_t(u, v) = \frac{d}{dt}B_t(u, v), \quad u, v \in V \tag{10.14}$$

and

$$\ddot{A}_t(u, v) = \frac{d^2}{dt^2}A_t(u, v), \ \ddot{B}_t(u, v) = \frac{d^2}{dt^2}B_t(u, v), \quad u, v \in V \tag{10.15}$$

as in Theorem 9.1 in §9.1. Then we show C^1 and C^2 rearrangements of eigenvalues explicitly, if these bilinear forms are continuous in t (Theorem 10.3 and Theorem 10.6 in §10.2).

We give the algorithm explicitly, as the transversal rearrangemet, Definition 10.3 in §10.6. Consequently, no rearrangement is necessary to establish C^1 or C^2 smoothness of eigenvalues in the region where their multiplicities are constant. Also elementary symmetric functions made by possible mutiple eigenvalues are always C^1 or C^2.

Our argument is executed in H^1 category without requiring any further elliptic regularities. Hence the Lipschitz continuity of $\partial\Omega$ is sufficient to ensure all the results on (10.8) under the general perturbation of domains.

10.2 Unilateral Derivatives and Rearrangements

Here we confirm Rellich's theorem [Rellich (1953)] on C^1 category, the continuous differentiability in t of rearranged eigenvalues. We show this rearrangement explicitly, to reach existence, characterization, and continuity of the second derivatives (Definition 10.3 in §10.7).

In more details, first, the existence of $\dot{A}_t$ and $\dot{B}_t$ in (10.14) implies that of the first unilateral derivatives in (10.8),

$$\dot{\lambda}_j^{\pm}(t) = \lim_{h \to \pm 0} \frac{1}{h}(\lambda_j(t+h) - \lambda_j(t)) \tag{10.16}$$

for each j and t. These derivatives, furthermore, are characterized by the other finite dimensional eigenvalue problems in accordance with the multiplicity of $\lambda_j(t)$ (Theorem 10.9 in §10.6). Second, if these $\dot{A}_t$ and $\dot{B}_t$ are continuous in t and if

$$\lambda_{k-1}(t) < \lambda_k(t) \leq \cdots \leq \lambda_{k+m-1}(t) < \lambda_{k+m}(t) \tag{10.17}$$

holds for $t \in I = (-\varepsilon_0, \varepsilon_0)$, there are C^1 curves

$$\tilde{C}_j, \ k \leq j \leq k+m-1,$$

made by at most countably many rearrangements of the C^0 curves

$$C_j = \{\lambda_j(t) \mid t \in I\}, \quad k \leq j \leq k+m-1.$$

(Theorem 10.11 in §10.7).

Here we notice two properties of the unilateral derivatives $\dot{\lambda}_j^{\pm}$. First, there arises that

$$\dot{\lambda}_j^+(t) = \dot{\lambda}_{2\ell+n-1-j}^-(t), \quad \ell \leq j \leq \ell+n-1$$

if

$$\lambda_{\ell-1}(t) < \lambda_\ell(t) = \cdots = \lambda_{\ell+n-1}(t) < \lambda_{\ell+n}(t) \tag{10.18}$$

holds for $\ell \geq k$, $\ell + n \leq m$ (Theorem 10.9 in §10.6). Second, the unilateral derivative $\dot{\lambda}_j^{\pm}$ are provided with the unilateral continuity as in

$$\lim_{h \to \pm 0} \dot{\lambda}_j^{\pm}(t+h) = \dot{\lambda}_j^{\pm}(t)$$

if both $\dot{A}_t$ and $\dot{B}_t$ are continuous in $t \in I$ (Theorem 10.10 in §10.6).

As for the second derivatives of eigenvalues, we assume the existence of $\ddot{A}_t$ and $\ddot{B}_t$ in (10.15) besides $\dot{A}_t$ and $\dot{B}_t$. Then each $j = 1, 2, \cdots$ admits the existence of

$$\ddot{\lambda}_j^{\pm}(t) = \lim_{h \to \pm 0} \frac{2}{h^2}(\lambda_j(t+h) - \lambda_j(t) - h\dot{\lambda}_j^{\pm}(t)). \qquad (10.19)$$

These limits are again unilateral and characterized by the other eigenvalue problem in a finite dimensional space (Theorem 10.12 in §10.8). The unilateral continuity of these $\ddot{\lambda}_j^{\pm}(t)$ is then assured under the continuity of $\ddot{A}_t$ and $\ddot{B}_t$ in t, similarly. These properties induce C^2 smoothness of $\tilde{C}_j$, $k \leq j \leq k+m-1$, once their C^1 smoothness is achieved (Theorem 10.15 in §10.8).

10.3 Characterization of Derivatives

Here we state the results on the first and the second derivatives of $\lambda_j = \lambda_j(t)$ in (10.8). Fix $t \in I$, and assume (10.18) for $\ell = k$ and $n = m$ for $k, m = 1, 2, \cdots$ with the convention $\lambda_0(t) = -\infty$:

$$\lambda_{k-1}(t) < \lambda_k(t) = \cdots = \lambda_{k+m-1}(t) < \lambda_{k+m}(t). \qquad (10.20)$$

Put

$$\lambda \equiv \lambda_k(t) = \cdots = \lambda_{k+m-1}(t), \qquad (10.21)$$

and let Y_t^{λ}, $\dim Y_t^{\lambda} = m$, be the eigenspace corresponding to this eigenvalue λ.

Theorem 10.1. *If $\{T_{t'}\}$ is differentiable at $t' = t$, there exist the unilateral limits*

$$\dot{\lambda}_j^{\pm} = \lim_{h \to \pm 0} \frac{1}{h}(\lambda_j(t+h) - \lambda), \quad k \leq j \leq k+m-1,$$

which satisfies

$$\nu_j \equiv \dot{\lambda}_j^+ = \dot{\lambda}_{2k+m-1-j}^-, \quad k \leq j \leq k+m-1. \qquad (10.22)$$

This ν_j is the q-th eigenvalue of the matrix

$$G_t^{\lambda} = \left(E_t^{\lambda}(\tilde{\phi}_i, \tilde{\phi}_j)\right)_{1 \leq i,j \leq m} \qquad (10.23)$$

for $q = j - k + 1$, where $\{\tilde{\phi}_j \mid 1 \leq j \leq m\}$ is a basis of Y_t^{λ}, satisfying

$$B_t(\tilde{\phi}_i, \tilde{\phi}_j) = \delta_{ij}, \quad 1 \leq i, j \leq m \qquad (10.24)$$

and

$$E_t^{\lambda} = \dot{A}_t - \lambda \dot{B}_t$$

for $\dot{A}_t$ and $\dot{B}_t$ defined by (10.12), (10.13), and (10.14).

Remark 10.1. If $\langle \phi_j \mid 1 \le j \le m \rangle$ is the other basis of Y_t^λ satisfying

$$B_t(\phi_i, \phi_j) = \delta_{ij}, \quad 1 \le i, j \le m, \tag{10.25}$$

it holds that

$$\phi_j = \sum_{i=1}^{m} q_j^i \tilde\phi_i, \ 1 \le j \le m$$

with the orthogonal $m \times m$ matrix $Q = (q_j^i)$. Hence ν_j, $k \le j \le k+m-1$, in (10.22) is determined, indpendent of the choice of $\langle \tilde\phi_j \mid 1 \le j \le m \rangle$.

Remark 10.2. Under (10.21), it holds that

$$\dot\lambda_k^+(t) \le \cdots \le \dot\lambda_{k+m-1}^+(t), \quad \dot\lambda_k^-(t) \ge \cdots \ge \dot\lambda_{k+m-1}^-(t).$$

A direct consequence of Theorem 10.1 is the existence of the unilateral derivatives $\dot\lambda_j^\pm(t)$ in (10.16) for any $t \in I$ and $j = 1, 2, \cdots$, if $\{T_t\}$ is differentiable in I. Then we obtain the following theorem.

Theorem 10.2. *If $\{T_t\}$ is continuously differentiable in $t \in I$, the functions $\dot\lambda_j^\pm = \dot\lambda_j^\pm(t)$ are unilaterally continuous, so that it holds that*

$$\lim_{h \to \pm 0} \dot\lambda_j^\pm(t+h) = \dot\lambda_j^\pm(t)$$

for any t and j.

Theorem 10.1 and Theorem 10.2 are proven in §10.6. Then we show the following theorem in §10.7.

Theorem 10.3 ([Rellich (1953)]). *Let $\{T_t\}$ be continuously differentiable in t, and assume*

$$\lambda_{k-1}(t) < \lambda_k(t) \le \cdots \le \lambda_{k+m-1}(t) < \lambda_{k+m}(t), \quad \forall t \in I$$

for some $k, m = 1, 2, \cdots$. Let

$$C_j = \{\lambda_j(t) \mid t \in I\}, \quad k \le j \le k+m-1$$

be C^0 curves. Then, there exist C^1 curves denoted by $\tilde C_i$, $k \le i \le k+m-1$, made by a rearrangement of

$$\{C_j \mid k \le j \le k+m-1\}$$

at most countably many times.

Turning to the second derivatives, we fix $t \in I$ again, and assume that $\{T_{t'}\}$ is twice differentiable at $t' = t$. Suppose (10.20), put λ as in (10.21), and let $k \le \ell_+ < r_+ \le k + m$ be such that

$$\dot{\lambda}^+_{\ell_+ - 1} < \lambda'_+ \equiv \dot{\lambda}^+_{\ell_+} = \cdots = \dot{\lambda}^+_{r_+ - 1} < \dot{\lambda}^+_{r_+} \tag{10.26}$$

in Theorem 10.1. To state the finite dimensional eigenvalue problem characterizing the second unilateral derivatives of $\lambda_j(t')$ at $t' = t$ for $\ell \le j \le r-1$, we introduce the following definition, recalling the eigenspace Y^λ_t corresponding to the eigenvaluie λ in (10.21).

Definition 10.2. Let

$$R : X = L^2(\Omega) \to Y^\lambda_t$$

be the orthogonal projection with respect to $B_t(\cdot, \cdot)$ and $P = I - R$, where

$$I : X \to X$$

is the identity operator. Let, furthermore, A_t, B_t, $\dot{A}_t$, $\dot{B}_t$, $\ddot{A}_t$, and $\ddot{B}_t$ be as in (10.12), (10.13), (10.14), and (10.15) in §10.1. Then we define

$$w = \gamma(u) \in PV, \quad u \in V$$

by

$$C_t(w, v) = -\dot{C}^{\lambda,\lambda'}_t(u, v), \quad \forall v \in PV \tag{10.27}$$

for $\lambda' = \lambda'_+$, where

$$C_t = A_t - \lambda B_t, \quad \dot{C}^{\lambda,\lambda'}_t = \dot{A}_t - \lambda \dot{B}_t - \lambda' B_t.$$

We put also

$$F^{\lambda,\lambda'}_t(u, v) = (\ddot{A}_t - \lambda \ddot{B}_t - 2\lambda' \dot{B}_t)(u, v) - 2C_t(\gamma(u), \gamma(v)), \quad u, v \in V.$$

Remark 10.3. To confirm the well-definedness of

$$\gamma(u) \in PV \subset V$$

for each $u \in V$, let $Q : L^2(\Omega) \to Z^k_t$ be the orthogonal projection with respect to $B_t(\cdot, \cdot)$, where Z^k_t denotes the finite dimensional subspace of $L^2(\Omega)$ generated by the eigenfunctions corresponding to the eigenvalues $\lambda_1(t), \cdots, \lambda_{k-1}(t)$. Let, furthermore, $V_0 = QV$ and $V_1 = (I - Q)RV$. First, there is a unique $w_0 \in V_0$ satisfying

$$C_t(w_0, v) = -\dot{C}^{\lambda,\lambda'}_t(u, v), \quad \forall v \in V_0 \tag{10.28}$$

because C_t is negative definite on $V_0 \times V_0$. Second, there is also a unique $w_1 \in V_1$ satisfying

$$C_t(w_1, v) = -\dot{C}_t^{\lambda,\lambda'}(u, v), \quad \forall v \in V_1 \tag{10.29}$$

because C_t is positive definite on $V_1 \times V_1$. Then we obtain (10.27) for $w = w_0 + w_1$. In fact, there arises that

$$C_t(w_0, v) = C_t(w_0, Qv)$$

and

$$C_t(w_1, v) = C_t(w_1, (I - Q)v)$$

for any $v \in V$, and hence

$$\begin{aligned} C_t(w, v) &= C_t(w_0, v) + C_t(w_1, v) \\ &= C_t(w_0, Qv) + C_t(w_1, (I - Q)v) \\ &= -\dot{C}_t^{\lambda,\lambda'}(u, Qv) - \dot{C}_t^{\lambda,\lambda'}(u, (I - Q)v) \\ &= -\dot{C}_t(u, v), \quad \forall v \in RV. \end{aligned}$$

We thus obtain $w \in RV$ satisfying (10.27). If $w \in RV$ is a solution to (10.27), conversely, then $w_0 = Qw \in V_0$ and $w_1 = (I - Q)w \in V_1$ solve (10.28) and (10.29), respectively. Then there arises the uniqueness of such $w = w_0 + w_1$ because these $w_0 \in V_0$ and $w_1 \in V_1$ are unique.

We have also

$$Y_t^\lambda = \langle \tilde{\phi}_j \mid k \leq j \leq k + m - 1 \rangle$$

with (10.24).

Theorem 10.4. *Under the above situation of (10.20) and (10.26), there exist*

$$\lambda_{j+}'' = \lim_{h \to +0} \frac{2}{h^2}(\lambda_j(t + h) - \lambda - h\lambda_+'), \quad \ell_+ \leq j \leq r_+ - 1. \tag{10.30}$$

This λ_{j+}'' is the q-th eigenvalue of the matrix

$$H_{t,\ell,r}^{\lambda,\lambda'} = \left(F_t^{\lambda,\lambda'}(\tilde{\phi}_i, \tilde{\phi}_j) \right)_{\ell \leq i,j \leq r-1} \tag{10.31}$$

for

$$\lambda' = \lambda_+', \quad \ell = \ell_+, \quad r = r_+$$

and $q = j - \ell + 1$.

Remark 10.4. If it holds that

$$\dot{\lambda}^-_{\ell_- - 1}(t) > \lambda'_- \equiv \dot{\lambda}^-_{\ell_-}(t) = \cdots = \dot{\lambda}^-_{r_- - 1}(t) > \dot{\lambda}^-_{r_-}(t)$$

for $k \le \ell_- < r_- \le k + m$, similarly, there arises the existence of

$$\tilde{\lambda}''_{j-} = \lim_{h \to -0} \frac{2}{h^2}(\lambda_j(t + h) - \lambda - h\lambda'_-), \quad \ell_- \le j \le r_- - 1$$

and this $\tilde{\lambda}''_{j-}$ is the q-the eigenvalue of the matrix $H^{\lambda,\lambda'}_{t,\ell,r}$ for

$$\lambda' = \lambda'_-, \quad \ell = \ell_-, \quad r = r_-$$

and $q = r - j$.

Remark 10.5. As in Remark 10.1 on the first derivative, the above

$$\lambda''_{j\pm}, \quad \ell_\pm \le j \le r_\pm - 1,$$

are independent of the choice of the ortho-normal system

$$\langle \tilde{\phi}_j \mid \ell_\pm \le j \le r_\pm - 1 \rangle.$$

Remark 10.6. Theorem 10.1 and Theorem 10.4 imply also the existence of the unilateral limits $\ddot{\lambda}^\pm_j(t)$ in (10.19) for any t and j if $\{T_t\}$ is 2-differentiable in $t \in I$, that is,

$$\ddot{\lambda}^\pm_j(t) = \lim_{h \to \pm 0} \frac{2}{h^2}(\lambda_j(t + h) - \lambda_j(t) - h\dot{\lambda}^\pm_j(t)).$$

Remark 10.7. By Liouville's theorem on general perturbation of domains studied in Chapter 8, the matrix G^λ_t in (10.23) is represented by the surface integrals of $\tilde{\phi}_j$, $1 \le j \le m$. This property is confirmed by [Jimbo–Ushikoshi (2015)] for the Stokes operator with Dirichlet condition under the normal deformation in §8.1. Similarly, the matrix $H^{\lambda,\lambda'}_t$ in (10.31) is represented by the surface integrals of $\tilde{\phi}_j$ and $\gamma(\tilde{\phi}_j)$, $1 \le j \le m$. Concerning the first eigenvalue, λ_1, there arises that $Z^k_t = \{0\}$ in Remark 10.3, which results in the positivity of C_t on PV, besides the simplicity of λ_1. This fact on the first eigenvalue is noticed by [Garabedian–Schiffer (1952–53)] for (10.1) in §10.1 with $\gamma_1 = \emptyset$ in two space dimension under the first order perturbation of the domain in t, so that

$$T_t = I + tS$$

in (8.1) of §8.1. There, the harmonic concavity of λ_1 with respect to the domain perturbation is assured.

Then we show the following theorems in §10.8.

Theorem 10.5. *If $\{T_t\}$ is continuously 2-differentiable in t, then $\ddot{\lambda}_j^{\pm} = \ddot{\lambda}_j^{\pm}(t)$ are unilaterally continuous, so that it holds that*

$$\lim_{h \to \pm 0} \ddot{\lambda}_j^{\pm}(t+h) = \ddot{\lambda}_j^{\pm}(t)$$

for any $t \in I$ and $j = 1, 2, \cdots$.

Theorem 10.6. *If $\{T_t\}$ is continuously 2-differentiable in t, the C^1 curves $\tilde{C}_j$, $k \le j \le k + m - 1$, in Theorem 10.3 is C^2.*

10.4   Reduction to the Abstract Theory

For the moment, we fix t and treat the bi-Lipschitz homeomorphism $T = T_t : \Omega \to \Omega_t = T_t\Omega$. Let

$$\tilde{\Omega} = \Omega_t, \quad f(y) = g(x), \quad y = Tx,$$

and confirm the chain rule for this transformation of variables, that is,

$$\nabla g = (DT)^T \nabla f, \quad dy = (\det DT) dx$$

for

$$\nabla g = (\frac{\partial g}{\partial x_1}, \cdots, \frac{\partial g}{\partial x_n})^T, \quad \nabla f = (\frac{\partial f}{\partial y_1}, \cdots, \frac{\partial f}{\partial y_n})^T,$$

and

$$DT = \begin{pmatrix} \frac{\partial y_1}{\partial x_1} & \cdots & \frac{\partial y_1}{\partial x_n} \\ \cdot & & \cdot \\ \frac{\partial y_n}{\partial x_1} & \cdots & \frac{\partial y_n}{\partial x_n} \end{pmatrix}.$$

Putting $\tilde{\gamma}_i = \gamma_{it} = T\gamma_i$ for $i = 0, 1$, we take the eigenvalue problem

$$-\Delta u = \lambda u \text{ in } \tilde{\Omega}, \quad u = 0 \text{ on } \tilde{\gamma}_0, \quad \frac{\partial u}{\partial \nu} = 0 \text{ on } \tilde{\gamma}_1, \qquad (10.32)$$

that is, (10.8) for $T = T_t$. Let

$$\tilde{V} = \{v \in H^1(\tilde{\Omega}) \mid v|_{\tilde{\gamma}_0} = 0\},$$

and introduce the weak form of (10.32),

$$u \in \tilde{V}, \quad \tilde{A}(u, v) = \lambda \tilde{B}(u, v), \quad \forall v \in \tilde{V}, \qquad (10.33)$$

where

$$\tilde{A}(u, v) = \int_{\tilde{\Omega}} \nabla_y u \cdot \nabla_y v \, dy, \quad \tilde{B}(u, v) = \int_{\tilde{\Omega}} uv \, dy. \qquad (10.34)$$

Given $\phi \in V$, we put

$$\psi(y) = \phi(x), \quad y = Tx.$$

Then it holds that

$$\phi \in V \iff \psi \in \tilde{V}$$

for $V \subset H^1(\Omega)$ defined by (10.3). Writing

$$U(x) = u(y), \ V(x) = v(y), \quad y = Tx, \qquad (10.35)$$

we obtain

$$
\begin{aligned}
\nabla_y u \cdot \nabla_y v &= [(DT)^{-1T}\nabla_x U] \cdot [(DT)^{-1T}\nabla_x V] \\
&= (\nabla_x U)^T (DT)^{-1}(DT)^{-1T}(\nabla_x V) \\
&= (\nabla_x U)^T Q(\nabla_x V) = Q[\nabla_x U, \nabla_x V]
\end{aligned}
$$

for $Q = (DT)^{-1}(DT)^{-1T}$, where F^T denotes the transpose of the matrix F. Then, (10.33) means

$$\int_\Omega Q[\nabla_x U, \nabla_x V]a \ dx = \lambda \int_\Omega UVa \ dx$$

for $a = \det DT$. The condition of normalization

$$\int_{\tilde{\Omega}} u^2 \ dy = 1$$

is also transformed into

$$\int_\Omega U^2 a \ dx = 1.$$

Under the family $\{T_t\}$ of homeomorphisms, therefore, the weak form (10.9) of (10.8) in §10.1, is equivalent to (10.11) for $V \subset H^1(\Omega)$ defined by (10.3) and B_t, A_t defined by (10.12)–(10.13) in §10.2. Here we confirm the following lemma.

Lemma 10.1. *The j-th eigenvalue of (10.8) is equal to that of (10.11).*

Proof. For the moment, let $\tilde{\lambda}_j(t)$ be the j-th eigenvalues of (10.8) and let $\lambda_j(t)$ be that of (10.11) for (10.12) and (10.13). By the mini-max principle (10.5), it holds that

$$\tilde{\lambda}_j(t) = \min_{\tilde{L}_j} \ \max_{v \in \tilde{L}_j \setminus \{0\}} \ \tilde{R}_t[v] = \max_{\tilde{W}_j} \ \min_{v \in \tilde{W}_j \setminus \{0\}} \ \tilde{R}_t[v], \qquad (10.36)$$

where $V_t \subset H^1(\Omega_t)$ is defined by (10.10),

$$\tilde{R}_t[v] = \frac{\tilde{A}_t(v,v)}{\tilde{B}_t(v,v)}$$

for

$$\tilde{A}_t(u,v) = \int_{\Omega_t} \nabla u \cdot \nabla v \, dx, \quad \tilde{B}_t(u,v) = \int_{\Omega_t} uv \, dx,$$

and $\{\tilde{L}_j\}$ and $\{\tilde{W}_j\}$ be the families of all subspaces of V_t with dimension and codimension j and $j-1$, respectively.

It holds also that

$$\lambda_j(t) = \min_{L_j} \max_{v \in L_j \setminus \{0\}} R_t[v] = \max_{W_j} \min_{v \in W_j \setminus \{0\}} R_t[v], \tag{10.37}$$

for

$$R_t[v] = \frac{A_t(v,v)}{B_t(v,v)}, \tag{10.38}$$

and $\{L_j\}$ and $\{W_j\}$ denote the families of all subspaces of V with dimension and codimension j and $j-1$, respectively.

If the set L is a j-dimensional subspace of V there is

$$\{\phi_\ell \in L \mid 1 \le \ell \le j\}$$

such that

$$\int_\Omega \phi_\ell \phi_{\ell'} \, dx = \delta_{\ell\ell'}$$

and

$$\phi = \sum_{\ell=1}^j c_\ell \phi_\ell, \quad c_\ell = \int_\Omega \phi \phi_\ell \, dx$$

for any $\phi \in L$, which implies

$$\psi = \sum_{\ell=1}^j c_\ell \psi_\ell$$

for $\psi = \phi \circ T_t^{-1}$ and $\psi_\ell = \phi \circ T_t^{-1}$. Hence we obtain $\dim \tilde{L}_t \le \dim L$ for

$$\tilde{L}_t = \{\phi \circ T_t^{-1} \mid \phi \in L\}.$$

The reverse inequality follows similarly, and hence it holds that

$$\dim \tilde{L}_t = \dim L = j.$$

Since $T_t : \Omega \to \Omega_t$ is a bi-Lipschitz homeomorphism, furthermore, $L \subset V$ if and only if $\tilde{L}_t \subset V_t$. We thus obtain

$$\tilde{\lambda}_j(t) = \lambda_j(t) \tag{10.39}$$

by (10.36)–(10.37). $\qquad\square$

We are ready to develop an abstract theory, writing L^2 norm in $X = L^2(\Omega)$ as $|\cdot|_X$. With V in (10.3), we recall that $\|\cdot\|_V$ denotes the norm in V and that the inclusion $V \hookrightarrow X$ is compact. It holds also that

$$|v|_X \le K\|v\|_V, \quad v \in V \tag{10.40}$$

for $K = 1$.

Henceforth, C denotes the generic positive constant. The above $A_t : V \times V \to \mathbf{R}$ and $B_t : X \times X \to \mathbf{R}$ for $t \in I$ are symmetric bilinear forms, satisfying

$$\begin{aligned} |A_t(u,v)| &\le C\|u\|_V\|v\|_V, & u,v \in V \\ A_t(u,u) &\ge \delta\|u\|_V^2, & u \in V \end{aligned} \tag{10.41}$$

and

$$\begin{aligned} |B_t(u,v)| &\le C|u|_X|v|_X, & u,v \in X \\ B_t(u,u) &\ge \delta|u|_X^2, & u \in X \end{aligned} \tag{10.42}$$

for some $\delta > 0$. Then the eigenvalues of (10.11) are denoted by

$$0 < \lambda_1(t) \le \lambda_2(t) \le \cdots \to +\infty.$$

The weak and the strong convergences of $\{u_j\} \subset Y$ to $u \in Y$ for $Y = X$ or $Y = V$ are, furthermore, indicated by

$$\mathrm{w} - \lim_{j\to\infty} u_j = u \text{ in } Y, \quad \mathrm{s} - \lim_{j\to\infty} u_j = u \text{ in } Y.$$

10.5   Continuity of Eigenvalues and Eigenspaces

Let $t \in I$ be fixed. We begin with the following theorem valid under the abstract setting in the previous section.

Theorem 10.7. *The conditions*

$$\lim_{h\to 0} \sup_{\|u\|_V, \|v\|_V \le 1} |A_{t+h}(u,v) - A_t(u,v)| = 0$$

$$\lim_{h\to 0} \sup_{|u|_X, |v|_X \le 1} |B_{t+h}(u,v) - B_t(u,v)| = 0, \tag{10.43}$$

imply

$$\lim_{h\to 0} \lambda_j(t+h) = \lambda_j(t) \tag{10.44}$$

for any $j = 1, 2, \cdots$.

Proof. We note that the j-th eigenvalue of (10.8) is given by the mini-max principle as in (10.37), for the Rayleigh quotient $R_t[v]$ defined by (10.38).

Given t, let

$$\alpha(h) = \sup_{\|u\|_V, \|v\|_V \le 1} |(A_{t+h} - A_t)(u, v)|$$

$$\beta(h) = \sup_{|u|_X, |v|_X \le 1} |(B_{t+h} - B_t)(u, v)|. \tag{10.45}$$

We obtain

$$(A_{t+h} - A_t)(v, v) \ge -\alpha(h)\|v\|_V^2 \ge -\frac{\alpha(h)}{\delta} A_t(v, v)$$

and

$$(A_{t+h} - A_t)(v, v) \le \alpha(h)\|v\|_V^2 \le \frac{\alpha(h)}{\delta} A_t(v, v)$$

by (10.41) and (10.43), which implies

$$(1 - \delta^{-1}\alpha(h))A_t(v, v) \le A_{t+h}(v, v) \le (1 + \delta^{-1}\alpha(h))A_t(v, v).$$

Similarly, there arises that

$$(1 - \delta^{-1}\beta(h))B_t(v, v) \le B_{t+h}(v, v) \le (1 + \delta^{-1}\beta(h))B_t(v, v)$$

for any $v \in V$.

Then it follows that

$$(1 - o(1))R_t[v] \le R_{t+h}[v] \le (1 + o(1))R_t[v]$$

uniformly in $v \in V \setminus \{0\}$, and hence

$$(1 - o(1))\lambda_j(t) \le \lambda_j(t + h) \le (1 + o(1))\lambda_j(t)$$

by (10.43). Thus we obtain (4.144). □

Let $u_j(t) \in V$ be the eigenfunction of (10.11) corresponding to the eigenvalue $\lambda_j(t)$:

$$B_t(u_j(t), u_{j'}(t)) = \delta_{jj'}$$
$$A_t(u_j(t), v) = \lambda_j(t)B_t(u_j(t), v), \quad \forall v \in V. \tag{10.46}$$

Fix t, assume (10.18), and define λ by (10.21). Although this multiplicity m is not stable under the perturbation of t, we obtain the following theorem concerning the continuity of eigenspaces with respect to t.

Let

$$Y_t^\lambda = \langle u_j(t) \mid k \le j \le k + m - 1 \rangle \tag{10.47}$$

be the subspace of X generated by the above eigenfunctions $u_j(t)$ for $k \leq j \leq k + m - 1$.

Lemma 10.2. *Under the above situation, any $h_\ell \to 0$ admits a subsequence, denoted by the same symbol, such that the limits*

$$s - \lim_{\ell \to \infty} u_j(t + h_\ell) = \phi_j \in Y_t^\lambda \text{ in } V, \quad k \leq j \leq k + m - 1 \qquad (10.48)$$

exists. In particular it holds that

$$B_t(\phi_j, \phi_{j'}) = \delta_{jj'}, \quad k \leq j, j' \leq k + m - 1. \qquad (10.49)$$

Proof. We have

$$\lim_{h \to 0} \lambda_j(t + h) = \lambda_j(t), \quad k \leq j \leq k + m - 1 \qquad (10.50)$$

by Theorem 10.7. It holds also that

$$\|u_j(t')\|_V \leq C, \quad k \leq j \leq k + m - 1, \ |t'| < \varepsilon_0 \qquad (10.51)$$

by (10.40)–(10.42) and (10.46), as in

$$\delta \|u_j(t')\|_V^2 \leq A_{t'}(u_j(t'), u_j(t'))$$
$$= \lambda_j(t') B_t(u_j(t'), u_j(t')) = \lambda_j(t').$$

Given $h_\ell \to 0$, therefore, we have a subsequence, denoted by the same symbol, which admits the weak limits

$$\text{w} - \lim_{\ell \to \infty} u_j(t + h_\ell) = \phi_j \text{ in } V, \quad k \leq j \leq k + m - 1, \qquad (10.52)$$

for some $\phi_j \in V$. From (10.46) it follows that

$$A_t(\phi_j, v) = \lambda_j(t) B_t(\phi_j, v), \quad \forall v \in V$$

and hence

$$\phi_j \in Y_t^\lambda, \quad k \leq j \leq k + m - 1.$$

Since $V \hookrightarrow X$ is compact, the weak convergence (10.52) implies the strong convergence

$$s - \lim_{\ell \to \infty} u_j(t + h_\ell) = \phi_j \text{ in } X, \qquad (10.53)$$

and hence (10.49). Now we improve this weak convergence (10.52) to the strong convergence (10.48) in V, using (10.43).

For this purpose, we put

$$v = u_j(t + h_\ell) - \phi_j$$

and recall (10.21) and (10.50). Then there arises that

$$
\begin{aligned}
\delta\|v\|_V^2 &\leq A_{t+h_\ell}(u_j(t+h_\ell) - \phi_j, v) \\
&= A_{t+h_\ell}(u_j(t+h_\ell), v) - (A_{t+h_\ell} - A_t)(\phi_j, v) - A_t(\phi_j, v) \\
&= \lambda_j(t+h_\ell)B_{t+h_\ell}(u_j(t+h_\ell), v) - (A_{t+h_\ell} - A_t)(\phi_j, v) - \lambda_j(t)B_t(\phi_j, v) \\
&= (\lambda_j(t+h_\ell) - \lambda_j(t))B_{t+h_\ell}(u_j(t+h_\ell), v) \\
&\quad + \lambda_j(t)(B_{t+h_\ell} - B_t)(u_j(t+h_\ell), v) + \lambda_j(t)B_t(u_j(t+h_\ell) - \phi_j, v) \\
&\quad - (A_{t+h_\ell} - A_t)(\phi_j, v) \\
&\leq C|\lambda_j(t+h_\ell) - \lambda_j(t)| \cdot |u_j(t+h_\ell)|_X \cdot K\|v\|_V \\
&\quad + \lambda_j(t) \cdot \beta(h_\ell) \cdot |u_j(t+h_\ell)|_X \cdot K\|v\|_V \\
&\quad + \lambda_j(t) \cdot C|u_j(t+h_\ell) - \phi_j|_X \cdot K\|v\|_V + \alpha(h_\ell)\|\phi_j\|_V \cdot \|v\|_V,
\end{aligned}
$$

and hence

$$
\begin{aligned}
\|v\|_V &\leq C\{|\lambda_j(t+h_\ell) - \lambda_j(t)| + \beta(h_\ell) + |u_j(t+h_\ell) - \phi_j|_X + \alpha(h_\ell)\} \\
&= o(1)
\end{aligned}
$$

by (10.43) and (10.53). $\qquad\qquad\qquad\qquad\qquad\qquad\qquad\qquad\square$

Remark 10.8. If $m = 1$, the eigenfunction $u_j(t)$ in (10.46) is uniquely determined up to the multiplication of ± 1, which implies $\phi_j = \pm u_j(t)$. In the other case of $m \geq 2$, the eigenfunction which attains (10.39) does not satisfy this property. Hence we have

$$
\phi_j = \sum_{i=1}^m q_j^i u_i(t)
$$

for $Q = (q_j^i)$ satisfying $Q^T = Q^{-1}$ in Theorem 10.2. In other words, the eigenfunction $u_j(t)$ corresponding to $\lambda_j(t)$ has more varieties than ± 1 multiplication, although the eigenspace Y_t^λ is determined. By this ambiguity the limit ϕ_j in (10.48) depends on the sequence $h_\ell \to 0$, which makes the argument below to be complicated.

10.6 First Derivatives

If $\{T_t\}$ is differentiable in the setting of §10.1, we can put

$$
\dot{A}_t(u, v) = \int_\Omega \dot{Q}_t[\nabla u, \nabla v]a_t + Q_t[\nabla u, \nabla v]\dot{a}_t \, dx, \quad u, v \in V
$$

and

$$\dot{B}_t(u, v) = \int_\Omega uv\dot{a}_t \, dx, \quad u, v \in X.$$

These $\dot{A}_t : V \times V \to \mathbf{R}$ and $\dot{B}_t : X \times X \to \mathbf{R}$ in (10.12) are bilinear forms satisfying

$$|\dot{A}_t(u, v)| \le C\|u\|_V\|v\|_V, \quad u, v \in V$$
$$|\dot{B}_t(u, v)| \le C|u|_X|v|_X, \quad u, v \in X \tag{10.54}$$

and

$$\lim_{h \to 0} \frac{1}{h} \sup_{\|u\|_V, \|v\|_V \le 1} \left|\left(A_{t+h} - A_t - h\dot{A}_t\right)(u, v)\right| = 0$$

$$\lim_{h \to 0} \frac{1}{h} \sup_{|u|_X, |v|_X \le 1} \left|\left(B_{t+h} - B_t - h\dot{B}_t\right)(u, v)\right| = 0. \tag{10.55}$$

Theorem 10.1 in §10.3, therefore, is reduced to the following abstract theorem. In this theorem, the assumption made by Theorem 10.7 in §10.5 is valid, and therefore, there arises that (10.44).

Theorem 10.8. *Let X, V be Hilbert spaces over $\mathbf{R}$, with compact embedding $V \hookrightarrow X$. Let*

$$A_t : V \times V \to \mathbf{R}, \quad B_t : X \times X \to \mathbf{R}$$

be symmetric bilinear forms satisfying (10.41)–(10.42) for any $t \in I$. Given t, assume the existence of the bilinear forms

$$\dot{A}_t : V \times V \to \mathbb{R}, \quad \dot{B}_t : X \times X \to \mathbb{R}$$

satisfying (10.54)–(10.55). Assume, finally, (10.20) in §10.3 for $\lambda_j(t)$ defined by (10.37)–(10.38). Then the conclusion of Theorem 10.1 holds.

Remark 10.9. In a formal argument, we write (10.11) as

$$u_t \in V, \ B_t(u_t, u_t) = 1$$
$$A_t(u_t, v) = \lambda_t B_t(u_t, v), \ v \in V. \tag{10.56}$$

First, taking a formal differentiation in t in this equality, we get

$$\dot{A}_t(u_t, v) + A_t(\dot{u}_t, v)$$
$$= \dot{\lambda}_t B_t(u_t, v) + \lambda_t \dot{B}_t(u_t, v) + \lambda_t B_t(\dot{u}_t, v), \ \forall v \in V. \tag{10.57}$$

Putting $v = u_t$, we obtain

$$\dot{A}_t(u_t, u_t) + A_t(\dot{u}_t, u_t) = \dot{\lambda}_t + \lambda_t \dot{B}_t(u_t, u_t) + \lambda_t B_t(\dot{u}_t, u_t) \tag{10.58}$$

by

$$B_t(u_t, u_t) = 1. \tag{10.59}$$

To eliminate $\dot{u}_t$ in (10.58), second, we use (10.59) to deduce

$$\dot{B}_t(u_t, u_t) + 2B_t(\dot{u}_t, u_t) = 0. \tag{10.60}$$

From

$$\lambda_t = \lambda_t B_t(u_t, u_t) = A_t(u_t, u_t)$$

it is derived also that

$$\dot{\lambda}_t = \dot{A}_t(u_t, u_t) + 2A_t(\dot{u}_t, u_t) \tag{10.61}$$

and then (10.58) is replaced by

$$\dot{A}_t(u_t, u_t) + \frac{1}{2}\dot{\lambda}_t - \frac{1}{2}\dot{A}_t(u_t, u_t) = \dot{\lambda}_t + \lambda_t\dot{B}_t(u_t, u_t) - \frac{1}{2}\lambda_t\dot{B}_t(u_t, u_t),$$

or,

$$\dot{\lambda}_t = \dot{A}_t(u_t, u_t) - \lambda_t\dot{B}_t(u_t, u_t). \tag{10.62}$$

Remark 10.10. As is noticed in Remark 10.8 in §5.9, if the eigenspace

$$Y_t^\lambda = \langle u_j(t) \mid k \le j \le k + m - 1 \rangle$$

corresponding to the eigenvalue $\lambda = \lambda_j(t)$, $k \le j \le k+m-1$, to (10.11), is one-dimensional as in $m = 1$, the eigenfunction u_t in (10.56) is unique up to the multiplication of ± 1, and this ambiguity is canceled in (10.62). This property is valid even if $m \ge 2$ as in Remark 10.1 in §10.3.

Turning to the rigorous proof, we use the following lemma, recalling $u_j(t) \in V$, Y_t^λ, and ϕ_j in (10.46), (10.47), and (10.48) in §10.5, respectively.

Lemma 10.3. *Under the assumption of Theorem 10.8, any $h_\ell \to 0$ admits a subsequence, denoted by the same symbol, such that*

$$\lim_{\ell\to\infty} \frac{1}{h_\ell}\{\lambda_j(t+h_\ell) - \lambda_j(t)\} = \dot{A}_t(\phi_j, \phi_j) - \lambda_j(t)\dot{B}_t(\phi_j, \phi_j) \tag{10.63}$$

for $k \le j \le k + m - 1$. It holds also that

$$\dot{A}_t(\phi_j, \phi_{j'}) - \lambda_j(t)\dot{B}_t(\phi_j, \phi_{j'}) = 0, \quad k \le j \ne j' \le k + m - 1. \tag{10.64}$$

Proof. Let $k \leq j, j' \leq k + m - 1$. Since

$$A_{t+h}(u_j(t+h) - \phi_j, u_{j'}(t+h) - \phi_{j'}) = A_{t+h}(u_j(t+h), u_{j'}(t+h))$$
$$-A_{t+h}(u_j(t+h), \phi_{j'}) - A_{t+h}(\phi_j, u_{j'}(t+h)) + A_{t+h}(\phi_j, \phi_{j'})$$

and

$$A_t(u_j(t+h) - \phi_j, u_{j'}(t+h) - \phi_{j'}) = A_t(u_j(t+h), u_{j'}(t+h))$$
$$-A_t(u_j(t+h), \phi_{j'}) - A_t(\phi_j, u_{j'}(t+h)) + A_t(\phi_j, \phi_{j'}),$$

it holds that

$$h(A_{t+h} - A_t) \left(\frac{u_j(t+h) - \phi_j}{h}, \frac{u_{j'}(t+h) - \phi_{j'}}{h} \right)$$

$$= \frac{1}{h}(A_{t+h} - A_t)(u_j(t+h) - \phi_j, u_{j'}(t+h) - \phi_{j'})$$

$$= \frac{1}{h}(A_{t+h} - A_t)(u_j(t+h), u_{j'}(t+h))$$

$$+ \frac{1}{h}(A_{t+h} - A_t)(\phi_j, \phi_{j'}) - \frac{1}{h}(A_{t+h} - A_t)(u_j(t+h), \phi_{j'})$$

$$- \frac{1}{h}(A_{t+h} - A_t)(\phi_j, u_{j'}(t+h)). \tag{10.65}$$

In (10.65), first, we obtain

$$\frac{1}{h}(A_{t+h} - A_t)(u_j(t+h), u_{j'}(t+h)) = \dot{A}_t(u_j(t+h), u_{j'}(t+h)) + o(1)$$

as $h \to 0$ by (10.51) and (10.55). Then (10.48) in §10.5 implies

$$\lim_{\ell \to \infty} \frac{1}{h_\ell}(A_{t+h_\ell} - A_t)(u_j(t+h_\ell), u_{j'}(t+h_\ell))$$

$$= \lim_{\ell \to \infty} \frac{1}{h_\ell}(A_{t+h_\ell} - A_t)(\phi_j, \phi_{j'}) = \dot{A}_t(\phi_j, \phi_{j'}). \tag{10.66}$$

Second, it holds that

$$A_{t+h}(u_j(t+h), \phi_{j'}) = \lambda_j(t+h)B_{t+h}(u_j(t+h), \phi_{j'})$$
$$A_t(u_j(t+h), \phi_{j'}) = \lambda_j(t)B_t(u_j(t+h), \phi_{j'})$$

by $\phi_{j'} \in Y_\lambda$, which implies

$$\frac{1}{h_\ell}(A_{t+h_\ell} - A_t)(u_j(t+h_\ell), \phi_{j'})$$

$$= \frac{1}{h_\ell}(\lambda_j(t+h_\ell)B_{t+h_\ell}(u_j(t+h_\ell), \phi_{j'}) - \lambda_j(t)B_t(u_j(t+h_\ell), \phi_{j'}))$$

$$= \frac{1}{h_\ell}(\lambda_j(t+h_\ell) - \lambda_j(t))B_{t+h_\ell}(u_j(t+h_\ell), \phi_{j'})$$

$$+ \frac{1}{h_\ell}\lambda_j(t)(B_{t+h_\ell} - B_t)(u_j(t+h_\ell), \phi_{j'})$$

$$= \frac{1}{h_\ell}(\lambda_j(t+h_\ell) - \lambda_j(t))(\delta_{jj'} + o(1)) + \lambda_j(t)\dot{B}_t(\phi_j, \phi_{j'}) + o(1)$$

by
$$B_{t+h_\ell}(u_j(t+h_\ell), \phi_{j'}) = B_{t+h_\ell}(\phi_j, \phi_{j'}) + o(1)$$
$$= B_t(\phi_j, \phi_{j'}) + o(1) = \delta_{jj'} + o(1)$$

and (10.55). Similarly, it follows that

$$\frac{1}{h_\ell}(A_{t+h_\ell} - A_t)(\phi_j, u_{j'}(t+h_\ell)) = \frac{1}{h_\ell}(\lambda_j(t+h_\ell) - \lambda_j(t))(\delta_{jj'} + o(1))$$
$$+\lambda_j(t)\dot{B}_t(\phi_j, \phi_{j'}) + o(1). \tag{10.67}$$

Finally, we obtain

$$\left| h_\ell(A_{t+h_\ell} - A_t)(\frac{u_j(t+h_\ell) - \phi_j}{h_\ell}, \frac{u_{j'}(t+h_\ell) - \phi_{j'}}{h_\ell}) \right|$$
$$\leq C h_\ell^2 \|\frac{u_j(t+h_\ell) - \phi_j}{h_\ell}\|_V \cdot \|\frac{u_{j'}(t+h_\ell) - \phi_{j'}}{h_\ell}\|_V$$
$$= C\|u_j(t+h_\ell) - \phi_j\|_V \cdot \|u_{j'}(t+h_\ell) - \phi_{j'}\|_V = o(1) \tag{10.68}$$

by (10.48) and (10.55). Then equalities (10.63)–(10.64) follow from (10.66)–(10.68) as

$$0 = 2\dot{A}_t(\phi_j, \phi_{j'}) - \frac{2}{h_\ell}(\lambda_j(t+h_\ell) - \lambda_j(t))(\delta_{jj'} + o(1))$$
$$-2\lambda_j(t)\dot{B}_t(\phi_j, \phi_{j'}) + o(1), \quad \ell \to \infty.$$

$\square$

.

Below we confirm that the process of taking subsequence in the previous lemma is not necessary, if $h_\ell \to 0$ is unilateral as in $h_\ell \to +0$ or $h_\ell \to -0$. Theorem 10.1 is thus reduced to the following theorem.

Theorem 10.9. *Under the assumption of Theorem 10.8, the unilateral limits*

$$\dot{\lambda}_j^\pm(t) = \lim_{h \to \pm 0} \frac{1}{h}\{\lambda_j(t+h) - \lambda_j(t)\} \tag{10.69}$$

exist, and it holds that

$$\dot{\lambda}_j^+(t) = \mu_{j-k+1}, \quad \dot{\lambda}_j^-(t) = \mu_{k+m-j}, \quad k \leq j \leq k+m-1. \tag{10.70}$$

Here, μ_q, $1 \leq q \leq m$, is the q-th eigenvalue of

$$u \in Y_t^\lambda, \quad E_t^\lambda(u,v) = \mu B_t(u,v), \quad \forall v \in Y_t^\lambda, \tag{10.71}$$

where Y_t^λ is the m-dimensional eigenspace of (10.11) in §10.1 corresponding to the eigenvalue λ of (10.21) in §10.3 defined by (10.47) in §10.5, and

$$E_t^\lambda = \dot{A}_t - \lambda\dot{B}_t. \tag{10.72}$$

In particular, it holds that

$$\dot{\lambda}_j^+(t) = \dot{\lambda}_{2k+m-1-j}^-(t), \quad k \leq j \leq k+m-1.$$

Proof. Since Y_t^λ is m-dimensional, the eigenvalue problem (10.71) admits m-eigenvalues denoted by

$$\mu_1 \leq \cdots \leq \mu_m.$$

By Lemma 10.3, on the other hand, any $h_\ell \to 0$ takes a subsequence, denoted by the same symbol, satisfying (10.63)–(10.64) for some

$$\phi_j \in Y_t^\lambda, \quad k \leq j \leq k + m - 1,$$

with (10.49) in §10.5.

This lemma ensures also the existence of

$$\tilde{\mu}_j = \lim_{\ell \to \infty} \frac{1}{h_\ell}(\lambda_j(t + h_\ell) - \lambda_j(t)), \tag{10.73}$$

and the equalities

$$E_t^\lambda(\phi_j, \phi_{j'}) = \delta_{jj'}\tilde{\mu}_j, \quad k \leq j, j' \leq k + m - 1. \tag{10.74}$$

We thus obtain

$$\phi_j \in Y_\lambda, \quad B_t(\phi_j, \phi_j) = 1$$
$$E_t^\lambda(\phi_j, v) = \tilde{\mu}_j B_t(\phi_j, v), \ \forall v \in Y_\lambda,$$

and therefore, $\mu = \tilde{\mu}_j$ is an eigenvalue of (10.71).

If $h_\ell \to +0$, there arises that

$$\tilde{\mu}_k \leq \cdots \leq \tilde{\mu}_{k+m-1}$$

by

$$\lambda_k(t + h) \leq \cdots \leq \lambda_{k+m-1}(t + h),$$

and hence

$$\tilde{\mu}_j = \mu_{j-k+1}, \quad k \leq j \leq k + m - 1.$$

Then we obtain the result because the value $\tilde{\mu}_j$ in (10.73) is independent of the sequence $h_\ell \to +0$, which is given arbitrarily.

In the other case of $h_\ell \to -0$, we obtain

$$\tilde{\mu}_j = \mu_{k+m-j}, \quad k \leq j \leq k + m - 1,$$

and the result follows similarly. $\qquad\square$

Remark 10.11. If the assumption of Theorem 10.8 holds for any $t \in I$, there are

$$\dot{\lambda}_j^{\pm}(t) = \lim_{h \to \pm 0} \frac{1}{h}(\lambda_j(t; h) - \lambda_j(t))$$

for any t and j. Henceforth, we assume this condition.

Theorem 10.2 in §10.3 is reduced to the following abstract theorem.

Theorem 10.10. *Let the assumption of Theorem 10.8 hold in I. Fix $t \in I$, and assume*

$$\lim_{h \to 0} \sup_{\|u\|_V, \|v\|_V \leq 1} \left| \dot{A}_{t+h}(u, v) - \dot{A}_t(u, v) \right| = 0$$

$$\lim_{h \to 0} \sup_{|u|_X, |v|_X \leq 1} \left| \dot{B}_{t+h}(u, v) - \dot{B}_t(u, v) \right| = 0. \tag{10.75}$$

Then, it follows that

$$\lim_{h \to \pm 0} \dot{\lambda}_j^{\pm}(t + h) = \dot{\lambda}_j^{\pm}(t).$$

Proof. Assume (10.20) in §10.3 and take $k \leq j \leq k + m - 1$. Since the assumption of Theorem 10.8 holds in I, any $t' \in I$ admits $u_j(t') \in V$ such that

$$B_{t'}(u_j(t'), u_j(t')) = 1$$
$$A_{t'}(u_j(t'), u_j(t')) = \lambda_j(t') B_{t'}(u_j(t'), u_j(t')) \tag{10.76}$$

and

$$\dot{\lambda}_j^+(t') = \dot{A}_{t'}(u_j(t'), u_j(t')) - \lambda_j(t') \dot{B}_{t'}(u_j(t'), u_j(t')). \tag{10.77}$$

by Lemma 10.3 and Theorem 10.9.

Given $t \in I$ with (10.75), we take $h_\ell \to +0$ and $u_j(t')$ in (10.76) for $t' = t + h_\ell$. Hence there is a subsequence denoted by the same symbol such that (10.48) with $\phi_j \in V$. Then it holds that

$$\dot{\lambda}_j^+(t) = \dot{A}_t(\phi_j, \phi_j) - \lambda_j(t) \dot{B}_t(\phi_j, \phi_j).$$

We thus obtain

$$\lim_{\ell \to \infty} \dot{\lambda}_j^+(t + h_\ell) = \lim_{\ell \to \infty} \{ \dot{A}_{t+h_\ell}(u_j(t + h_\ell), u_j(t + h_\ell))$$
$$- \lambda_j(t + h_\ell) \dot{B}_{t+h_\ell}(u_j(t + h_\ell), u_j(t + h_\ell)) \}$$
$$= \dot{A}_t(\phi_j, \phi_j) - \lambda_j(t) \dot{B}_t(\phi_j, \phi_j) = \dot{\lambda}_j^+(t)$$

by (10.77), and hence

$$\lim_{h \to +0} \dot{\lambda}_j^+(t + h) = \dot{\lambda}_j^+(t)$$

because $h_\ell \to +0$ is arbitrary.

The proof of

$$\lim_{h \to -0} \dot{\lambda}_j^-(t + h) = \dot{\lambda}_j^-(t)$$

is similar. $\square$

Remark 10.12. The limits (10.69) are unilaterally locally uniform in $t \in I$. In fact, if there are $t_k \downarrow t_0 \in I$, $\delta > 0$, and $h_\ell \downarrow 0$ such that

$$\left| \frac{1}{h_\ell} (\lambda_j(t_k + h_\ell) - \lambda_j(t_k)) - \dot{\lambda}_j^+(t_k) \right| \geq \delta,$$

for example, we obtain

$$\left| \frac{1}{h_\ell} (\lambda_j(t_0 + h_\ell) - \lambda_j(t_0)) - \dot{\lambda}_j^+(t_0) \right| \geq \delta,$$

with $k \to \infty$, and then a contradiction with $\ell \to \infty$.

10.7   Rearrangement of Eigenvalues

For simplicity we introduce the following notations to prove Theorem 10.3 in §10.3. Continue to put $I = (-\varepsilon_0, \varepsilon_0)$, and let $f_j \in C^0(I)$, $1 \leq j \leq m$, satisfy

$$f_1(t) \leq \cdots \leq f_m(t), \quad t \in I. \tag{10.78}$$

Assume the existence of the unilateral limits

$$\dot{f}_j^\pm(t) = \lim_{h \to \pm 0} \frac{1}{h} (f_j(t + h) - f_j(t)) \tag{10.79}$$

with their unilateral continuity

$$\lim_{h \to \pm 0} \dot{f}_j(t + h) = \dot{f}_j^\pm(t) \tag{10.80}$$

for any j and t. Assume, finally,

$$\dot{f}_j^+(t) = \dot{f}_{2k+n-j-1}^-(t), \quad k \leq j \leq k + n - 1, \tag{10.81}$$

provided that

$$f_{k-1}(t) < f_k(t) = \cdots = f_{k+n-1}(t) < f_{k+n}(t), \tag{10.82}$$

where $1 \leq n \leq m$, $1 \leq k \leq m - n + 1$, and $t \in I$. In (10.82) we understand

$$f_0(t) = -\infty, \quad f_{m+1}(t) = +\infty.$$

We call

$$K = K_{k,n}(t) = \{k, \cdots, k + n - 1\}$$

an *n-cluster* at t with entry k if (10.82) arises, and also

$$p(K) = \max\{j \mid k \leq j \leq k + [\tfrac{n}{2}] - 1, \ \dot{f}_j^+(t) < \dot{f}_{2k+n-1-j}^+(t)\} - k + 1$$

its *p-value*. Here, we understand $p(K) = 0$ if

$$\dot{f}_k^+(t) = \dot{f}_{k+n-1}^+(t),$$

noting

$$k \le i \le j \le k+n-1 \quad \Rightarrow \quad \dot{f}_i^+(t) \le \dot{f}_j^+(t).$$

We construct a rearrangement of C^0-curves

$$C_j = \{f_j(t) \mid t \in I\}, \ 1 \le j \le m,$$

denoted by $\tilde{C}_j$, $1 \le j \le m$, so that are C^1 in $t \in I$. This rearrangement is done only on

$$I_1 = \{t \in I \mid \text{there exists a cluster } K \text{ at } t \text{ such that } p(K) \ge 1\}.$$

To introduce this rearrangement, we note the following facts in advance. First, given $2 \le n \le m$, let

$$I_1^n = \{t \in I \mid \text{there exists an } n\text{-cluster } K \text{ at } t \text{ such that } p(K) \ge 1\}.$$

If $t \in I_1^n$ and $K = K_{k,n}(t)$ satisfies $p(K) \ge 1$, it holds that

$$f_{k+n-1}(t') > f_k(t'), \quad 0 < |t' - t| \ll 1. \tag{10.83}$$

Hence this t is an isolated point of I_1^n. In particular, each I_1^n, $2 \le n \le m$, is at most countable, and hence so is I_1 by

$$I_1 = \bigcup_{n=2}^m I_1^n.$$

Second, given $t \in I_1$ and $1 \le j \le m$, if $j \in K = K_{k,n}(t)$ holds for some K in $p(K) \ge 1$, this K is unique.

Definition 10.3. The curves $\tilde{C}_j$, $1 \le j \le m$, are called the transversal rearrangement of C_j, $1 \le j \le m$, if the following operations are done.

(1) If $t \in I_1$, $1 \le j \le m$, and $j \in K$ hold for $K = K_{k,n}(t)$ with $p = p(K) \ge 1$, the curve C_j for $k \le j \le k+p-1$ and $k-n-p \le j \le k-n-1$ on the right, is connected to $C_{2k+n-j-1}$ on the left at t.
(2) No rearrangements to C_j, $1 \le j \le m$, are made otherwise.

The curves $\tilde{C}_j$, $1 \le j \le m$, are uniquely constructed from C_j, $1 \le j \le m$, by this transversal rearrangement. From the results in the previous section, Theorem 10.3 is a consequence of the following theorem.

Theorem 10.11. *Under the above situation, the C^0 curves $\tilde{C}_j$, $1 \le j \le m$, made by the transversal rearrangement of C_j, $1 \le j \le m$, are C^1.*

Proof. This theorem is obvious if $m = 1$. Now we show it by an induction on m, assuming the assertion up to $m - 1$.

Take $t_0 \in I \setminus \overline{I_1}$, and make the transversal rearrangement of C_j, $1 \le j \le q$, toward left and right diections. Let t_ℓ, $\ell = 1, 2, \cdots$, be the successive points of I_1 in the left direction:

$$t_\ell \in I_1, \ t_{\ell-1} > t_\ell, \ (t_\ell, t_{\ell-1}) \cap I_1 = \emptyset, \quad \ell = 1, 2, \cdots.$$

If $\{t_\ell\}$ is finite, these C_j's are successfully rearranged to C^1 curves on $(-\varepsilon_0, t_0)$. If not, there is

$$t_* = \lim_{\ell \to \infty} t_\ell \in [-\varepsilon_0, t_0).$$

Then the case $t_* = -\varepsilon_0$ ensures the same conclusion.

Letting $t_* > -\varepsilon_0$, we show that $\{\tilde{C}_j \mid 1 \le j \le m\}$ are C^1 curves on $(t_* - \delta, t_* + \delta)$ for $0 < \delta \ll 1$. Once this fact is proven, we can repeat this process up to $t = -\varepsilon_0$. Turning to the right direction then, we conclude that $\tilde{C}_j$, $1 \le j \le m$, are C^1 curves on $I = (-\varepsilon_0, \varepsilon_0)$ and the proof is complete.

To establish this propery we distinguish the cases $t_* \in I_1$ and $t_* \notin I_1$. If $t_* \in I_1$, first, the above assertion follows from the assumption of induction. In fact, each $K = K_{k,n}(t_*)$ with $p(K) \ge 1$ admits (10.83), while

$$\tilde{f}_j(t) = \begin{cases} f_j(t), & 0 < t - t_* \ll 1 \\ f_{2k+n-j-1}(t), & 0 < t_* - t \ll 1 \end{cases}, \quad j = k, \ j = k + n - 1$$

are C^1 around $t = t_*$. Then we apply the assumption of induction to C_j, $k + 1 \le j \le k + n - 2$, to get n-C^1 curves in $(t_* - \delta, t_* + \delta)$ made by C_j, $k \le j \le k + n - 1$. Operating this process to any $K = K_{k,n}(t_*)$ with $p(K) \ge 1$ at $t = t_*$, we get C^1 curves $\tilde{C}_j$, $1 \le j \le m$, in $(t_* - \delta, t_* + \delta)$ by this transversal rearrangement of C_j, $1 \le j \le m$, at $t = t_*$.

If $t_* \notin I_1$, we take the cluster decomposition of $\{1, \cdots, m\}$ at $t = t_*$, that is,

$$1 = k_1 < k_1 + n_1 = k_2 < \cdots < k_{s-1} + n_{s-1} = k_s < k_s + n_s = m$$

satisfying

$$\bigcup_{r=1}^{s} K_{k_r, n_r}(t_*) = \{1, \cdots, m\}.$$

Since

$$p(K_{k_r, n_r}(t_*)) = 0, \quad 1 \le r \le s$$

holds by the assumption, there are a_r, $1 \le r \le s$, such that

$$\dot{f}_j^+(t_*) = a_r, \quad \forall j \in K_{k_r, n_r}.$$

We obtain, on the other hand,

$$f_{k_{r-1}}(t) < f_{k_r}(t), \quad |t - t_*| \ll 1, \ 1 \le r \le s+1$$

under the agreement

$$f_{k_0}(t) = -\infty, \quad f_{k_{s+1}}(t) = +\infty,$$

and hence $\tilde{C}_j$, $1 \le j \le m$, are C^1 on $[t_*, t_* + \delta)$ for $0 < \delta \ll 1$. If t_* is not a right accumulating point of I_1, which means the existence of $0 < \delta \ll 1$ such that

$$I_1 \cap (t_* - \delta, t_*] = \{t_*\},$$

therefore, these $\tilde{C}_j$, $1 \le j \le m$, made by the transversal rearrangement of C_j, $1 \le j \le m$, are C^1 on $(t_* - \delta, t_* + \delta)$.

In the other case that t_* is a right accumulating point of I_1, these $\tilde{C}_j$, $1 \le j \le m$, are C^1 on $(t_* - \delta, t_*]$, similarly. Then it holds that

$$\dot{f}_j^-(t_*) = a_r, \quad \forall j \in K_{k_r, n_r} = K_{k_r, n_r}(t_*)$$

by $p(K_{k_r, n_r}) = 0$, and hence $\tilde{C}_j$, $1 \le j \le m$, are C^1 on $(t_* - \delta, t_* + \delta)$. $\square$

10.8 Second Derivatives

If $T_t : \Omega \to \Omega_t$ is 2-differentiable, we have the other bilinear forms $\ddot{A}_t : V \times V \to \mathbf{R}$ and $\ddot{B}_t : X \times X \to \mathbf{R}$ satisfying

$$|\ddot{A}_t(u, v)| \le C\|u\|_V\|v\|_v, \quad u, v \in V$$
$$|\ddot{B}_t(u, v)| \le C|u|_X|v|_X, \quad u, v \in X \tag{10.84}$$

uniformly in t and

$$\lim_{h \to 0} \frac{1}{h^2} \sup_{\|u\|_V, \|v\|_V \le 1} \left|\left(A_{t+h} - A_t - h\dot{A}_t - \frac{h^2}{2}\ddot{A}_t\right)(u, v)\right| = 0$$

$$\lim_{h \to 0} \frac{1}{h^2} \sup_{|u|_X, |v|_X \le 1} \left|\left(B_{t+h} - B_t - h\dot{B}_t - \frac{h^2}{2}\ddot{B}_t\right)(u, v)\right| = 0 \tag{10.85}$$

for each t. Hence Theorem 10.4 is reduced to the following abstract theorem.

Theorem 10.12. *In Theorem 10.8 in §10.6, assume, furthermore, (10.84)–(10.85). Then the conclusion of Theorem 10.4 in §10.3 holds.*

Remark 10.13. Differently from Theorem 9.1 in §9.1, the assumption and the conclusion of this theorem require the Taylor expansions of the second order at fixed $t \in I$. Later in Theorem 10.15 below, we discuss the differentiability of $\tilde{\lambda}'_j(t)$ defined by Theorem 10.11 in §10.8

Remark 10.14. Here we develop a formal argument as in Remark 10.9 in §10.6 concerning the first derivative. Assuming (10.62), first, we deduce

$$\ddot{\lambda}_t = \ddot{A}_t(u_t, u_t) + 2\dot{A}_t(\dot{u}_t, u_t) - \dot{\lambda}_t \dot{B}_t(u_t, u_t) - \lambda_t \ddot{B}_t(u_t, u_t) - 2\lambda_t \dot{B}_t(\dot{u}_t, u_t)$$
$$= 2(\dot{A}_t - \lambda_t \dot{B}_t)(\dot{u}_t, u_t) + D_t(u_t, u_t) \qquad (10.86)$$

for

$$D_t(u, v) = \ddot{A}_t(u, v) - \dot{\lambda}_t \dot{B}_t(u, v) - \lambda_t \ddot{B}_t(u, v), \quad u, v \in V. \qquad (10.87)$$

Putting $v = \dot{u}_t$ in (10.57) in §10.6, second, we reach

$$(\dot{A}_t - \lambda_t \dot{B}_t)(u_t, \dot{u}_t) = -(A_t - \lambda_t B_t)(\dot{u}_t, \dot{u}_t) + \dot{\lambda}_t B_t(u_t, \dot{u}_t). \qquad (10.88)$$

Then, (10.59), (10.86), and (10.88) imply

$$\ddot{\lambda}_t = -2(A_t - \lambda_t B_t)(z_*, z_*) + 2\dot{\lambda}_t B_t(u_t, z_*) + D_t(u_t, u_t)$$
$$= -2(A_t - \lambda_t B_t)(z_*, z_*) - \dot{\lambda}_t \dot{B}_t(u_t, u_t) + D_t(u_t, u_t)$$
$$= -2(A_t - \lambda_t B_t)(z_*, z_*) + \ddot{A}_t(u_t, u_t) - 2\dot{\lambda}_t \dot{B}_t(u_t, u_t)$$
$$\quad - \lambda_t \ddot{B}_t(u_t, u_t) \qquad (10.89)$$

for $z_* = \dot{u}_t$.

Remark 10.15. We observe that $\dot{u}_t \in V$ is not uniquely determined by (10.57) in §10.6, which is derived formally. It has, more precisely, the ambiguity of addition of an element in Y_t^λ. This ambiguity, however, cancels in (10.89) by equality (10.90) below.

We now develop rigorous argument valid even to (10.17) in §10.2. Define λ by (10.21) in §10.3 and recall $Y_t^\lambda = \langle u_j(t) \mid k \leq j \leq k + m - 1 \rangle$ for $u_j(t)$ satisfying (10.46) in §10.5. Let, also,

$$C_{t'}^j = A_{t'} - \lambda_j(t')B_{t'}, \quad k \leq j \leq k + m - 1, \ t' \in I.$$

Then we obtain

$$C_t^j = A_t - \lambda B_t \equiv C_t$$

and

$$C_t(u, v) = 0, \quad \forall(u, v) \in Y_t^\lambda \times V. \qquad (10.90)$$

By Lemma 10.3 in §10.6, given $h_\ell \to 0$, which may not be of definite sign, we have a subsequence, denoted by the same symbol, satisfying (10.48) in §10.5,

$$\mathrm{s} - \lim_{\ell \to \infty} u_j(t + h_\ell) = \phi_j \in Y_t^\lambda \text{ in } V, \quad k \le j \le k + m - 1.$$

There exists

$$\dot{\lambda}_j^* = \lim_{\ell \to \infty} \frac{1}{h_\ell}(\lambda_j(t + h_\ell) - \lambda_j(t)), \quad k \le j \le k + m - 1 \tag{10.91}$$

passing to a subsequence, with the equality

$$\dot{\lambda}_j^* \delta_{jj'} = \dot{A}_t(\phi_j, \phi_{j'}) - \lambda \dot{B}_t(\phi_j, \phi_{j'}), \quad k \le j, j' \le k + m - 1.$$

Remark 10.16. If we take the subsequence of $h_\ell \to 0$ to be unilateral, the limit $\dot{\lambda}_j^*$ in (10.91) exists and is either $\dot{\lambda}_j^+(t)$ or $\dot{\lambda}_j^-(t)$.

Let

$$\dot{C}_t^{*j} = \dot{A}_t - \dot{\lambda}_j^* B_t - \lambda \dot{B}_t.$$

It holds that

$$\lim_{\ell \to \infty} \sup_{\|u\|_V \le 1, \|v\|_V \le 1} \left| \frac{1}{h_\ell}(C_{t+h_\ell}^j - C_t)(u, v) - \dot{C}_t^{*j}(u, v) \right| = 0$$

for $k \le j \le k + m - 1$ by (10.55) in §10.6, and also

$$\dot{C}_t^{*j}(u, v) = 0, \quad \forall u, v \in Y_\lambda, \ k \le j \le k + m - 1 \tag{10.92}$$

by Lemma 10.3. Put

$$z_\ell^j = \frac{1}{h_\ell}(u_j(t + h_\ell) - \phi_j).$$

Lemma 10.4. *It holds that*

$$\lim_{\ell \to \infty} C_t(z_\ell^j, v) = -\dot{C}_t^{*j}(\phi_j, v), \quad \forall v \in V. \tag{10.93}$$

Proof. Given $v \in V$, we obtain

$$C_t(z_\ell^j, v) = \frac{1}{h_\ell}C_t(u_j(t + h_\ell) - \phi_j, v) = \frac{1}{h_\ell}C_t(u_j(t + h_\ell), v)$$

$$= -\frac{1}{h_\ell}(C_{t+h_\ell}^j - C_t)(u_j(t + h_\ell), v) = -\dot{C}_t^{*j}(u_j(t + h_\ell), v) + o(1)$$

by (10.51) in §10.5 and (10.90). Then (10.93) follows from (10.48) in §10.5. $\square$

Recall that

$$R : X \to Y_t^\lambda = \langle u_j(t) \mid k \le j \le k + m - 1 \rangle$$

is the orthogonal projection with respect to $B_t(\cdot, \cdot)$, and $P = I - R$. Hence there is a unique $z_*^j \in PV$ satisfying

$$C_t(z_*^j, v) = -\dot{C}_t^{*j}(\phi_j, v), \quad \forall v \in PV. \tag{10.94}$$

Remark 10.17. Equality (10.94) ensures

$$\gamma_{\lambda_j^*}(\phi_j) = z_*^j, \quad k \le j \le k + m - 1$$

under the notation of Definition 10.2 in §10.3.

Lemma 10.5. *It holds that*

$$w - \lim_{\ell \to \infty} P z_\ell^j = z_*^j \quad in \ V, \quad k \le j \le k + m - 1. \tag{10.95}$$

Proof. Lemma 10.4 ensures that $\{P z_\ell^j\}$ converges weakly in PV, and hence is bounded there:

$$\|P z_\ell^j\|_V \le C.$$

Then, passing to a subsequence denoted by the same symbol, there is $\tilde{z}_j \in PV$ such that

$$w - \lim_{\ell \to \infty} P z_\ell^j = \tilde{z}_j \quad in \ V,$$

which satisfies

$$C_t(\tilde{z}_j, v) = -\dot{C}_t^{*j}(\phi_j, v), \quad \forall v \in V \tag{10.96}$$

by Lemma 10.4. Since such $\tilde{z}_j \in PV$ is unique, we obtain the result with $\tilde{z}_j = z_*^j$. $\qquad\square$

Remark 10.18. Since (10.96) holds with $\tilde{z}_j = z_*^j$, this $z_j^* \in PV$ defined by (10.94) satisfies

$$C_t(z_*^j, v) = -\dot{C}_t^{*j}(\phi_j, v), \quad \forall v \in V.$$

Remark 10.19. Generally, the inequality

$$\|z_\ell^j\|_V \le C$$

is not expected, which causes the other difficulty to the proof of Theorem 10.12. In fact, if $m = 1$ and $\phi_j = u_j(t)$, for example, this property means

$$|B_t(u_j(t), w_{h_\ell}^j)| \le C, \quad w_h^j = \frac{1}{h}(u_j(t + h) - u_j(t)).$$

In the formal argument, we have actually (10.59) in §10.6, which, however, does not assure the actual convergence

$$\lim_{h\to 0} B_t(u_j(t), w_h^j) = -\frac{1}{2}\dot{B}_t(u_j(t), u_j(t)).$$ (10.97)

In fact, the equality

$$1 = B_{t+h}(u_j(t+h), u_j(t+h)) = B_t(u_j(t), u_j(t))$$

just implies

$$0 = \frac{1}{h}\{B_{t+h}(u_j(t+h), u_j(t+h)) - B_t(u_j(t), u_j(t))\}$$

$$= \frac{1}{h}(B_{t+h} - B_t)(u_j(t+h), u_j(t+h))$$

$$\quad + \frac{1}{h}\{B_t(u_j(t+h), u_j(t+h)) - B_t(u_j(t), u_j(t))\}$$

$$= \dot{B}_t(u_j(t), u_j(t)) + o(1) + \frac{1}{h}B_t(u_j(t+h) + u_j(t), u_j(t+h) - u_j(t))$$

$$= \dot{B}_t(u_j(t), u_j(t)) + 2B_t\left(\frac{u_j(t+h) + u_j(t)}{2}, w_h^j\right) + o(1)$$

and hence

$$\lim_{h\to 0} B_t\left(\frac{u_j(t+h) + u_j(t)}{2}, w_h^j\right) = -\frac{1}{2}\dot{B}_t(u_j(t), u_j(t))$$ (10.98)

differently from (10.97). Here, the condition $\|w_h^j\|_X = O(1)$ is necessary to conclude

$$B_t(u_j(t+h), w_h^j) = B_t(u_j(t), w_h^j) + o(1)$$

in the left-hand side of (10.98) from $\phi_j = u_j(t)$ in (10.48). Our purpose, however, was to assure

$$\|w_h^j\|_V = O(1),$$

which is reduced to $\|w_h^j\|_X = O(1)$ by $\|P_h w_h^j\|_V = O(1)$. This is a circular reasoning.

Lemma 10.6. *It holds that*

$$s - \lim_{\ell\to\infty} Pz_\ell^j = z_*^j \quad in \ V, \qquad k \le j \le k+m-1.$$

Proof. Since $V \hookrightarrow X$ is compact, tehre arises that

$$s - \lim_{\ell\to\infty} Pz_\ell^j = z_*^j \quad in \ X$$ (10.99)

in the previous lemma. Then we obtain

$$
\begin{aligned}
\delta\|Pz_\ell^j - z_*^j\|_V^2 &\le A_t(Pz_\ell^j - z_*^j, Pz_\ell^j - z_*^j) \\
&= C_t(Pz_\ell^j - z_*^j, Pz_\ell^j - z_*^j) + o(1) \\
&= C_t(Pz_\ell^j, Pz_\ell^j - z_*^j) + o(1) = C_t(z_\ell^j, Pz_\ell^j - z_*^j) + o(1)
\end{aligned}
$$

by (10.95) and (10.99).

Since $\phi_j \in Y_t^\lambda$ it holds that

$$
\begin{aligned}
C_t(z_\ell^j, Pz_\ell^j - z_*^j) &= \frac{1}{h_\ell} C_t(u_j(t + h_\ell), Pz_\ell^j - z_*^j) \\
&= \frac{1}{h_\ell}(C_t - C_{t+h_\ell}^j)(u_j(t + h_\ell), Pz_\ell^j - z_*^j) \\
&= -\dot{C}_t^{*j}(u_j(t + h_\ell), Pz_\ell^j - z_*^j) + o(1) \\
&= -\dot{C}_t^{*j}(\phi_j, Pz_\ell^j - z_*^j) + o(1) = o(1)
\end{aligned}
$$

by (10.48) in §10.5 and (10.95), because

$$
v \in V \ \mapsto\ \dot{C}_t^{*j}(\phi_j, v) \in \mathbf{R}
$$

is a bounded linear mapping. Then the result follows as

$$
\lim_{\ell \to \infty} \|Pz_\ell^j - z_*^j\|_V = 0.
$$

$\square$

Lemma 10.7. *There exists*

$$
\ddot{\lambda}_j^* \equiv \lim_{\ell \to \infty} \frac{2}{h_\ell^2}(\lambda_j(t + h_\ell) - \lambda_j(t) - h_\ell \dot{\lambda}_j^*), \quad k \le j \le k + 1 - m \quad (10.100)
$$

with

$$
\ddot{\lambda}_j^* = (\ddot{A}_t - \lambda\ddot{B}_t - 2\dot{\lambda}_j^* \dot{B}_t)(\phi_j, \phi_j) - 2C_t(z_*^j, z_*^j).
$$

Proof. By (10.92) and Lemma 10.6, we have

$$
\begin{aligned}
C_t(z_\ell^j, z_\ell^j) &= C_t(Pz_\ell^j, Pz_\ell^j) = C_t(z_*^j, z_*^j) + o(1) \\
&= C_t(Pz_\ell^j, z_*^j) + o(1) = C_t(z_\ell^j, z_*^j) + o(1). \quad (10.101)
\end{aligned}
$$

It holds that

$$
\begin{aligned}
C_t(z_\ell^j, z_*^j) &= \frac{1}{h_\ell} C_t(u_j(t + h_\ell) - \phi_j, z_*^j) = \frac{1}{h_\ell} C_t(u_j(t + h_\ell), z_*^j) \\
&= \frac{1}{h_\ell}(C_t - C_{t+h_\ell}^j)(u_j(t + h_\ell), z_*^j) = -\dot{C}_t^{*j}(u_j(t + h_\ell), z_*^j) + o(1) \\
&= -\dot{C}_t^{*j}(\phi_j, z_*^j) + o(1)
\end{aligned}
$$

by (10.90) and $\phi_j \in Y_t^\lambda$, which implies

$$C_t(z_*^j, z_*^j) = -\dot{C}_t^{*j}(\phi_j, z_*^j) + o(1) \qquad (10.102)$$

by (10.101). It holds also that

$$\dot{C}_t^{*j}(\phi_j, z_*^j) = \frac{1}{h_\ell}\dot{C}_t^{*j}(\phi_j, P(u_j(t + h_\ell) - \phi_j)) + o(1)$$
$$= \frac{1}{h_\ell}\dot{C}_t^{*j}(\phi_j, u_j(t + h_\ell) - \phi_j) + o(1)$$
$$= \frac{1}{h_\ell}\dot{C}_t^{*j}(\phi_j, u_j(t + h_\ell)) + o(1) \qquad (10.103)$$

by (10.92) and $\phi_j \in Y_t^\lambda$.

Here, we use the asymptotics

$$C_{t+h_\ell}^j(\phi_j, u_j(t + h_\ell)) = C_t^j(\phi_j, u_j(t + h_\ell)) + h_\ell \cdot \dot{C}_t^{*j}(\phi_j, u_j(t + h_\ell))$$
$$+ \frac{1}{2}h_\ell^2 \cdot \ddot{C}_{t,\ell}^{*j}(\phi_j, u_j(t + h_\ell)) + o(h_\ell^2)$$

for

$$\ddot{C}_{t,\ell}^{*j} = \ddot{A}_t - \lambda\ddot{B}_t - 2\dot{\lambda}_j^*\dot{B}_t - \frac{2}{h_\ell^2}(\lambda_j(t + h_\ell) - \lambda_j(t) - h\dot{\lambda}_j^*)B_t, \qquad (10.104)$$

derived from (10.55) in §10.6 and (10.85). Since

$$C_{t+h_\ell}^j(\phi_j, u_j(t + h_\ell)) = C_t^j(\phi_j, u_j(t + h_\ell)) = 0$$

holds by (10.90), we obtain

$$C_t^j(z_*^j, z_*^j) = -\dot{C}_t^{*j}(\phi_j, z_*^j) + o(1) = -\frac{1}{h_\ell}\dot{C}_\ell^{*j}(\phi_j, u_j(t + h_\ell)) + o(1)$$
$$= \frac{1}{2}\ddot{C}_{t,\ell}^{*j}(\phi_j, u_j(t + h_\ell)) + o(1) = \frac{1}{2}\ddot{C}_{t,\ell}^{*j}(\phi_j, \phi_j) + o(1)$$
$$= \frac{1}{2}(\ddot{A}_t - \lambda\ddot{B}_t - 2\dot{\lambda}_j^*\dot{B}_t)(\phi_j, \phi_j) - \frac{1}{h_\ell^2}(\lambda_j(t + h_\ell) - \lambda_j(t) - h_\ell\dot{\lambda}_j^*) + o(1)$$

by (10.102)–(10.103) and $B_t(\phi_j, \phi_j) = 1$. Then it follows that

$$\lim_{\ell\to\infty} \frac{2}{h_\ell^2}(\lambda_j(t + h_\ell) - \lambda_j(t) - h_\ell\dot{\lambda}_j^*)$$
$$= (\ddot{A}_t - \lambda\ddot{B}_t - 2\dot{\lambda}_j^*\dot{B}_t)(\phi_j, \phi_j) - 2C_t(z_*^j, z_*^j),$$

and the proof is complete. $\qquad \square$

Lemma 10.8. *If* $\dot{\lambda}_* \equiv \dot{\lambda}_j^* = \dot{\lambda}_{j'}^*$ *arises for some* $k \le j \ne j' \le k + m - 1$, *then it holds that*

$$(\ddot{A}_t - \lambda\ddot{B}_t - 2\dot{\lambda}_*\dot{B}_t)(\phi_j, \phi_{j'}) = 2C_t(z_*^j, z_*^{j'}).$$

Proof. As in the previous lemma we obtain

$$C_t(z_*^j, z_*^{j'}) = C_t(Pz_\ell^j, z_*^{j'}) + o(1) = C_t(z_\ell^j, z_*^{j'}) + o(1)$$

$$= \frac{1}{h_\ell} C_t(u_j(t+h_\ell) - \phi_j, z_*^{j'}) + o(1) = \frac{1}{h_\ell} C_t(u_j(t+h_\ell), z_*^{j'}) + o(1)$$

$$= \frac{1}{h_\ell}(C_t - C_{t+h_\ell})(u_j(t+h), z_*^{j'}) + o(1)$$

$$= -\dot{C}_t^{*j}(u_j(t+h_\ell), z_*^{j'}) + o(1) = -\dot{C}_{t*}^j(\phi_j, z_*^{j'}) + o(1)$$

$$= -\frac{1}{h_\ell}\dot{C}_t^{*j}(\phi_j, P(u_{j'}(t+h_\ell) - \phi_{j'})) + o(1)$$

$$= -\frac{1}{h_\ell}\dot{C}_t^{*j}(\phi_j, u_{j'}(t+h_\ell) - \phi_{j'}) + o(1)$$

$$= -\frac{1}{h_\ell}\dot{C}_t^{*j}(\phi_j, u_{j'}(t+h_\ell)) + o(1)$$

by $\phi_j, \phi_{j'} \in Y_t^\lambda$. Then it holds that

$$C_{t+h_\ell}^j(\phi_j, u_{j'}(t+h_\ell)) = C_t(\phi_j, u_{j'}(t+h_\ell)) + h_\ell \cdot \dot{C}_t^{*j}(\phi_j, u_{j'}(t+h_\ell))$$

$$+\frac{1}{2}h_\ell^2 \cdot \ddot{C}_{t,\ell}^{*j}(\phi_j, u_{j'}(t+h_\ell)) + o(h_\ell^2)$$

with

$$C_t(\phi_j, u_{j'}(t+h_\ell)) = 0$$

and

$$C_{t+h_\ell}^j(\phi_j, u_{j'}(t+h_\ell)) = (C_{t+h_\ell}^j - C_{t+h_\ell}^{j'})(\phi_j, u_{j'}(t+h_\ell))$$

$$= (\lambda_{j'}(t+h_\ell) - \lambda_j(t+h_\ell))B_{t+h_\ell}(\phi_j, u_{j'}(t+h_\ell))$$

by (10.90), to conclude

$$C_t(z_*^j, z_*^{j'}) = \frac{1}{h_\ell^2}(\lambda_j(t+h_\ell) - \lambda_{j'}(t+h_\ell))B_{t+h_\ell}(\phi_j, u_{j'}(t+h_\ell))$$

$$+\frac{1}{2}\ddot{C}_{t,\ell}^{*j}(\phi_j, u_{j'}(t+h_\ell)) + o(1). \tag{10.105}$$

Then we use

$$\lambda_j(t+h_\ell) = \lambda + h_\ell\dot{\lambda}_j^* + \frac{h_\ell^2}{2}\ddot{\lambda}_j^* + o(h_\ell^2)$$

$$\lambda_{j'}(t+h_\ell) = \lambda + h_\ell\dot{\lambda}_{j'}^* + \frac{h_\ell^2}{2}\ddot{\lambda}_{j'}^* + o(h_\ell^2)$$

with $\dot{\lambda}_j^* = \dot{\lambda}_{j'}^*$, to deduce

$$\lim_{\ell\to\infty}\frac{1}{h_\ell^2}(\lambda_j(t+h_\ell) - \lambda_{j'}(t+h_\ell))B_{t+h_\ell}(\phi_j, u_{j'}(t+h_\ell))$$

$$= \frac{1}{2}(\ddot{\lambda}_j^* + \ddot{\lambda}_{j'}^*)B_t(\phi_j, \phi_{j'}) = 0.$$

Then the result follows from (10.104), (10.105), and the previous lemma.   □

Recall $F_t^{\lambda,\lambda'}$ of Definition 10.2 in §10.3.

Lemma 10.9. *Define* $\tilde{\mu}_k \leq \cdots \leq \tilde{\mu}_{k+m-1}$ *by*

$$\{\tilde{\mu}_j \mid k \leq j \leq k+m-1\} = \{\dot{\lambda}_j^* \mid k \leq j \leq k+m-1\},$$

and assume $k \leq \ell < r \leq k+m$ *be such that*

$$\tilde{\mu}_{\ell-1} < \tilde{\mu} \equiv \tilde{\mu}_\ell = \cdots = \tilde{\mu}_{r-1} < \tilde{\mu}_r.$$

under the agreement of

$$\tilde{\mu}_{k-1} = -\infty, \quad \tilde{\mu}_{k+m} = +\infty.$$

Then, $\sigma = \ddot{\lambda}_j^*$, $\ell \leq j \leq r-1$, *is an eigenvalue of*

$$u \in Y_{\lambda,t}^{\ell,r}, \quad F_t^{\lambda,\tilde{\mu}}(u,v) = \sigma B_t(u,v), \ \forall v \in Y_{\lambda,t}^{\ell,r},$$

for

$$Y_{\lambda,t}^{\ell,r} = \langle u_j(t) \mid \ell \leq j \leq r-1 \rangle \subset Y_t^\lambda.$$

Proof. Lemmas 10.7 and 10.8 imply

$$F_t^{\lambda,\tilde{\mu}}(\phi_j,\phi_{j'}) = \delta_{jj'} \ddot{\lambda}_j^*, \quad \ell \leq j, j' \leq r-1,$$

and hence the result follows from

$$Y_{\lambda,t}^{\ell,r} = \langle \phi_j \mid \ell \leq j \leq r-1 \rangle, \quad B_t(\phi_j,\phi_{j'}) = \delta_{jj'}.$$

$\square$

Theorem 10.12 is now reduced to the following theorem.

Theorem 10.13. *Fix* $t \in I$, *and assume (10.20) in §10.3. Put* λ *as in (10.21) and let*

$$k \leq \ell_+ < r_+ \leq k+m$$

be such that

$$\dot{\lambda}_{\ell_+-1}^+(t) < \lambda'_+ \equiv \dot{\lambda}_{\ell_+}^+(t) = \cdots = \dot{\lambda}_{r_+-1}^+(t) < \dot{\lambda}_{r_+}^+(t). \tag{10.106}$$

Then there exists

$$\lambda''_{j+} = \lim_{h\to+0} \frac{2}{h^2}(\lambda_j(t+h) - \lambda - h\lambda'_+), \quad \ell_+ \leq j \leq r_+ - 1. \tag{10.107}$$

It holds, furthermore, that

$$\lambda''_{j+} = \sigma_{j-\ell_++1}^+, \quad \ell_+ \leq j \leq r_+ - 1,$$

where σ_q^+, $1 \leq q \leq r_+ - \ell_+$, is the q-th eigenvalue of

$$u \in Y_\lambda^{\ell,r}, \quad F_t^{\lambda,\lambda'}(u,v) = \sigma B_t(u,v), \quad \forall v \in Y_\lambda^{\ell,r} \tag{10.108}$$

for

$$\lambda' = \lambda_+', \quad \ell = \ell_+, \quad r = r_+.$$

If

$$\dot{\lambda}_{\ell_- - 1}^-(t) > \lambda' \equiv \dot{\lambda}_{\ell_-}^-(t) = \cdots = \dot{\lambda}_{r_- - 1}^-(t) > \dot{\lambda}_{r_-}^-(t),$$

similarly, there arises that

$$\lambda_{j-}'' = \lim_{h \to -0} \frac{2}{h^2}(\lambda_j(t+h) - \lambda - h\lambda'), \quad \ell \leq j \leq r-1$$

with

$$\lambda_{j-}'' = \sigma_{j-r_-}^-,$$

where σ_q^-, $1 \leq q \leq r_- - \ell_-$, is the q-th eigenvalue of (10.108) for

$$\lambda' = \lambda_-', \quad \ell = \ell_-, \quad r = r_-.$$

Proof. In the previous lemma, we obtain

$$\ddot{\lambda}_\ell^* \leq \cdots \leq \ddot{\lambda}_{r-1}^*.$$

Hence the result follows similarly to Lemma 10.9. □

Theorem 10.5 in §10.3 is reduced to the following abstract theorem. The proof is similar to that of Theorem 10.10 in §10.6.

Theorem 10.14. *Let the assumption of Theorem 10.12 hold for any t. Fix $t \in I$, and assume, furthermore,*

$$\lim_{h \to 0} \sup_{\|u\|_V, \|v\|_V \leq 1} \left| \ddot{A}_{t+h}(u,v) - \ddot{A}_t(u,v) \right| = 0$$

$$\lim_{h \to 0} \sup_{|u|_X, |v|_X \leq 1} \left| \ddot{B}_{t+h}(u,v) - \ddot{B}_t(u,v) \right| = 0. \tag{10.109}$$

Then it holds that

$$\lim_{h \to \pm 0} \ddot{\lambda}_j^\pm(t+h) = \ddot{\lambda}_j^\pm(t), \quad \ell_\pm \leq j \leq r_\pm - 1.$$

Remark 10.20. By the above theorems and Remark 10.16, the limits (10.19),

$$\ddot{\lambda}_j^\pm(t) = \lim_{h \to \pm 0} \frac{2}{h^2}(\lambda_j(t+h) - \lambda_j(t) - h\dot{\lambda}_j^\pm(t))$$

exist for any j, provided that the conditions (10.41), (10.42) in §10.4, (10.54), (10.55) in §10.6, (10.84), and (10.85) in this section hold. If these conditions hold for any $t \in I$, these limits are locally unilaterally uniform in $t \in I$, similarly to Remark 10.12 in §10.6 concerning the first derivative.

Finally, Theorem 10.6 in §10.3 is reduced to the following abstract theorem.

Theorem 10.15. *If (10.109) is valid to any t in the previous theorem, C^1 curves $\tilde{C}_j$, $1 \leq j \leq m$, in Theorem 10.3 in §10.3 are C^2.*

Proof. Define $\tilde{\lambda}_j(t)$ by

$$\tilde{C}_j = \{\tilde{\lambda}_j(t) \mid t \in I\}, \ 1 \leq j \leq m.$$

From the proof of Theorem 10.11, Theorem 10.13 and Theorem 10.14 guarantee the existence of

$$\tilde{\lambda}_j''(t) = \lim_{h \to 0} \frac{1}{h^2}(\tilde{\lambda}_j(t+h) - \tilde{\lambda}_j(t) - h\tilde{\lambda}_j'(t)) \tag{10.110}$$

together with its continuity in t,

$$\lim_{h \to \pm 0} \tilde{\lambda}_j''(t+h) = \tilde{\lambda}_j''(t)$$

for any t and j. This convergence (10.110), furthermore, is locally uniform in $t \in I$ by Remark 10.20.

Then it follows that

$$\tilde{\lambda}_j(t+h) = \tilde{\lambda}_j(t) + h\tilde{\lambda}_j'(t) + \frac{h^2}{2}\tilde{\lambda}_j''(t) + o(h^2)$$

$$\tilde{\lambda}_j(t) = \tilde{\lambda}_j(t+h) - h\tilde{\lambda}_j'(t+h) + \frac{h^2}{2}\tilde{\lambda}_j''(t+h) + o(h^2),$$

as $h \to 0$, which implies

$$\lim_{h \to 0} \frac{1}{h}(\tilde{\lambda}_j'(t+h) - \tilde{\lambda}_j'(t)) = \lim_{h \to 0} \frac{1}{2}(\tilde{\lambda}_j''(t+h) + \tilde{\lambda}_j''(t)) = \tilde{\lambda}_j''(t)$$

for any $t \in I$. Hence these $\tilde{C}_j$, $1 \leq j \leq m$, are C^2. $\qquad\square$

Bibliography

Adams, R.A. (1978) *Sobolev Spaces*, Academic Press, Boston.

Alikakos, N.D. (1979), L^p *bounds of solutions of reaction-diffusion equations*, Comm. Partial Differential Equations 4, 827–868.

Azegami, H. (2020), *Shape Optimization Problems*, Springer, Singapore, 2020.

Bartz, J., Struwe, M., and Ye, R. (1994), *A new approach to the Ricci flow on* $\S^2$, Ann. Scuola Norm. Sup. Pisa Cl. Sci. IV, 21 (1994) 475–482.

Biler, P., (1998), *Local and and global stability of some parabolic systems modelling chemotaxis*, Adv. Math. Sci. Appl. 8, 715–743.

Biler, P., Hilhorst, D., and Nadzieja, T. (1994), *Existence and nonexistence of solutions for a model of gravitational interaction of particles. II.*, Colloq. Math. 67, 297–308.

Brenner, S.C. and Scott, L.R. (2007), *The Mathematical Theory of Finite Element Methods*, third edition, Springer-Verlag, New York.

Brezis, H. and Merle, F. (1991) *Uniform estimates and blow-up behavior for solutions of* $-\Delta u = V(x)e^u$ *in two dimensions.* Comm. Partial Differential Equations 16, 1223–1253.

Brezis, H. and Strauss, W. (1973) *Semi-linear second-order elliptic equations in* L^1. J. Math. Soc. Japan 25, 565–590.

Caglioti, E., Lions, P.-L., Marchioro, C., and Privilenti, M. (1992), *A special class of stationary flows for two-dimensional Euler equations: A statistical mechanics description.* Comm. Math. Phys. 143, 501–525.

Caglioti, E., Lions, P.-L., Marchioro, C., and Pulvirenti, M. (1995), *A special class of stationary flows for two-dimensional Euler equations: A statistical mechanics description, Part II*, Comm. Math. Phys. 174, 229–260.

Chan, S.Y.A., Chen, C.C., and Lin, C.S. (2003) *Extremal functions for a mean field equations in two dimensions*, In; Lectures on Partial Differential Equations (S.-Y.A. Chang, C.-S. Lin, and S.-T. Yau, eds.), International Press, New York, pp. 61–93.

Chang, S.-Y.A and Yang, P.-C. (1987), *Prescribing Gaussian curvature on* S^2, Acta Math. 159, 215–259.

Chang, S.-Y.A and Yang, P.-C. (1988), *Conformal deformation of metrics on* S^2, J. Differential Geometry 27, 259–296.

Chanillo. S and Kiessling, M. (1994), *Rotational symmetry of solutions of some nonlinear problems in statistical mechanics and in geometry*, Comm. Math. Phys. 160, 217–238.

Chavanis, P.H. (2004), *Generalized kinetic equations and collapse of self-gravitating Langevin particles in D dimensions*, Banach Center Publ. 66, 79–101.

Chavanis, P.H. (2008), *Two-dimensional Brownian vortices*, Physica A 387, 6917–6942.

Chavanis, P.H. and Sire, C. (2004), *Anormalous diffusion and collapse of self-gravitating Langevin particles in D dimensions*, Phys. Rev. E 69, 016116.

Chen, W. and Li, C. (1991), *Prescribing Gaussian curvatures on surfaces with conical sigularities*, J. Geom. Anal. 1, 615–622.

Cheng, K.S. and Lin, C.S. (1997), *On the asymptotic behavior of solutions to the conformal Gaussian curvature equations in* $\mathbf{R}^2$, Math. Anal. 308, 119–139.

Chill, R (2006), *On the Lojasiewicz-Simon gradient inequaity on Hilbert spaces*, Proceedings of 5th European-Magkrebian Workshop on Semigroup Theory, Evolution Equations and Applications, Jendoubi, M.A. ed., 25–36.

Chow, B. (1991), *The Ricci flow on the 2-sphere*, J. Differential Geometry, 33 (1991), 325–334.

Chow, B. and Knopf, D. (2004), *The Ricci Flow: An Introduction*, American Mathematical Society, Providence RI.

Chow, S.-H. and Hale, J.K. (1982), *Methods of Bifurcation Theory*, Springer-Verlag, New York.

Chueh, K., Conley, C., and Smoller, J. (1977), *Positively invariant regions for systems of nonlinear diffusion equations*, Indiana Univ. Math. J. 26, 373–392.

Crandall, M.G. and T. Liggett, T. (1971), *Generation of semi-groups of nonlinear transformations on general Banach spaces*, Amer. J. Math. 93, 265–293.

Dauge, M. (1988), *Elliptic Boundary Value Problems on Corner Domains*, L.N. in Math. 1341, Springer-Verlag, Berlin.

de Figueiredo, D.G., Lions, P.L. and Nussbaum, R.D. (1982), *A priori estimates and existence of positive solutions of semilinear elliptic equations*, J. Math. Pure Appl. 61, 41–63.

Eyink, G.L., Spohn H., and Chen, W. (1993), *Negative-temperature states and large-scale, long-lived vortices in two-dimensional turbulence*, J. Stat. Phys. 70, 833–886.

Fellner, K., Latos, E., and Suzuki, T. (2019), *Large-time asymptotics of a public goods game model with diffusion*, Monat. Matem. 190, 101–121.

Fellner, K., Morgan, J., and Tang, B.Q. (2021), *Uniform-in-time bounds for quadratic reaction-diffusion systems with mass dissipation in higher dimensions*, Discrete Conti. Dyn. Syst. S. 12, 635-651.

Fontana, L. (1993), *Sharp borderline Sobolev inequalities on comact Riemannian manifolds*, Comment. Math. Helvetici 68, 415–454.

Friedman, A. (1982), *Variational Principles and Free-Boundary Problems*, Wiley, New York.

Flanders, H. (1963), *Differential Forms with Applications to the Physical Sciences*, Academic Press, New York.

Fujita, H., Saito, N., and Suzuki, T. (2001), *Operator Theory and Numerical Methods*, Elsevier, Amsterdam.

Gajewski, H. and Zacharias, K. (1998), *Global behaviour of a reaction-diffusion system modelling chemotaxis*, Math. Nachr. 195, 77–114.

Garabedian P.R., and Schiffer, M. (1952-53), *Convexity of domain functionals*, J. Ann. Math. 2, 281–368.

Geselowitz, D.G. (1967), *On bioelastic potentials in an inhomogeneous volume conductor*, Biophys. J. 7, 1–11.

Geselowitz, D.G. (1970), *On the magnetic field generated outside on inhomogeneous volume conductor by integral current sources*, IEEE Tras. Magn. MAG-6, 911–925,

Gidas, B., Ni, W.-M., and Nirenberg, L. (1979), *Symmetry and related properties via the maximum principle*, Comm. Math. Phys. 68, 209–598.

Girault. V. and Raviart, P.-A. (1986), *Finite Element Methods for Navier-Stokes Equations*, Springer-Verlag, Berlin.

Grisvard, P. (1985), *Elliptic Problems in Nonsmooth Domains*, Pitman, Boston.

Grisvard, P. and Iooss, G. (1975), *Problémes aux limites unilatéraux dans des domaines non réguliers*, Publ. Séminaries Mathématiques, Univ. Rennes 9, 1–26.

Grossi, M. and Takahashi, F. (2010), *Nonexistence of multi-bubble solutions to some elliptic equations on convex domains*, J. Funct. Anal. 259, 904–917.

Hamilton, R.S. (1982) *Three manifolds with positive Ricci curvature*, J. Differential Geometry 17 (1982) 255–306.

Hamilton, R.S. (1988) *The Ricci flow on surfaces*, Contem. Math. 71, Amer. Math. Soc. Providence, pp. 237–262.

Haraux, A and Jendoubi, M.A. (2001), *Decay estimates to equilibrium for some evolution equations with an analytic nonlinearity*, Asymptotic Analysis 26, 21–36.

Hargus, M. and Iooss, G. (2011), *Local Bifurcations, Center Manifolds, and Normal Forms in Infinite-Dimensional Dynamical Systems*, Springer-Verlag.

Henry, D. (1981), *Geometric Theory of Semilinear Parabolic Equations*, Lecture Notes in Math. 840, Springer-Verlag.

Herrero, M.A. and Velázquez, J.J.L. (1996), *Singularity patterns in a chemotaxis model*, Math. Ann. 306, 583–623.

Holm, D.D., Schmah, T., and Stoica, C. (2009), *Geometry Mechanics and Symmetry*, Oxford University Press, New York.

Hong, C.W. (1987), *A best constant and the Gaussian curvature*, Proc. Amer. Math. Soc. 97, 737–747.

Huang, S.Z. (2006), *Gradient Inequalities*, Amer. Math. Soc., Providence.

Huston, V. and Schmitt, K. (1992), *Permanence and the dynamics of biological systems*, Math. Biosc. 111, 1-71.

Jäger, W. and Luckhaus, S. (1992), *On explosions of solution to a system of partial differential equations modelling chemotaxis*, Trans. Amer. Math. Soc. 329, 819–824.

Jimbo, S. and Ushikoshi, E. *Hadamard variational formula for the multiple eigenvalues of the Stokes operator with the Dirichlet bundary conditions*, Far East J. Math. Sci. 98 (2015) 713-739.

Joyce, G and Montgomery, D. (1973), *Negative temperature states for the two-dimensional guiding-centre plasma*, J. Plasma Phys. 10, 107–121.

Kadlec, J. (1964), *On the regularity of the solution of the Poisson problem on a domain with boundary locally similar to the boundary of a convex open set*, Czechoslovak Math. J. 14, 386–393.

Kamenomostskaja, S.L. (1961), *On the Stefan probem* (in Russian), Mat. Sbornik 53, 488–514.

Kanel, J.I. (1990) *Solvability in the large of a system of reaction-diffusion equations with the balance condition*, Differential Equations 26, 331–339.

Kanou, M., Sato, T., and Watanabe, K. (2013), *Interface retularity of the solutions for the rotation free and divergence free system in Euclidian space*, Tokyo J Math. 36, 473–482.

Kanou, M., Sato, T., and Watanabe, K. (2016), *Interface regularity of the solutions to the systems on Riemannian manifold*, Tokyo J. Math. 39, 83–100.

Karali, G., Suzuki, T., and Yamada, Y. (2013), *Global-in-time behavior of the solution to a Gierer-Meinhardt system*, Discrete Contin. Dyn. Syst. Ser. A 33, 2885–2900.

Kato, T. (1976), *Perturbation Theory for Linear Operators*, second edition, Springer-Verlag, Berlin.

Kavallaris, N.I., and Suzuki, T. (2010), *An analytic approach to the normalized Ricci flow-like equation*, Nonlinear Analysis TMA 72, 2300–2317.

Kavallaris, N.I., and T. Suzuki, T. (2015), *An analytic approach to the normalized Ricci flow-like equation: revisited*, Appl. Math. Letters 44, 30–33.

Kavallaris, N.I., and T. Suzuki, T. (2022), *Gradient inequality and convergence to steady-states of the normalized Ricci flow*, Nonlinear Analysis TMA 221, 112906.

Kobayashi, M., Suzuki, T., and Yamada, Y. (2019), *Lotka-Volterra systems with periodic orbits*, Funkcial. Ekvac., 62, 129–155.

Kobayashi, T. Suzui, T., and Watanabe, K. (2003), *Interface regularity for Maxwell and Stokes system*, Osaka J. Math. 40, 925–943.

Kondrat'ev, V.A. (1967), *Boundary problems for elliptic equations in domains with conical or angular points*, Trans. Mosc. Math. Soc. 16, 227–313.

Kufner, A., John, O., and Fučik, S. (1977), *Function Spaces*, Academia, Prague.

Kurokiba, M. and Ogawa, T. (2003), *Finite time blow-up of the solution for a nonlinear parabolic equation of drift-diffusion type*, Differential Integral Equations 16, 427–452.

Ladyženskaya, O.A., Solonikov, V.A., and Ural'ceva, N.N. (1968), *Linear and Quasi-linear Equations of Parbolic Type*, Amer. Math. Soc., Providence.

Latos, E. Suzuki, T., and Yamada, Y. (2012), *Transient and asymptotic dynamics of a prey-predator system with diffusion*, Math. Meth. Appl. Sci. 35, 1101–1109.

Li, Y.Y. (1999), *Harnack type inequality: the method of moving planes*, Comm. Math. Phys. 200, 421–444.

Li, Y.Y. and Shafrir, I. (1994), *Blow-up analysis for solutions of* $-\Delta u = Ve^u$ *in dimension two*, Indiana Univ. Math. J. 43, 1255–1270.

Lin, C.S. (2000), *Uniqueness of solutions to the mean field equations for the spherical Onsager vortex*, Arch. Rational Mech. Anal. 153, 153–176.

Lin, C.S. (2007), *An expository survey on recent development of mean field equation, Discrete and Contin. Dyn. Syst. Ser. A* 19, 387–410.

Lin, C.S. and Lucia, M. (2006), *Uniqueness of solutions for a mean field equation on torus*, J. Differential Equations 229, 172–185.

Lions, P.L. (1984), *The concentration compactness principle in the calculus of variation. The locally compact case. Part I.*, Ann. Inst. H. Poincaré, Anal. Non Linéaire 1, 109–145.

Lions, P.L. (1997), *On Euler Equations and Statistical Physics*, Cattedra Galileiana, Pisa.

Łojasiewicz, S. (1963), *Une propriété topologique des sous-ensembles analytiques réels*, Colloques internationaux du C.N.R.S #117, Les équations aux dérivées partielles.

Lotka, A.J. (1925), *Elements of Physical Biology*, William and Wilkins, London.

Luckhaus, S., Sugiyama, Y., and Velázquez, J.J.L. (2012), *Measure valued solutions of the 2D Keller-Segel system*, Arch. Rational Mech. Anal.. 26, 31–80.

Masuda, K. and Takahashi, K. (1994), *Asymptotic behavior of solutions of reaction-diffusion systems of Lotka-Volterra type*, Differential Integral Equations 7, 1041–1053.

Meyers, N.G. (1963), *An L^p-estimate for the gradient of solutions of second order elliptic divergence equations*, Ann. Scoul. Norm. Sup. Pisa III 17, 189–206.

Miyanishi, Y. and Suzuki, T. (2017), *Eigenvalues and eigenfunctions of double layer potentials*, Trans. Amer. Math. Soc. 369, 8037–8059.

Mizoguchi, N. (2020), *Refined asymptotic behavior of blowup solutions to a simplified chemotaxis system*, Comm. Pure Appl. Math. 75, 1870-1886.

Mizohata, S. (1973), *The Theory of Partial Differential Equations*, Cambridge Univ. Press, Cambridge.

Montgomery, T. and Joyce, G. (1974), *Statistical mechanics of negative temperature*, Phys. Fluids 17, 1139–1145.

Morita, S. (2001), *Geometry of Differential Forms*, Amer. Math. Soc., Providence, RI.

Moser, J. (1971), *A sharp form of an inequality of N. Trudinger*, Indiana Univ. Math. J. 20, 1077–1092.

Nagai, T. (1994), *Blow-up of radially symmetric solutions to a chemotaxis system*, Adv. Math. Sci. Appl. 5, 581–601.

Nagai, T., Senba, T., and Suzuki, T. (2001), *Concentration behavior of blow-up solutions for a simplified system of chemotaxis*, Kokyuroku RIMS 1181, 140–176.

Nagai, T., Senba, T., and Yoshida, K. (1997), *Applications of the Trudinger-Moser inequality to a parabolic system of chemotaxis*, Funkcial. Ekvac. 20, 411–433.

Nagasaki, K. and Suzuki, T. (1990a), *Asymptotic analysis for two-dimensional ellip-tic eigenvalue problem with an exponential non-linearity*, Asymptotic Analysis 3, 173–188.

Nagumo, M. (1942), *Uber die Lge der Integralkurven gewohnlicher Differentialgleichungen*, Proc. Phys.-Math. Soc. Japan 24, 551–559.

Naito, Y. and Suzuki, T. (2008), *Self-similarity in chemotaxis systems*, Colloquium Mathematics 111, 11–34.

Nečas, J. (1967), *Le Methodes Directres en Théorie des Equations Elliptiques*, Masson, Paris.

Nečas, J. (1983), *Introduction to the Theory of Nonlinear Elliptic Equations*, Teubner, Leipzig.

Ohtsuka, H., *An approach to regularize the vortex model*, Proc. 5th East Asia PDE Conference, (H.J. Choe, C.-S. Lin, T. Suzuki and J. Wei, eds) Gakuto International Series on Mathematical Science and Applications, Gakkotosyo, Tokyo, pp. 245-264.

Ohtsuka, H. Senba, T., and Suzuki, T. (2007), *Blowup in infinite time in the simplifid system of chemotaxis*, Adv. Math. Sci. Appl. 17, 445–472.

Oleinik, O.A. (1960), *A method of solution of the general Stefan problem* (in Russian), Soviet Math. Dokl. 1, 1350–1354.

Onofri, E. (1982), *On the positivity of the effective action in a theory of random surfaces*, Comm. Math. Phys. 86, 321–326.

Onsager, L. (1949), *Statistical hydrodynamics*, Suppl. Nuovo Cimento 6, 279–287.

Perelman, G. (2002), *Ricci flow with surgery on three-manifolds*, arXiv:math. DG/0303109.

Perelman, G. (2003), *Finite extinction time for the solutions to the Ricci flow on certain three-manifolds*, arXiv:math. DG/0307245.

Pierre, M. (2010), *Global existence in reaction-diffusion systems with control of mass: a survey*, Milan J. Math. 78, 417–455.

Pierre, M., Suzuki, T., and Yamada, Y. (2019) *Dissipative reaction diffusion systems with quadratic growth*, Indiana Univ. Math. J. 68, 291–322.

Pohozaev, S.I. (1965), *Eigenfunction of the equation $\Delta u + \lambda f(u) = 0$*, Soviet Math. Dokl. 6, 1408–1411.

Pointin, Y.B. and Lundgren, T.S. (1976), *Statistical mechanics of two- dimensional vortices in a bounded container*, Phys. Fluid 19, 1459–1470.

Rellich, F. (1953), *Perturbation Theory of Eigenvalue Problems*, Lecture Notes, New York Univ.

Sato, T. and Suzuki, T. (2023), *Morse indices of the solution to the inhomogeneous elliptic equation with exponentially dominated nonlinearities* Ann. Mat. Pura Applica, 202, 551–599.

Sawada, S. and Suzuki, T. (2008), *Derivation of the Equilibrium Mean Field Equations of Point Vortex System and Vortex Filament System* Theor. Appl. Mechanics Japan 56, 285–290.

Sawada, S. and Suzuki, T. (2017), *Relaxation theory for point vortices*, Vortex Structures in Fluid Dynamic Propblems (Perez-de-Tejada, H. ed.) INTECH 2017, Chapter 11, pp. 205–224.

Senba, T. (2007), *Type II blowup of solutions to a simplified Keller-Segel system in two dimensions, Nonlinear Analysis* 66, 1817–1839.

Senba, T. and Suzuki, T. (2000), *Some structures of the solution set for a stationary system of chemotaxis, Adv. Math. Sci. Appl.* 10, 191–224.

Senba, T. and Suzuki, T. (2001), *Chemotactic collapse in a parabolic-elliptic system of chemotaxis, Adv. Differential Equations* 6, 21–50.

Senba, T. and Suzuki, T. (2002a), *Weak solutions to a parabolic-elliptic system of chemotaxis, J. Funct. Anal.* 191, 17–51.

Senba, T. and Suzuki, T. (2002b), *Time global solutions to a parabolic-elliptic system modelling chemotaxis, Asymptotic Analysis* 32, 63–89.

Senba, T. and Suzuki, T. (2003), *Blowup behavior of solutions to rescaled Jäger-Luckhaus system, Adv. Differential Equations* 8, 787–820.

Shafrir, I. (1992), *A* sup + inf *inequality for the equation* $-\Delta u = V e^u$, *C.R. Acad. Sci. Paris, Série I* 315, 159–164.

Simader, C.G. (1972), *On Dirichlet's Boundary Value Problem*, Lecture Notes in Math. 268, Springer-Verlag, Berlin.

Simon, L. (1983), *Asymptotics for a class of nonlinear evolution equations with applications to geometric problems, Ann. Math.* 118, 525–571.

Sire, C. and Chavanis, P.-H. (2002), *Thermodynamics and collapse of self-gravitating Brownian particles in D dimensions, Phys. Rev. E* 66, 046133.

Struwe, M (2002). *Curvature flows on surfaces, Ann. Scuola Norm. Sup. Pisa Cl. Sci.* (5) 1, 247–274.

Suzuki, T. (1982), *Full-discrete finite element approximation of evolution equation* $u_t + A(t)u = 0$ *of parabolic type, J. Fac. Sci. Univ. Tokyo, Sec IA* 29, 195–240.

Suzuki, T. (1992), *Global analysis for a two-dimensional elliptic eigenvalue problem with exponential nonlinearity, Ann. Inst. Henri Poincaré, Analyse Non Linéaire* 9, 367–398.

Suzuki, T. (2005). *Free Energy and Self-Interacting Particles*, Birkhäuser, Boston.

Suzuki, T. (2013), *Exclusion of boundary blowup for 2D chemotaxis system provided with Dirichlet boundary condition, J. Math. Pure Appl.* 100, 347–367.

Suzuki, T. (2015b), *Blowup in infinite time for 2D Smoluchowski-Poisson equation, Differential Integral Equations* 28, 601–630.

Suzuki, T. (2015). *Mean Field Theories and Dual Variation*, second edition, Atlantis Press, Paris, France.

Suzuki, T. (2018). *Chemotaxis, Reaction, Network*, World Scientific Publishing, Singapore.

Suzuki, T. (2020). *Semilinear Elliptic Equations, Classical and Modern Theories*, Walter de Gruyter GmbH, Berlin, 2020.

Suzuki, T. (2021), *Interface vanishing for nonstationary Maxwell equation, Adv. Math. Sci. Appl.* 30, 555–570.

Suzuki, T. (2022). *Applied Analysis, Mathematics for Science, Technology, Engineering*, third edition, World Scientific Publishing, Singpore.

Suzuki, T. (2022), *Local and global behavior of solutions to 2D-elliptic equation with exponentially-dominated nonlinearities, Asymptotic Analysis* 128, 465–494.

Suzuki, T. and Takahashi, R. (2009), *Degenerate parabolic equation with critical exponent derived from the kinetic theory, I. Gnereation of the weak solution*, Adv. Differential Equations 14, 503–524.

Suzuki, T. and Takahashi, R. (2009), *Degenerate parabolic equation with critical exponent derived from the kinetic theory, II. Blowup threshold*, Differential Integral Equations 14, 1153–1172.

Suzuki, T. and Takahashi, R. (2010), *Degenerate parabolic equation with critical exponent derived from the kinetic theory, IV. Structure of the blowup set*, Differential Integral Equations 15, 223–250.

Suzuki, T. and Takahashi, R. (2012), *Degenerate parabolic equation with critical exponent derived from the kinetic theory, III. ε-regularity*, Differential Integral Equations 25, 223–250.

Suzuki, T. and Tasaki, S. (2010), *Stationary solutions to a thermoelastic system on shape memory materials*, Nonlinearity 23, 2623-2656.

Suzuki, T. and Tsuchiya, T. (2005) *Convergence analysis of trial free boundary methods for the two-dimensional filtration problem*, Numer. Math. 100, 537–564.

Suzuki, T. and Tsuchiya, T. (2011), *Weak formulation of Hadamard variation applied to the filtration problem*, Japan. J. Indus. Appl. Math. 28, 327–350.

Suzuki, T. and Tsuchiya, T. (2016), *First and second Hadamard variational formulae of the Green function for general domain perturbations*, J. Math. Soc. Japan, 68, 1389–1419.

Suzuki, T. and Tsuchiya, T. (2023a), *Liouville's formulae and Hadamard variation with respect to general domain perturbations*, J. Math. Soc. Japan, 75, 1025–1053.

Suzuki, T. and Tsuchiya, T. (2023b), *Hadamard variation of eigenvalues with respect to general domain perturbations*, J. Math. Soc. Japan, submitted.

Suzuki, T. and Watanabe, K. (2023) *Interface vanishing of $d - \delta$ systems in Euclidean spaces*, in preparation.

Suzuki, T. and Yamada, Y. (2013), *A Lotka-Volterra system with diffusion*, Nonlinear Analysis and Interdisciplinary Sciences (Aiki, T., Fukao, T., Kenmochi, N., Niezgódka, M., and Ôtani, M. eds.), Gakuto International Series on Mathematical Sciences and Applications, *Gakkotosyo* 36, 215-236.

Suzuki, T. and Yamada, Y. (2015), *Global-in-time behavior of Lotka-Volterra system with diffusion - skew symmetric case*, Indiana Univ. Math. J. 64, 181-216.

Trudinger, N.S. (1967), *On imbedding into Orlicz space and some applications*, J. Math. Mech. 17, 473–484.

Vázquez, J.L., *Smoothing and Decay Estimates for Nonlinear Diffusion equations: Equations of Porous Medium Type*, Oxford Univ. Press, Oxford.

Volterra, V. (1926), *Variazioni e fluttuazioni del numero d'individui in specie animali conviventi*, Mem. Acad. Licen. 2, 31–113.

Yagi, A. (2010), *Abstract Parabolic Evolution Equations and their Applications*, Springer Verlag, Berlin.

Index